Human Existence in the 21st Century

Reality, Greed, Religion and War at the End of Civilization

by

Antoine Averie

This book is dedicated to all the babies & children of Earth who deserve a peaceful planet from which to explore and inhabit a star system...

Chapters & Sections

Human Existence in the 21st Century

"An Extraterrestrial View"

A Paradox of Possibilities

The extremes are stunning...When Earth is viewed as a single organizational structure there are patterns that emerge from the activities of over seven billion humans that are indicative of the multiple outcomes possible in this random experiment of life on a single world...It's nearly certain the next three hundred years will determine the structure and future of life on Earth for the next 1,000 years with predicted populations of 10 to 13 billion humans, a continuing growth in technology and scientific discovery, ongoing political upheaval, unending territorial posturing and mass migration on a violent, primitive world that grows more connected even as it grows more chaotic...In this century there are over 100 satellites circling this planet, multiple spacecrafts exploring our star system and at least one spacecraft orbiting every planet but Neptune and Uranus...Multiple missions to Mars verify the human interest in this world that will be the first to be visited and less than sixty years after the launch of the world's first satellite humans have caught up with a distant comet and landed a complex machine on it's surface...Industrious and creative, curious and intelligent, these are incredible achievements with every mission a verification of human progress over the last 100 years and a growing global awareness of the physical laws that define this reality...Each success a verification of the ability of self-aware, sentient intelligence to reprocess the raw materials of a planet into complex tools with extraordinary purpose...

During the same time-frame in an exploding global civilization without constraint, humans have fought multiple wars killing millions of innocent humans, created weapons capable of destroying a planet, enslaved entire populations to steal their resources, polluted the atmosphere in ways that will influence our climate for thousands of years, destroyed the habitat of uncounted species and impacted Earth's oceans in a way that could challenge the survival of every creature on this world...Today there are thousands of females forced into genital mutilation, billions of humans with no toilets or clean water, millions of babies and children who go hungry every day while conflicts escalate in Africa and the Middle-East in a continuous cycle of violence without end, a stunning global ignorance of the framework of reality that supports human existence and the real possibilities of life without ethnic discrimination, gender bias, racial inequality and war...*How can there be these two extremes?*

Humanity is just beginning to understand and address the problems and solutions necessary for a sentient species to manage a planet and survive for millions of years in a single star system and every year more people grow aware of the possibilities of a connected Earth with a global civilization based on science and equality rather than territory, dogma, greed, superstition and fear...But every year the chaos and fragmentation grow more, not less, and a growing indifference to the situational realities of billions of humans is creating a world with sharply divided lines drawn with weapons, walls and money as primitive instincts older than the human species dominate the mindsets of some of the most powerful people on Earth creating a paradigm of inequality and territoriality that threatens the existence of every society, every nation and every species on this planet...How does this change? *How can a species with 7.4 billion individuals reach a consensual understanding of the nature of human reality and do what is necessary to survive?* How do we increase Earth's IQ to create a majority population with intelligence and awareness that understands causality and can change this primitive ignorance that threatens human existence? There is only one way and it will take a long time...An intense focus on initial brain growth during the first three years of life for every baby born on Earth...Adequate nutrition, secure attachment, nurturance, sleep, positive stimulation and play...Essential human interaction during the first 1,000 days of life to create a neural density critical in the development of "reality intelligence", a deep inter-connective network of internal communication within the brain that develops primarily in the first 3 years of life and determines every individual's understanding of their own existence...Neural density critical in the development of a natural morality, an instinctual, intuitive empathy and a connective intelligence that can understand human reality and the nature of human existence for a lifetime...

This is the message that reverberates throughout this book and it would change this world if a majority of humans understood that after tens of thousands of years without the necessary knowledge or technology to understand and control human existence *we are finally here*...In less than one hundred years humans have discovered and understood the physical laws and processes necessary to manage a world, the power of the atom, quantum actions that make technology possible, space travel, multiple alternative energy sources, techniques to map the human brain and a growing medical knowledge that will eventually cure all disease...Finally we've learned to control electromagnetism that will be part of human existence until the end and we've learned how to travel through space developing speed of light networks everywhere on Earth to enable anyone, anywhere to communicate and process information while wiring the entire planet with technology that will grow more and more complex and will be an integral part of human existence for millions of years...A critical time-point in the history of a species, this century is the beginning of a transition to a global civilization inhabiting a star system, an inclusive civilization capable of surviving for millions of years while providing education, health, freedom and equal standing for every human on Earth...A thousand-year process of change...

Or the 21st century could be the beginning of a long decline into an abyss that will take centuries to recover from and challenge the survival of every creature on Earth...It's unknown how the unchecked processes that continue today will challenge human survival over the next 100-300 years...Fossil fuel extraction increases daily despite a collective knowledge of consequence and the climate will change dramatically for a very, very long time while human resources and production capabilities are being stretched by a growing world population with no programs or policies anywhere to address this in a serious way...Any intelligent species would have to control reproduction and in this century humans have many ways to prevent pregnancies in every female on Earth but the problem is not technology or safety, it's widespread ignorance, religious dogma, a lack of reality intelligence and a domination of women by primitive men and cultures to prevent education and choice...Earth can comfortably support maybe five billion humans and there has to be some limit on population until we can terraform Mars or Venus...This is not a moral question...No intelligence wants to destroy life...With consensual awareness and modern medicine humans could voluntarily control their birth rate in multiple ways using data, cooperation and effective techniques *but this will not happen*...In this century Earth's IQ is too low for self-determination and sex is one of the strongest instincts in all the creatures on this world...

It's easy to imagine an intelligent species maintaining a mean population of around 5 or 6 billion (+ or – 100 million) for thousands of years with a planet-wide realization of the need for balance but in this century humans are nowhere close to this and a near future with 13 billion humans competing for resources will be interesting as the impact of a crowded Earth with inequality is becoming increasingly clear...Small-minded governments continue to divide populations into categories and discriminate for advantage while popular uprisings, civil war and widespread terrorism promise violence with unknown consequences...The detonation of just a few nuclear weapons by terrorists or nations will challenge the entire world and the release of contagions and biological agents could have a deleterious global effect on the progress of an entire species...The collapse of the 1st World Civilization is not only possible but very probable given the forces that dominate in this century with short-sighted policies and the growing danger of inaction...An implosion of the Middle-East (happening in this decade), a sea level rise of five to seven feet, an aggressive China determined to rule lands and seas in southeast Asia, an unknown and increasingly dysfunctional North Korea, a growing chaos in Pakistan with nuclear weapons aimed at India, severe drought, unpredictable climate, extreme weather, collapsing food chains, poverty, famine and starvation on an overcrowded world where over 80% of the inhabitants are unaware that humans exist on a large rock with no salvation...We will not find a heaven to save us from ourselves and we won't suddenly find ourselves next to another planet to resupply our resources...

To those of us who understand this reality, maybe one or two billion humans spread around the globe without unity or a single voice, the *murders* (500,000+ a year), *greed* (billions of dollars obtained illegally every year at the expense of the majority of Earth's population), *war* (65 nations and more than 600 militia/guerrilla groups involved in conflicts), *poverty* (3 billion humans existing on less than $2.50 a day), *gender discrimination* (40% of women suffer physical or sexual violence, 14 million girls are forced into marriage every year and the WHO reports 125 million females alive in 2016 are victims of genital mutilation), *ethnic discrimination* (billions of people on a daily basis affecting income, freedom and health) and *religion* (over 80% of Earth's population are affiliated with some religious organization who insist every event is "God's Will" and the way to "salvation" is to pray and donate more money), reflect behavior and actions so primitive and ignorant that the possibility of a harmonious, intelligent Earth with awareness and equality for every person is almost impossible to imagine...Even the most powerful humans on Earth are unable to understand existence with the president of the United States "praying" to an imaginary God for guidance in a nation that is increasingly religious and increasingly violent with over 300 million guns legal to carry almost anywhere at anytime...

National polls found 40% of Americans believe Earth was created just 6,000 to 10,000 years ago (Gallup 2014), 57% believe religion can solve all of today's 'problems' (Gallup 2014) and 25% did not know the Earth orbits the Sun (National Science Foundation 2014)

And before every formal political session in the United States Congress lawmakers invoke this ancient ritual of 'prayer' in an ongoing attempt to communicate with this deity asking for 'wisdom' and 'guidance' as they consider and enact legislation affecting millions of humans...After this ritual they generally ignore the needs of millions of poor, sick and elderly humans to focus on the corporate powers who donate money to their campaigns and work on the electoral strategy necessary to insure their future employment...The historical record in this century reflecting the actions of these political bodies in creating a world recession with reckless legislation, inadequate regulation and corporate collusion coupled with nonsensical military aggression and war on a scale that has caused the deaths of hundreds of thousands of innocent people casts doubt on the 'wisdom' and 'judgment' of this imaginary creature they believe is guiding them...

This behavior by leaders who control the largest arsenal in the world and huge wealth is both disappointing and more than a little alarming...In this century there are thousands of solutions proposed which could make huge differences and it's amazing how intelligent people with solutions to many of humanity's problems still expect corporations and political players controlled by corruption and the free flow of money to listen and suddenly understand what must be done...It will never happen that way because most of these "leaders" cannot understand what's necessary to change these primitive rituals and competitive instincts that enrich a few at the expense of many...Earth needs new generations of intelligent humans who recognize the nature of our existence to create a majority with the power to change the status-quo and use new technology, emerging science and creative solutions to manage human habitat, reduce emissions from every power plant on Earth, eliminate internal combustion machines for electrical alternatives, stop the endless production of weapons available to almost anyone on Earth and address population control, the necessity to recycle everything, hunger, poverty and disease...With a global majority ignorant of consequence and unsympathetic to the fate of everyone outside of their "tribe", it's a race against time for this civilization which will almost certainly collapse but hope for the next civilization if there is a focus on babies, inclusion and sustainability on the only world known that supports human existence...

3

INTRODUCTION

In an indifferent universe there is a possibility that a species will evolve through a series of random occurrences and unexpected events over a vast period of time until that species reaches a level of awareness and intelligence that enables them to recognize the reality they exist in and understand the next step necessary to determine the intellectual and evolutionary path of future generations...This is that time in the history of human existence...

Defining a World

The history of the human species is retrospect...No matter that what happened in the past hour is as enduring a history as what happened 100,000 years ago, history is retrospect with events considered historical only after years of connections, patterns of behavior, cause, effect and insight by scholars and pundits to determine the measurable effect of human activities on human activities...Many people over thousands of years have contributed to the knowledge and insight that defines human history but only recently has a framework defining eons, ages, epochs, dynasties, millennium, centuries and decades been recognized by most of the world as a standard in an emerging puzzle that is continually updated with new archaeological discoveries and connections still being found in the oceans and continents that create the outermost structure of this planet...In the analysis of history, a millennium (1,000 years) is often too long to directly link causality and a decade (10 years) too short for definitive conclusions, but a century (100 years) is a period of time containing multiple generations whose actions and decisions have a measurable effect in regional and world events and is a template for the analysis of various histories stretching back 6,000 years (60 centuries)...Within this framework Earth has existed some 45 million centuries with human divergence from primates beginning just 50,000 centuries ago...

This book is not a history book...It is about now...In a 'modern age' and with a modern calendar that's been accepted for just a few centuries, 'now' is defined as the 21st century, 2,000 years since the acknowledged birth of Jesus Christ, a man who lived in the ancient Roman Empire with an uncertain existence (no evidence) and an undeniable impact...At this point in the history of human existence with an accepted technology defining time to an accuracy of nanoseconds, it's unlikely there will be another major revision of structured time but historical milestones are critical in defining the development of intellect with a global acceptance of universal time zones and mapping the planet with longitude and latitude, representative first steps in developing a global knowledge and cooperation with the awareness and ability to manage a single planet with an efficiency that will ensure the survival of vital species and habitat for millions of years...Vital for success is technology to manipulate a proven reality, numerical analysis of the millions of long-term and short-term variables that define human reality and a global consensus of cause and effect with a knowledgeable, manageable population who understand the nature of their existence...In the 21st century the human species is nowhere close to this...In 1968 a picture of Earth from the first manned spacecraft to circle our moon impacted millions of humans in a moment of clarity with what seemed at the time was a global realization of the singularity of Earth and a fragile existence not guaranteed by any existential force, but what has happened in the nearly fifty years since that time has done little to protect Earth's fragility with global pollution, unchecked population growth, endless war and a growing inequality in a fractured and chaotic world civilization with an uncertain future...This book is a snapshot of that civilization, at this moment, in this time...

The diversity and nuance identifying nations and populations in the 21st century add a unique and challenging complexity to in-depth analysis and a comprehensive understanding of the behavior, demographics and solutions on a world of 7.4 billion individuals, but what is most stunning in a global civilization that can now be analyzed and understood as a single complex organizational structure is an *instinctive territorial obsession* in the human animal that is older than our species, an instinct that has been a dominant determinant in every war ever fought and in every civilization that has flourished and declined...An instinct that still dominates in this century with almost every nation on Earth possessing a standing military whose requirements far outweigh every other aspect of the society it protects from invasion and conquest, a vigilance still believed necessary on 21st century Earth...

When these military forces or militias are controlled by humans with a limited awareness, an absence of empathy, a paranoid, delusional view of reality and a deep-seated primitive fear of everyone and everything, the result is greed and corruption on unimaginable scales, persecution of anyone and everyone imagined as a threat and the horror of death, rape, mutilation, torture and genocide initiated by human monsters with no understanding of human existence...Idi Amin, Hitler, Pol Pot, Stalin, Kim Jong-ll, Kim il Sung, Lenin, Leopold ll, Tito, Charles Taylor, Pinochet, Franco, Mbasogo, Duvalier, Alturabi, Than Shwe and Sukarno from a past with thousands of names...Still holding power in 2016 are Mugabe, al-Assad, al-Bashir, Ali Khamenei and Idriss Deby, often allied with the west, increasingly wealthy as their populations starve and uncompromisingly brutal when challenged...Such monsters would not exist if the initial growth and development of all human infants were a priority in every culture on Earth, but that world will take many years...

Until that time these "leaders" exist with the compliance of nations worldwide and in many situations without fear of a world focused on profit and sovereignty who voice concern, morality and compassion but do little to save the billions of women and children who still suffer from chronic inaction...Humans are the endpoint of a billion years of evolutionary roulette and after thousands of years of ignorance and the deaths of millions, *the physical and intellectual reality of human existence is now known*...There is one reality and no amount of spin or denial can change that...A reality recognized by intelligence and dismissed by ignorance...This book was written to define this known reality and to explain why a self-aware sentient species with an incontrovertible scientific knowledge of the nature of their existence continue to engage in unending war and conflict while destroying the ecosystem that supports their existence...It's about the impact of initial brain growth on the awareness and intelligence of adult humans and why this species is in grave danger from their own technology...In the territorial global environment of this 21st century, situational realities in every nation on Earth continually change but the physical and spatial nature of human existence is proven, definable and undeniable...

An accurate representation of the size of active military forces and the expenditures dedicated to arming and growing these forces is almost impossible in a world with more than 195 sovereign nations and the cloak of secrecy every nation feels is necessary to gain an advantage...The numbers represented in this book are approximations derived from multiple sources including official publications profiling every nation on Earth and the statistics released from governments regarding the size and expense of their military forces which define a primitive world civilization fighting over territory, resources and ideology, unaware of the possibilities of a consensual awareness with the resources of a star system...

This book is an amalgamation of scientific research, historical evidence, empirical evidence, anecdotal evidence, observational evidence and data from the archives of knowledge collected by the human species over thousands of years...The numbers represented here to explain the diversity of human civilization and the consequence of human activity are relatively concrete when reflecting historical trends and measurable science but many measurements of human activity are variable when representing the ongoing and ever-changing realities of a global civilization with over seven billion humans...Multiple references have been researched, reviewed and updated to reflect the most accurate measurement of a world that cannot be frozen in time to be studied but inherent in this process is query and criticism from the exposure of such realities and the author welcomes any discourse that will shine a light on the possibilities of human existence and the challenge of survival in this century...

Although a connective idea (*reality intelligence*) is evident throughout the book, each chapter is a complete entity and this book can be read in any order one chooses...It's about the reality of human existence in three-dimensional space, a definable reality realized just recently in human history and a reality which threatens the survival of the only sentient, self-aware intelligence known to exist...It's about the future of life on a single world, a small rock in a vast emptiness containing trillions of rocks...It's about seven billion humans sharing one planet...It's about us...

Human Genome = Difference between any 2 humans on Earth is less than .1%...

Relative Reality

In *The Grand Design,* Steven Hawking and Leonard Mlodinow state that "there is no picture or theory independent concept of reality" and humans have no way of knowing if existence within this reality is contained within some other reality that is undetectable and unimaginable to life forms who exist on a single world...This is the great question of philosophy, religion and science and it's almost certain the answer will never be known...Only in the last 100 years have humans realized and verified the size of the "space" they exist in and the reality of this existence is staggering...An emptiness over 93 billion light years in diameter expanding at an ever-increasing rate with familiar atoms making up just 4.6% of everything and the remainder believed to be dark matter (24%) and dark energy (71.6%), modern scientific conjecture defining the nature of a universe that will exist for trillions of years...

What we do know is the reality we know we exist in has properties, patterns, rules and events that are measurable and predictable in enough ways for creatures like us to understand *what we are* even if there is no real chance of knowing *why we are*...Any intelligence existing in this universe who attempts to discover the properties and rules of their reality will discover the same universe we do, a universe with consistent and predictable laws and properties...The recent discovery and verification of quantum theory proves there is a great deal about reality not yet understood but this quantum world exists at a scale that has very little to do with how most humans think of existence and the continuing challenge of survival requires an acceptance that the complexity of this reality supports human existence or we would not be here...Within this known reality the focus of the human species must be on increasing intelligence to create a global awareness of the balance of nature that allows life to exist and to live within that balance...It is our only chance to survive...

This much is known...Humans are the dominant species on planet Earth...This is our world and we can survive as a species for millions of years if we can increase the intelligence of all humans to a point where there is a global consensual awareness of what is necessary for survival...Human history has just begun and we are the only intelligent life in a solar system with eight planets and 170 moons to explore, to inhabit and to use for resources vital to a continued existence...We are life that has become self-aware and after thousands of years we can finally understand the diversity and singularity of all humans and the possibilities of sentient intelligence...Both male and female have an equal ability for intellectual capability, curiosity, wisdom, insight, guile, wit, problem solving, determination, bravery, leadership, humor, genius, survival and skills which are not determined by race, gender or ethnicity and excepting for obvious limitations imposed by physical damage through genetic factors, random events, prenatal care or unforeseeable circumstances, *every human baby born healthy has an equal possibility to develop an intellect and awareness comparable to every other member of our species*...This is key...This means all humans have the potential to develop the intelligence and awareness necessary to recognize and understand the patterns of life on this planet which enable different species to live in a shared environment that's always changing through time...Over thousands of years humans will have to adapt to geological processes affecting the lives of millions, climate processes that will change habitat and threaten food production, threats from space which could be catastrophic without intervention and many situational realities of existence that require a consensual global awareness and intelligence able to respond to any threat...We exist because of the abundance of life that surrounds us and have evolved to a point where we can adapt to life better than life can adapt to us...Humans are the caretakers of this planet, the only sentient, self-aware inhabitants in this star system and no matter what the challenge, human activity *will* determine the future of all life on Earth...

How does a sentient intelligence, a young primitive species who are just now understanding the measurable reality that defines their existence, survive this exponential stage of growth? How can any government, any group or any policy change the focus or behavior of over seven billion creatures who exist without a consensual awareness or understanding of the impact and consequence of their collective actions?

Questions without answers...Even with the incredible explosion of science and technology in the past century and a collective global knowledge that has increased more in the past 400 years than in the previous 10,000 years, the habits, beliefs and dogma that have determined human activity for most of history do not disappear in just a few generations...This planet is changing and the existence of seven billion humans is the reason...Human activity is now linked to almost every process on Earth affecting biodiversity of species, weather on climatological scales, physical processes which shape the planet and accelerating innovation, new technology and progress in all the sciences, humanities and realities that represent a world civilization...An explosion of science, technology and over four billion people in just fifty years is bringing into focus a future with virtually unlimited possibilities and a future which could threaten the existence of life on the only planet known to have life, a single blue world in an indifferent universe, fragile, balanced and resilient...A cyclical ecosystem that supports carbon-based life...

Every human alive today is a member of the 1st World Civilization, a new age of instantaneous communication and spatial awareness anywhere and anytime in a global society fractured in uncounted ways as humanity struggles with questions never asked and dangers never faced...Isolated cultures, civilizations and histories are being exposed and challenged by new realities proven and predictable and a growing awareness of human existence on a single world that shines an unblinking light on the primitive customs, superstitions, beliefs and dogma that dominated and dictated human development for over 100,000 years...Adaptation and change are necessary for the continued evolution of awareness and intellect and it's essential a majority of people understand the nature of human existence if we are to survive in a reality increasingly obvious to a species beginning to understand the possibilities of a future lasting millions of years...Despite scientific discoveries that have swept away thousand year old shadows, there are still billions of humans who do not understand the reality of their own existence, who still worship gods and demons, who believe in predetermined fate, who kill other people without cause because they're "different" and who do not recognize the balance of the world that supports their existence, unable to recognize all humans are fundamentally the same...This must change if our species is to continue, but how? Is it possible to increase the intelligence and awareness of an entire species?

The simple answer is *yes*...In this century many researchers studying human intelligence, human development and human behavior are beginning to understand that *the key to universal intelligence in all humans is rooted in the earliest stages of life*, the most essential stage of growth that will determine a baseline intelligence critical in an individual's ability to recognize reality...There are many studies that demonstrate over and over the positive impact of secure attachment, nutrition and sleep on developing babies and in the past two decades, with new technology and techniques to image and measure brain growth, there is a growing body of evidence showing a measured effect of enhanced brain growth in infants nurtured in secure environments with close, continuous contact, adequate nutrition, abundant sleep and positive stimulation...*The human brain develops more in the first thousand days after birth than at any other time during a human life* and this critical stage of development determines a lifetime of awareness with dense neural growth early in life that is essential for the brain to communicate with itself...Internal communication between the ancient, primitive areas of the brain and a highly developed frontal cortex that determines every individual's ability to understand and recognize the reality of their physical existence and the conceptual reality of being a human...A brain with a dense connectivity to understand empathy, morality and reality while continually processing billions of bits of information with a three-dimensional experience to understand the nature of it's own existence...

The only thing holding back the human species is millions of years of evolution...It's not a little thing...The million year old brain is fully formed when born but the intellectual brain takes another three years...A focus on initial brain growth in every baby born will change the world, a thousand days of undivided attention on the needs of every small human...This is the only way to develop "reality intelligence", a deep intuitive awareness determined by neural density and neural connectivity that is essential in enabling every brain to communicate with itself...

Human Reality

The diversity of humans throughout history and on Earth today is undisputed with the myriad of cultural beliefs and practices, superstition and dogma, wisdom and common sense, written text, word of mouth, human trial and error, and the religious adherence and religious structure intertwined with early growth and societal expectations that have been instrumental in defining the way humans in every society view and practice human development after birth...Science is only now starting to recognize the critical importance of initial brain growth in infants and it's unknown how generations of humans in isolated cultures and societies have understood and recognized the impact of environmental factors, secure attachment and positive human interaction on developing babies or if any of these cultures recognized how influential this period of growth is in determining an intuitive intellect and awareness when the infant human reaches maturity and becomes an adult...Every human lives a singular existence innately focused on stability and survival and although we are isolated internally from every other human on the planet, there are recognizable patterns of thinking and reasoning that determine thoughts, emotions, behaviors and actions that reflect internal rationalizations, intelligence and every individual's awareness of the differing situational realities that define their existence...These internal mechanisms of communication and thought which determine behavior, personality, intelligence and beliefs have been a mystery for all of history and many are still a mystery, but new research that can measure brain growth and brain activity coupled with numerous studies based on historical, empirical, and anecdotal evidence is now revealing a complex causality relationship between the development of essential structure and dense neural growth in the first three years of life and the behavior, beliefs, intelligence, awareness and actions of every individual throughout a lifetime...

For many years instincts critical to human survival were believed to be mostly isolated from the intellectual centers of thought and reason, influencing a person's behavior and their actions when it was necessary but otherwise dormant in the modern "civilized" human...What is now being realized is how powerful these primitive 'instincts' are and how critical initial brain growth is in the formation of dense neural 'bridges' connecting every region of the brain to enable a fluidity of internal communication essential in the development of an *intelligence* that can recognize reality, a *morality* that is instinctive and does not need to be 'taught' and an *empathy* that recognizes the inter-connectivity of all life on this world...Stunning is an increasing body of evidence implicating the impact of external environmental influences in these first years of life as the primary determinant in the development of this connective neural growth critical to every human's understanding of reality and crucial in the early formation of the organizational structure essential for memory storage and brain function over a lifetime...

Variation within species and diversity across cultures define the populations of Earth, but for all the diversity in this world almost everyone can be included in one of three groups determined by each person's environment, viability, activities, influences and experiences in the first one thousand days after birth...Within these three groups representing the billions of people on Earth today are uncounted variations of how each individual understands reality, how they process reality and how they rationalize their existence but their intelligence, their cognitive awareness and their ability as adults to recognize and understand the reality of the world they exist in is essentially determined by age three...

Every human consciousness is shaped by life situations, early education, random chance, unexpected events, decisions both spurious and rational, ordered context and outside influences that combine to create a path in life with many unexpected and unknown possibilities, yet every person's thoughts, beliefs, empathy, morality, intelligence, cognitive awareness and actions are influenced and determined by the density of connective neural growth and the development of organizational structure that occurs in the first three years of life...

At one end of this spectrum are over a billion humans who are intelligent, capable, interactive, creative, non-violent, understand the reality we exist in and want to move the world toward a shared co-existence to create global stability, peace, innovation and a future with civilizations and leaders who understand the possibilities of a shared human existence lasting millions of years...

In the middle group are billions of humans with varying degrees of awareness, who exist in the world as they perceive it with questions of God, morality, justice and purpose, many people who are intelligent and many who are not, people who discriminate against others based on religious, ethnic and cultural beliefs, people who believe human existence is a temporary thing that will soon end because the human species is innately violent and doomed to self-destruction, people who believe that war is part of human nature and killing is just something we do, people who believe that greed and selfishness are inherent in our species and people who believe and support governments, dictators, monarchies, religions and ideologies who promise security and rewards that cannot be realized or simply do not exist...*Billions of people unable to understand the reality or the possibilities of their own existence...*

At the extreme end of this spectrum are millions of humans whose neural growth was severely retarded during infancy due to extreme neglect, lack of nutrition, brutality and violence, the absence of any love or affection and extreme negative experiences...These individuals who survive childhood and merge into populations can seem 'normal' in many ways when they become adults but they do not have any substantial kind of feedback loop or communication between the primitive and rational centers of the brain to reconcile fear, survival, aggression, sex, anxiety, longing, morality and reason...*Unstable and unpredictable, these humans can be very dangerous to other living creatures...*

If this sounds simplistic, it is not...Early experiences, especially profound negative experiences in the first three years of life, can severely retard the growth of neurons and the density of connective neural bridges in developing brains that are essential in determining intellectual capacity and reality awareness in the adult human...The earlier in life a baby has to survive severe negative experiences, the worst the damage, inhibiting the development of organizational structure and inhibiting neural growth during the most critical time of brain development...*There are millions of babies born every year on this planet that receive little nurturance, inadequate food and water, extremely disrupted sleep patterns, suffer sickness and distress on a daily basis and have no chance of developing the dense neural growth and organizational structure necessary to understand human reality...*These millions of babies will be the adult population of Earth one day and like billions of people today they will be blinded by religion, think that war and murder are justifiable, understand little of the fragility and the balance of a planet that supports their existence or the real possibilities of a human civilization with the resources of a star system who recognize equality and diversity in all humans...

If we want to change the world we have to start at day 0 for all babies born on Earth and find a way to love, feed and stimulate every human being during the first three years of life...

Political Reality

As many world populations grow younger and billions of humans in oppressive and stagnant situations believe any chance for a life of security and prosperity is becoming more unattainable, there will be growing dissent, uprisings, demands and violence that will be challenged by long-standing dictators, monarchs, autocrats, ruling families and corrupt democratic political coalitions as a threat to existing power structures and many of these governments will respond with violence...In a 2012 report released by Freedom House, 48 nations were identified where populations are oppressed and restricted with varying degrees of severity by 'elected' governments and coalitions not selected by the citizens in free and fair elections or in too many cases in power with no elections at all...In every nation it's about controlling the activities of the population and controlling the resources...

At one extreme is North Korea where individuality and freedom of information and movement are absolutely forbidden with harsh penalties and censorship of all things the government wants to control...At another extreme is Israel who want desperately for the world to believe they are a free and democratic society while passing laws to prevent demonstrations and boycotts by any individual or group criticizing anything "Israeli" and refusing to recognize the rights or needs of the populations in the territories occupied by Israel or even the rights of some of their own citizens determined by gender, race, ethnicity and heritage...Countries in the Middle East who are currently responding to internal demands for expanded freedoms with strategic restrictions and violence against their citizens include Saudi Arabia, Lebanon, Bahrain, The United Arab Emirates, Egypt and Syria, where a government and a military desperate to hold power and control over their own citizens have killed more than 350,000 people to keep al-Assad in power in a conflict that has displaced millions of refugees and millions of innocent children to surrounding nations resulting in extreme suffering, sickness, death and a growing desperation which threatens to increase chaos and destabilize the entire region...

*Syria in 2015...5.5 million children need assistance, life expectancy has decreased by 20 years in a single decade, 4.8 million humans displaced, humanitarian aid threatened and blocked, electricity infrastructure collapsing...*UN

On a planet where individual life and liberty is determined primarily by profit and greed, the democratic world watches Syria burn while protecting corporations and state interests from the global financial risk of "involvement"...Within this new human paradigm of interconnected global finance that enrich a small percentage of the world in ways recently unimaginable, oppressive governments, human monsters and despots survive and flourish while staying in power for many years, repressing citizens, starving children, enriching themselves, selling and buying energy, goods and services from military manufacturers and nations worldwide who "look the other way" as millions of humans suffer...Mugabe in Zimbabwe (35 years in power), Biya in Cameroon (40 years in power), al-Assad in Syria (16 years in power), Al-Bashir in Sudan (27 years in power), Afewerki in Eritrea (24 years in power), Hun Sen in Cambodia (31 years in power), Russia, China, Ethiopia, North Korea, Egypt, Rwanda, Gambia, Chad, Iran, Nicaragua, Ecuador and Mexico, a short list of the many countries with governments and rulers seeking more "control" over citizen rights and a focus on restricting all freedoms that might challenge the power of the sitting government...The United States, a nation who believes itself to be the gold standard of freedom and democracy, re-authorized parts of it's "Patriot Act" allowing the government to spy on anyone it deems 'necessary' and recently introduced the "USA Freedom Act" which puts some restrictions on large-scale "trolling" for information and some time limits for service providers holding individual records for investigators but activities continue with little oversight...The United Nations has become ineffectual as a cooperative enclave insuring nation sovereignty and the rights of all humans on Earth with an ineffective Security Council and a membership of 195 nations who prioritize profit and territorial concerns over justice and human rights...In 2016 Syria burns, the rich get richer, the poor get poorer, environmental concerns are increasingly secondary to the pursuit of energy at any cost and a growing military presence in almost every nation is a priority over education, housing, health-care and rational, logical planning for a sustainable future on a chaotic and crowded Earth...

In a world of seven billion, how is it possible to change the patterns of behavior in individuals, governments and corporations who control the world's resources but do not have the awareness or the intelligence to understand the situational reality of a single planet with an exploding population? It's not known at this time if enough humans with intelligence and awareness exist to build a world community with policies to rescue and enhance the growth of every baby born, policies that recognize the primitive aspects of greed and corruption, policies that recognize this planet as a cyclic, interactive ecosystem that's fragile, resilient and balanced in biological, geological and climatological ways to support human existence and policies that recognize human singularity and human diversity...

We are all the same species...

Gender, race and ethnicity are variations of the same creature and every single human on Earth is created and structured in the same way...Humans are the endpoint of a complex evolutionary process that took over three billion years and we are just now realizing how incredible this reality is and the possibilities of existence in our solar system...If there are not enough humans who understand reality and can effect political change across the planet in a large enough way to affect the structure of global practices, *if climate change continues* with little or no consensus among corporations and governments affecting world pollution of the atmosphere and the oceans, *if inequality continues to grow* impacting the poor everywhere so a billion wealthy humans can enjoy excess, *if religion and myth dictate human activity* rather than science, logic, awareness and compassion or *if nothing changes*, the 1st World Civilization will crumble...Real change in the short-term (100 years?) is difficult to envision but may be determinant on one largely unknown factor: *How many people alive in this century understand the reality of human existence and recognize the negative impact and destructive policies of this emerging world civilization?* If we could know a number representative of a growing, collective intelligence with an awareness of now and if we knew of individuals willing to challenge for political change, willing to run for office and willing to chart a path for progress based on science and reason instead of religion and ideology maybe we could effect real change and make a difference, but there is little hope...

World powers, corporations and the nations that protect and support these corporate structures will continue to burn fossil fuels until there is no more supply and they will burn them without regard for consequence...It's almost certain there will be a nuclear weapon exploded in the next fifty years through design or accident as thousands of these weapons still exist on the planet...Rising sea levels and climate change will affect global food production through drought and famine as millions are displaced by environmental destruction and the changes brought about by human activity...Inequality will grow and entire populations will grow increasingly restless, angry and resistant to reforms and policies meant to control them...*It's hard to find solutions when over half of the humans alive cannot understand the problems*...Without change the first world civilization will implode under the weight of billions of desperate humans and the fears of desperate nations who will build walls and barriers on their borders in a survivalist effort to stay sovereign and stable...Military's will grow with control of water, food, energy and territory affecting the lives of billions and it's unknown how a world without hope will respond, but this is clear: The social and political structure of planet Earth in only a few hundred years will be very different from the structure we see today...

A Future Uncertain

In the 21st century, the seven billion humans alive on Earth live in a global maelstrom where the strong and rich take from the weak and poor and thousands, maybe millions of babies born every year, are malnourished, mistreated or abandoned and ignored...Millions of ignorant humans justify violence based on religion, gender, race and ethnic identity while nations and the corporations protected by those nations focus on destructive energy policies and greater wealth without understanding or caring about the negative impact on vital species or the planet...There are generally no political systems among all the nations of Earth who are not corrupted and influenced by money and power...10% of the richest humans own 85% of the world's wealth and over half of the population of Earth own less than 1% of global wealth, a growing imbalance that's not sustainable...Human civilization is an amalgamation of primitive mindsets, ancient traditions, ignorance and religious dogma coupled with historic innovation, unparalleled advances in science, technology and quantum possibilities in a volatile mix that dominate and influence human behavior from terrorist enclaves in third world countries to the boardrooms of Wall Street...As nations and individuals on Earth develop incredible technologies in medicine, industry, science and communication that promise long life, a future of limitless energy and a shared existence unimagined, millions of humans go hungry, conflicts erupt on multiple continents, nations posture for power claiming disputed territory while corporations build weapons and governments threaten violence to defend against real and imagined danger...Despite what is indisputably the most productive, creative and progressive hundred years ever known, the human species is balanced on the edge of a knife...The planet's food chains are being disrupted in oceans and on land, atmospheric pollution threatens every human on Earth, millions of refugees flee from war and hunger to countries who are reluctant to help and the world is unwilling and unable to react in any definitive way because of the deep financial ties that twist and wind through every nation on Earth with self-serving governments, private corporations and industries of war profiting from the chaos they help create...This is human reality...Many voices try daily to convince others of their version of the truth but there is only one reality...Still, there is hope...

With knowledge comes awareness and with awareness comes possibilities...The future of Earth is still undetermined but the need for intelligent action is now...The next evolutionary step for humanity will not be a random event, a genetic variation or a mutation from space...It will be an evolutionary step we initiate...A global, high priority focus on the early development of human intelligence to create an awareness where everyone, with all of our human diversity, ethnicity, culture, skills, curiosity and unlimited levels of emotional involvement and attachment, everyone understands the physical and spatial reality of human existence...And this will happen only when every human baby is cared for physically and emotionally in the first three years of life to maximize initial brain development, a global challenge that will take hundreds if not thousands of years...Intelligent life seeks to know itself and inquiry, observation and experimentation are the tools we've developed to explore our reality...Who will argue with maximizing brain growth in the babies who will be the next explorers and thinkers to determine the future of humanity? The next 1,000 years will determine if our species can use the knowledge acquired over the last 10,000 years and do what's needed to balance our habitat and survive industrial activity, ethnic diversity, overpopulation and war...Over the next thousand years it will be possible to harness the energy of the sun, recognize the limitations of human population, explore the asteroids and moons for resources and establish a world cooperation and awareness that will grow and thrive for millions of years...This is the future of the human species if there is a focus worldwide on the needs of every baby born...It's the only way to create a world civilization with an awareness of equality, a knowledge of reality and an understanding of necessity...A clear majority of humans with an intuitive awareness of the challenge of survival and the possibilities of a shared existence lasting millions of years...These are small humans who will be adults in just 20 years and what is done in the first one thousand days of their life will determine their intelligence and awareness for a lifetime...

Earth's IQ seeks a priority standard of care for all the children of Earth during the first three years of life so they are never hungry without someone to feed them, never lonely without someone to hold them, never hurt without someone to fix them and never awake without someone aware...Every child born on Earth...The first thousand days...

AN EMERGING AWARENESS

All living organisms with a sensory system have an awareness of "something"...Instinctively or intellectually, every creature has some sensation of themselves and in most species a sense of external space extending out from the center of this awareness...For humans, mammals, reptiles and all species with a central nervous system, the center of awareness is the brain...This is the structure of life on this planet, a consistent physical organization across species with a brain located in a central cavity, eyes, ears, nose and mouth in close proximity and millions of nerve endings extending from the base of the brain to every surface point on the creature essential for recognizing their reality and surviving the external environment that surrounds them...This basic physical structure is similar for almost all species in the kingdom classification *Animalia,* a connection of life extending billions of years into the past...

The first living organisms on Earth were aware of their reality only to the extent of the sensors, filaments and appendages on the exterior of their outer surface...Some were aware of light and dark and some were not, a sensory reality extending in all directions of exceedingly small dimensions...A reality measured in inches, an extremely small *"bubble of awareness"* barely extending past the dimensions of a creature's own physical body...This was their world and it was *everything...*As life evolved and brains grew more complex these boundaries of sensory awareness grew larger depending on mobility and the ability of each species to recognize patterns in nature with an increasing neural capacity necessary to process and understand the environmental information delivered non-stop to the brain from the senses...

A chimpanzee can see the sun, the moon and the stars but has no understanding of this pattern of reality...All of it's senses are similar to our human senses but this animal, closer to humans (DNA) than any other creature on Earth, has a "bubble of awareness" that stops somewhere just beyond the top of the trees...Any concept of reality beyond that point does not exist for the chimp...Whales have an awareness that extends across thousands of miles of ocean and have shown this through migratory patterns and songs but we don't know if they are consciously aware of the southern oceans as a future destination while they are feeding in the northern oceans...Are they in some way aware of the entire boundaries of their reality or is it just instinct and habit that guide the whales from one part of the planet to another?...On Earth today, the vision, hearing, smell, touch and taste in many animals is equal to or better than the same senses that deliver information to the human brain but not one of the millions of species who share this planet come anywhere close to a human's understanding of existence...

In three billion years, life on Earth has developed from an awareness of light and dark, limited mobility and sensory filaments to a complex organism with an awareness extending *13.7 billion light years* in all directions...This is the human "bubble of awareness"...A reality that is recognized by just one species on Earth and an existence defined by human science, a process of questions, experiments and understanding that has continued since human ancestors asked what, how and why is 'fire'...Many realizations of scientific proofs were quick and obvious but other realities of human existence were believed to be one way for thousands of years until discovery and experiments proved otherwise...The flat earth, the fixed sky and the Earth centered solar system were ideas believed by most of the world only a thousand years ago...Very few humans believe these ideas today, just ten centuries later...

Imagine biological life on a single planet evolving over billions of years into a self-aware sentient species who recognized the structure of their existence, built the necessary tools and recorded the orbits of the stars circling their central galactic black hole...This is an incredible achievement for a creature only 2 meters tall who discovered the properties of electromagnetism less than 200 years ago...It is just the beginning of human possibilities and the limits of what humans can discover and develop is unimaginable...We don't yet know the capabilities of the human brain...We don't yet know the limits of reality...How clever, intelligent and resourceful our species needed to be over thousands of years to reach a time where it is possible to create a record of stellar movement at a distance of 27,000 light years to prove the existence of one of the strangest, least understood objects in the universe...

Physical Reality

Every human alive shares with all other humans an awareness of our basic three-dimensional reality...This awareness is what defines existence...All of us exist within similar developmental and organizational structures (bodies) essential for survival that let us experience our singular existence on this world...These biological bodies, a basic design and structure that is similar for all animals on Earth, have developed over millions of years as a protective carrier for the brain, the who of who you are...Imagine 7.4 billion brains without their bodies moving around on the planet, each one a three-pound mass of neurons sitting on top of a three-foot spinal column with millions of nerve strands and endings sensing and experiencing reality with just five ways to understand the nature of existence: sight, sound, touch, taste and smell...If a brain loses one or two of these senses it can be challenging to survive...If a brain loses all sensory input it is essentially dead...*This is life*...An intelligence, awareness and instinct developed over millions of years to survive and understand the environment we exist in and the reality of everything humans experience from their singular three-dimensional viewpoint on Earth...A primitive species with extraordinary technology...

TIME: 2017CE...10,000 years since the beginning of recorded human history...One million years since the controlled use of fire by human ancestors...65 million years since the last mass extinction...230 million years since Earth's greatest mass extinction...3 billion years since the first life appeared in the oceans...5 billion years since the birth of the Sun and the formation of the planets and 13.7 billion years since the inflationary event that created the Universe...2017CE...170 years since electricity was controlled in a meaningful way...157 years since oil was discovered in America...87 years since the large scale structure of the Universe was realized...60 years since the launch of the first space satellite and the beginning of the 1st World Civilization...Current Reality: Seven billion humans living and fighting over ideology, religion, land and resources...Time: 2017CE...Situation: Critical...

Humans are a single species existing on the surface of a large, solid spherical object called Earth, circling a huge sphere of thermonuclear fire called Sun...A sentient species of self-aware creatures who exist on the surface of Earth with many different types of plant and animal life, carbon-based life that is everywhere on this world...In 2017 there are more than 7,427,223,753 humans on Earth, a single "planet" that is part of a debris field extending out from the Sun (the center of local navigational space) to a distance of more than one light year...Six trillion miles (9.5 trillion kilometers) of rocks, comets, dust and gas circling the Sun with Earth and seven other planets...Three solid planets, Mercury, Venus, Mars and four large organized concentrations of gas, Jupiter, Saturn, Neptune and Uranus, giant gas planets at a greater distance from the Sun than the four rocky planets...Circling these planets are more than 170 moons, many of them explorable with resources that could sustain and power the human species for millions of years and between the fourth and fifth planet exist a grouping of millions of rocks (asteroid belt) with unknown resources...Our entire star system is accelerating through an expanding emptiness of unknown size and origin we call the Universe...

Earth, the third planet from the Sun, has one moon, Luna, the fifth largest moon among all the planets and the only extraterrestrial surface humans have visited...There are millions of objects in this debris field leftover from the formation of the Solar System, dust, comets and very distant objects not yet discovered (some maybe bigger than Earth) orbiting the Sun in a swirling, circular, mathematically predictable pattern that continues without cessation...Today, with increasingly complex tools, science, and the growth of technology, the speed and direction of anything our species can observe and measure can be predicted with increasing accuracy...In this definable reality, Earth has made over 4,400,000,000 orbits of the Sun, a middle-age star...*This predictability of motion and behavior means our Sun will remain stable for more than a billion years before this star begins to run out of fuel and Earth will no longer support life*...This is a very long time for a species less than a million years old...An existence unimaginable...A reality with many possibilities and a reality with many dangers...

If there is a vast intelligence within the boundaries of human awareness, it's unknown...If there's a vast intelligence *outside* the boundaries of observation and awareness, it's unknowable...*What exists on the other side of the "Big Bang"*? Today this is a question without an answer...A billion years from now it will still be a question without an answer...Within our observable, measurable reality there are no gods, but what may be possible is mostly unknown...The possibility for life elsewhere is staggering with the physical laws in the universe consistent everywhere and the same atoms and molecules that create everything on Earth abundant in the enormous gas clouds forming stars and solar systems...It's clear the same processes that created this solar system and supplied the needed ingredients for life are going on everywhere times a billion and it's almost certain intelligent life exists somewhere else in this universe but the human race is young and just recently aware of possibilities...*If we were to detect an alien transmission in the next million years, it would be soon*...The history of our species has only just begun and the challenge of survival is real, but after millions of years of evolutionary development we are the dominant intelligence that now inhabits Earth...The modern human...

Early Greeks including Pythagoras and Aristotle believed the Earth was round in 5ᵗʰ century BC and Eratosthenes of Cyrene, a librarian at the Ancient Library of Alexandria, calculated Earth's circumference around 240 BC...

Emerging in Africa only 200,000 years ago, humans (Homo sapiens) have been recording their existence for over 6,000 years with writing and far longer (30,000 years) in color cave paintings...The total population of Earth remained less than 10 million until 12,000 years ago and only 400 years ago, Earth's population was just 600 million people...*In the last 300 years, human population has increased by more than 6 billion* with an increase of four billion people in just the last 50 years and the global effects of this population growth are evident everywhere...Wars and conflict exist all over the planet in places with boundaries, military's and societies living with different ideologies...Most of the resources on this planet are controlled by a small percentage of people who own large corporate structures, vast wealth and influence governments with direct control over military's and security...Human females, once more numerous than the men on Earth, are systematically, historically and routinely dominated and abused across many known boundaries and cultures despite scientific evidence proving the female human and the male human are essentially identical...The ignorance fueling this discrimination will disappear one day as a growing awareness of the necessity of initial brain growth slowly increases intelligence and changes abhorrent behavior but that future is not today...Although humans control many things that could reverse ignorant policies dictating energy use, reorder a global civilization to eliminate poverty and recognize the essential actions necessary to manage a world, the human species is still young and still primitive...In the 21st century, war and conflict rage across the planet on multiple continents, millions of babies are born neglected and helpless in very negative environments and the humans who are unable to understand reality, possibly 80% of Earth's population, still worship myths and legends, pollute the environment on a global scale, destroy species vital to our continued existence and kill each other for ideas, power and resources...

Earth orbits the Sun at an average speed of 67,108 miles per hour...

Scientific Reality

It's been just 328 years since Issac Newton published the most complete explanation of reality known until that date, *The Principia*, a book that brought to focus the scientific progress of a thousand years in a structured explanation of human reality in physical space and time...Issac Newton explained *how* the Earth moved in three-dimensional space and the collective human understanding of reality took an enormous step forward...For the first time, *less than 400 years ago*, humans began to know the structure of their solar system and the reality of their physical existence, finally understanding in new ways the science that explained the mysteries of existence...The world people once believed was flat and endless was a sphere circling an enormous ball of fire with a regularity that was predictable in very precise ways...The "lights in the sky" whose motions were recorded for thousands of years were also spheres that orbited the Sun with the Earth...And the Moon, mysterious, frightening, and inspiring for thousands of years was just a large rock circling the Earth in a measurable, predictable way...These realizations had been the subject of speculation, observation, experimentation and discovery for a long time and various serious "proofs" of this observable reality were published in the 200 years before *The Principia,* but Issac Newton put it together with explanation...Newton proved with laws, mathematics, observation and experiment the *predictability* of our physical environment and the reality of human existence...After thousands of years of questions, humans began to know...

The situational reality of Earth's physical existence will remain unchanged for millions of years...

In the three centuries since Newton, a continuous progression of science and inquiry has shown beyond doubt the structure, the size and the substance of everything humans are a part of...The size of our three-dimensional reality is measurable and extends in all directions billions of light years...The chemistry and physics that determine our existence appear to be the same everywhere in the universe and the time-scale we exist in is determinant and predictable...The fractured history of humanity and the development of different species on this planet is knowable and evolution is now understood and studied everywhere on Earth...Humans alive today exist at the beginning of the 1st World Civilization, a time of technology, knowledge and awareness that define the reality of human existence and our collective scientific understanding of all things has grown more in the last 300 years than in the previous ten thousand years...We know *where* we are relative to everything else within our observable reality...We finally know *what* we are, combinations of elements (atoms) made inside stars and found everywhere we look in the universe...We know *how* we are, composite carbon based structures grown from a microscopic size based upon a DNA blueprint that determines the growth of every living thing on Earth but we don't know *why* we are because the question *why do humans exist?* is the same question as *why does the universe exist?* It's complicated...The origins of life are unknown and the why of life is still a mystery with the physical laws, elements and complexity required for biological life to exist, but intelligence seeks to know itself and defining human reality defines the intelligence of the human species...Until the last human dies, millions of years from now or maybe billions of years from now, we will continue to ask these questions and will continue to search for the answers...

What we do know is every human, every animal, every plant, everything that is life exists on a single sphere circling a ball of energy that is neither solid or gas...We know all life has similar origins and we know that all members of the human species are essentially the same...No matter what gender, no matter what color, no matter what ethnicity, no matter where born...*All humans are the same in the most elemental and fundamental way*...This is a crucial realization...This is a reality of life...This is science...So the question is: *Why do billions of people not understand this?*

Why are billions of human beings with essentially the same capacity for intellectual awareness unable to understand our three-dimensional reality and the nature of human existence in an intuitive and cognitive way?

Thirty-six years ago Carl Sagan presented *Cosmos,* a episodic series that was the most complete summation of reality ever seen...Eloquent, accurate and factual, it's estimated over 800 million people have watched this scientific exposition of human existence yet little in the world has changed...Why didn't 800 million humans, world leaders as well as any citizen of Earth, stop and recognize the reality of our collective existence and the balance of life and resources on a single world with a closed, cyclic ecosystem? Why didn't war, pollution, killing, hate and greed stop? Why are over six billion people unable to understand a proven, definable reality and unable to understand the needs and possibilities of the human species? In this 21st century, in a new normal of information technology and worldwide distribution, there are literally thousands of documentaries watched by millions of people that explain in exquisite detail the reality of life on Earth and the balance of resources on this one planet that allows sentient life to exist...Vast libraries explain the science and technology that verify human discovery and continue to confirm processes measurable in very precise ways that expose the ignorance and dogma of thousands of generations...*Still there are billions of people on Earth who cannot recognize, understand or accept what is obvious*...Many are intelligent, knowledgeable and skilled in multiple disciplines and they are still unable to assimilate, comprehend and understand human reality...Why? The answer is stunning and the impact on the human species is undeniable...The reason billions of people on Earth today do not understand reality and the nature of their own existence is because they *can't*...

Reality Intelligence

Reality Intelligence is a natural property of the brain that develops in the first three years of life determined by the density of neural growth connecting the internal structure of the brain...These dense neural pathways and the initial development of organizational structure are critical in determining a brain's ability to communicate with itself, a fluid and comprehensive communication necessary for an intuitive and cognitive understanding of reality...

Reality Intelligence in the simplest terms is an intuitive and cognitive awareness of the nature and reality of human existence...It is a level of intelligence and awareness determined by the connective neural density in the networks and organizational structure of the brain which develop primarily during the first 1,000 days of life, a density of growth that increases the fluidity and volume of communication within the brain and merges the rational, logical and most recently evolved regions of the brain with the primitive and ancient regions of the brain that communicate fear, anxiety, survival, hunger, longing, sex and aggression...Human science is not advanced enough to count individual neurons and connections (synapses) to understand this complex interaction in a direct way, but there are volumes of studies that show initial brain growth in infants is enhanced in a safe environment of secure attachment, good nutrition, abundant sleep and positive stimulation...Consequently, early brain growth and development is inhibited in infants who must survive in an environment of violence, fear, abandonment, neglect, indifference, brutality, insecurity and hunger, especially in the first year of life...What has not yet been realized or understood is *how critical this initial stage of brain growth is in determining the awareness and intelligence of a human being over a lifetime*, a permanent connective communication network within the brain that is a fundamental part of human intelligence and may be the determinant factor in the future of the human species...(Reality Intelligence is discussed in depth in Part 2)

The ability of every individual to understand human reality is determined in the first years of life...

It is a deep intelligence, a subconscious property of the brain determined by the development of permanent structure (essentially the internal wiring of the brain) in the first months and years after birth...When memory networks are fully formed around the age of 3, they are 'permanent' in a way that enables people to remember events from childhood in old age...Reality intelligence is much the same thing...A permanent internal structure of connective neural density, a framework influencing behavior and thought for the entirety of an individual's existence that enables the brain to communicate in a fluid way with all of it's differing regions, a diverse and dense network of internal communication that determines morality, empathy and intellectual awareness...It is essentially a feedback loop between the **cerebrum** (**cerebral cortex** and the **frontal lobe**, the processor of awareness where all decisions and all analysis take place) and the primitive but crucial brain structure that evolved over millions of years to react to external stimuli and activate reactionary survival instincts critical in the successful evolution of our species in dangerous and challenging environments...The cumulative effect of maximizing brain growth during infancy is an intuitive and cognitive awareness of human existence when mature and a natural morality and intuitive empathy that accepts the mystery of *'why we exist'* while recognizing that all humans are fundamentally the same in a reality we've only begun to understand and explore..

There are significant thresholds of neural growth and structure in every baby's brain that are critical in an individual's lifetime understanding of existence with the density of neural growth and structure primarily determined by how babies experience this early growth period in a world where they are helpless and vulnerable...

There are four essential areas of focus in the first three years that are crucial in maximizing brain growth...*Secure Attachment* = Love, nurturance, touching, holding and a feeling of security and safety...*Nutrition* = Adequate nutrients and calories that are critical for a developing brain...*Sleep* = As much as they want...*Positive Stimulation* = Play, positive experiences and no physical discipline...

Reality Intelligence = A deep permanent connective state of awareness...

It takes awareness and cooperation among all adults that interact with these small humans to help create a positive environment that will enhance neural growth and create an individual who will understand reality, morality and empathy but it's a difference that lasts a lifetime...An intuitive morality and complex understanding of the nature of life is inherent in children who are nurtured this way in infancy...A morality, an intelligence and an understanding of very basic realities that are the foundation of consciousness for a lifetime...An awareness of actuality that develops as we age and experience our life...An awareness that humans don't damage, abuse or kill other humans...An understanding of the balance of everything in our physical environment with a recognition and a curiosity of the patterns of life on this planet that enable different species to live in a finite, shared environment...Babies raised this way will grow up with an understanding of the balance and diversity of life on a single planet and an awareness of the cooperation necessary to create a civilization lasting millions of years...*If the brain does not reach this critical threshold of neural growth in the first three years due to malnutrition, abuse, severe negative life experiences or simply neglect and indifference, the individual will never recognize or understand the reality of their own existence, no matter how much proof, no matter how many facts, no matter how good the science, no matter how sound and logical the argument...*
This defines the majority of Earth's population in the 21ˢᵗ century...

The Future

What will be the nature of human civilization in a million years? A picture of the Andromeda galaxy photographed in infrared wavelengths in 2010 will be famous, familiar, historical in 10,000 years, in 100,000 years, even in 1,000,000 years and the galaxy will still be generally where it is now...A picture taken in the year 1,002,010 will look very much the same as the picture taken in 2010...Endless time in a vast emptiness...All science and observation envision a stable universe that will evolve and develop for billions of years and our planet exists in a highly predictable mathematical interplay with every other object in the solar system that will continue for the next billion years as it has for the last billion years...With intelligence, awareness and a little bit of luck, human civilization should last for many millions of years...We are young...In a million years humans will have explored all the moons in this solar system, sent robotic probes to the nearest stars and maybe even sent people on generational ships to other planets circling other stars...Humans willing to leave everything known and familiar to explore and inhabit a new star system...In another thousand years we might hear back from them...And when the Sun begins to change over a billion years from now, our species will identify survivable star systems, organize in large complex spaceships and establish civilizations around young distant G-class stars or even red dwarfs...*These possibilities are real...*We are the end product of the evolution of intelligence on Earth and have reached a stage of awareness where we can understand and control our reality...*No other dominant creature will evolve on this planet as long as humans exist...*This is our world and this is our solar system, with all the resources, all the challenges and all the rewards...

When asked the question "How long will humans exist?", many people predict the end of our species in only thousands of years, an indication of how unaware most people are to the possibilities of a dominant species...There is evidence of species in the evolutionary history of Earth who survived for thousands, even millions of years and we are the most intelligent species yet to evolve...It's conceivable the human animal as an individual and as a species has the physical capacity, the intelligence and the instinct to survive nearly any man-made disaster, biological, technological or environmental...Seven billion people with the collective knowledge to understand and survive nuclear war, biological agents, geological upheaval, environmental catastrophe and begin to build civilization again but we exist in an indifferent universe and there are possibilities which could exterminate all life on Earth...One is a cascading runaway climate event turning the Earth into Venus...It would take ignorant irresponsibility on a planet-wide scale to initiate such an event and let it continue in an uncontrollable way but it's unknown if there is a climatological "tipping point" where uncontrolled extreme conditions on Earth could become a catalyst for our destruction...Other possibilities in a violent universe include extremely rare events like runaway stars or gamma ray bursts capable of exterminating an entire planet, scenarios so rare and uncontrollable it's senseless to worry about such things, but there are very real dangers with almost universal agreement that the biggest "doomsday" threat to the human species is an impact event from a large object in this solar system...There is a long history of these catastrophic events and without human intervention it will happen again...

It's sobering to understand the scale of reality and realize how small, how special, how durable and how vulnerable Earth is...Humans exist because this star system resides in a quiet part of the galaxy and has a large planet (Jupiter) scientists believe is crucial in diverting impact asteroids and comets from possible collisions with Earth...There are millions of comets and asteroids moving in the debris field orbiting the Sun and many of these large bodies may pass through the inner solar system just once every 20,000 years...We've had the telescope for 400 years...To watch, identify, predict and interact with space debris threatening this world will be the biggest challenge facing the human race for the rest of our existence...Earth has a history of impact events causing global changes that can affect every process on the planet and scientists are still developing the technology to protect Earth from an impact event...It's a race against time...Geological and fossil records prove the random and catastrophic planet changing impacts from the past and it's a certainty Earth will face a huge threat in the future, maybe in 10 million years or maybe just 200 years...*For an intelligent species who understand the reality they exist in, setting up a defense network to monitor their space environment for millions of years would be the highest priority*...The human species is unique, resilient, intelligent, curious and adaptable and it would take a large unforeseeable event or series of events to exterminate our entire species...If we can increase global intelligence (Earth's IQ) to where a majority of humans understand this reality and the consequence of ignorance and inaction, a continued existence for millions of years is possible...

Maximizing initial brain growth in all babies will be the next evolutionary step for the human species...

The Reality of Now

The challenge of change in a world where a majority of people cannot understand reality or the possibilities of a peaceful co-existence is extreme and it's increasingly clear from the chaotic situational realities defining this world with multiple wars, murder on a global scale, increasing suicides, world poverty, widespread pollution and a growing ignorance of necessity, it's clear a majority of Earth's population in the 21st century will never accept the scientific explanations that define human existence, no matter how compelling the evidence or logical and sound the argument...*The amazing thing is it's not their fault*...Infants have little control over events or environmental impacts during their first 1,000 days of life and the critical development of dense neural pathways that determine their intelligence and awareness are dependent on how they experience and survive these first years of life...Sadly, in this primitive 'civilized' complex global society, a world of extreme wealth with a growing inequality that is stunning in it's ignorance of causality, even as new technology and new research is proving daily how important the beginning years of life are, there are millions of babies and children who continue to exist in terrible conditions of poverty and abuse, malnutrition, hopelessness, conflict and life situations over which they have no control...Even in 'developed nations' there are millions of babies born healthy in an endless variety of environmental situations by generations of adults who feed and clothe them but attach no importance to this critical period of brain development...Many caretakers believe intellectual development begins at age four or five in formal education settings with no emphasis or understanding of the first years that are crucial in developing awareness and intelligence...And the cultural norms, instructions, manuals, habits, the "do's" and the "don'ts" and the long-held beliefs passed down through multiple generations dictating how societies and individuals care for a human baby in the first three years of life are a menagerie of enlightened realizations and horror stories with 'daily beatings to build strength and character', 'isolation to build individuality' 'binding so they don't crawl like animals' and in many cultures a vastly different treatment of the male and female throughout a lifetime...

Human science has yet to develop techniques that allow an accurate count of neural density in individual brains and ethical considerations do not allow restricting nutrients, necessary sleep and stimulation to compare small infants for cause and effect, but there is abundant, compelling evidence confirming this process and one day science will map the human brain in exquisite detail and develop the tools to count individual neurons, identify important connective structure and begin to understand the causality relationship that defines intellectual processing from the earliest stages of life...This is not possible in this century but there is overwhelming evidence that implicates the first months and years of life in patterns of thought and activity that determine behavior over a lifetime...The signature of every serial killer known is abandonment, violence and extreme neglect in the first months after birth that can profoundly affect a life in a very short time...*It takes just weeks or months of extreme situational distress in infancy to negatively affect a human for a lifetime with a subconscious fear and confusion that never ends and a lack of empathy and understanding toward any human in any situation...*

Arguably, the most critical determinant in this equation is nutrition and there must be a global priority on adequate nutrition and essential nutrients for every child on Earth...Millions of children and babies in Africa and developing areas of the world go hungry daily and even in highly developed countries millions are undernourished due to political stalemate, indifferent polices and broken safety nets...In a 21st century with vast global wealth and technological wonders, a third of all young children in North Korea show signs of stunted growth due to food shortages and in some nations in sub-Saharan Africa it's estimated more than 50% of children under age five suffer chronic malnutrition in cultures, nations and societies where young children are secondary and expendable...What's invisible is the under-developed brains of these children that will never recover and will result in a population of survivors unable to understand reality, survivors easily influenced by others with extreme purposes and survivors who will create another generation of humans without an awareness of existence...

Even something many people believe to be harmless, *allowing a baby to cry itself to sleep,* can inhibit neural growth if done repeatedly in the first years of life...Four year old children understand why they are being left alone to go to sleep...An infant does not understand when they are left alone and no human responds to their cries...

*What is being understood as never before is the **permanency** of initial neural growth in the first three years that essentially determines the brain's ability to communicate with itself and understand existence for a lifetime...*

The care of every baby born and the development of *reality intelligence* in the first three years of life is the key to the future of the human species and is discussed in the second part of the book which speaks to the importance of early brain growth and development...First, the reality of human existence and the reality of planet Earth in the 21st century...

"I believe our future depends powerfully on how well we understand this Cosmos"...Carl Sagan

PLANET EARTH

WORLD ORGANIZATIONS

There is no World Government at this time in human history

The land mass of Earth is divided into over 195 territorial divisions designated as countries or nations with the boundaries, governing structure, population and history of each country the result of discovery, exploration, war, migration, cooperation, environmental forces and human events over the last 10,000 years of recorded history...Some countries and boundaries have existed for thousands of years and other countries have just recently been established...At this time in human history all the land on Earth is recognized, measured and managed by humans in some way...

ETHNICITY AND RACE
Gender, racial and ethnic discrimination are a transitional phase of a developing intelligence...

There are volumes of books, multiple theories and explanations defining race and ethnicity, its origins and its impact on human populations and they are all right in differing ways...The complexity of human development over the past 200,000 years while settling and exploring this planet is staggering and origins of ethnicity and of different races on Earth are simply variations of population growth over a long history as large and small groups of humans migrating across regions of Earth remained isolated for long periods of time due to a relatively low world population and the geographical barriers of water and land masses...These isolated populations (who existed over thousands of years with little contact or knowledge of other groups of humans) created hundred's of diverse civilizations with unique identities and a recognizable, undeniable human bond of "sameness", evidence that race and ethnicity (maybe the most divisive issues on Earth in this century) are a consequence of migration and dispersion at every level imaginable with multiple variations of ancestry, religion, history, language, nationality, physical appearance, cultural norms and a strong sense of identity with the humans in one's "tribe"...

Today, with a population of 7.4 billion and an incredible diversity that maybe was essential for the expansion and survival of our species in an unknown and hostile environment, it's increasingly clear that all humans beings are fundamentally the same...No matter what color, no matter what gender, no matter variations in physical features, no matter where one is born or what they believe, all humans are the same species...Still, governments everywhere on Earth define broad ethnic categories and many nations have strict separatist laws, different standards of inclusion and requirements determined solely by one's ethnicity and race...In an intelligent world with a global awareness of reality, ethnic variations would be recognized in the celebration and remembrance of cultures and histories which have defined the emergence of the human species in a myriad of different ways, valuable interesting histories of the differing realities experienced by multiple human societies while acquiring the knowledge necessary to create a global society but today there are still billions of people on this planet unable to understand what is proven...Any consideration of skin color as an indicator of awareness and intellect is primitive and ignorant with no scientific evidence that pigment variations determine intellectual ability...And any person who believes females are inferior to males in any way beyond physical dimensions are simply unable to understand a known reality...In a convergence of humanity unprecedented until now, every human is a member of an integrated global civilization that will continue to merge for thousands of years where it's possible, in some very distance future, everyone will be the same color with racial and ethnic profiling an ancient history of a species search for intellectual maturity...

With the ability in this century to photograph, measure and recognize this world from space and with the technology to monitor Earth in real time, all nations can now recognize and understand the boundaries of all nations...And despite the boundary disputes and ideological and political differences between countries, there is a sharing of trade and commerce across the entire world for the first time in human existence with an ability to communicate and participate in a global collection of historical divisions that all represent the intelligence and awareness of a single species...

In the 21st century there are over 23,000 world or intergovernmental organizations on Earth with varying degrees of inclusion, effectiveness and visibility...Many are subsets of bigger organizations and many are focused on financial prosperity and international support, existing just to exist...There is nothing that can be envisioned as a foundation for world governance with 195 individual nations firmly focused on territorial sovereignty and a national independence recognized by almost all nations to a degree where sovereign nations and governments will sit idle while illegitimate leaders and ruthless dictators murder their own citizens and steal their resources...This mindset grows stronger in this century and is responsible for thousands of innocent humans dying needlessly and millions of children and infants trying to survive in desperate ways while global financial inter-connectivity and corporate protections prevent multiple nations from stopping atrocities that grow easier for them to ignore...It is increasingly difficult to wage war on another nation and there are few protests and little intervention if a campaign of terror and violence is unleashed within the borders of a "sovereign" nation...

UNITED NATIONS

Currently 193 nations are recognized as members of this organization...Founded in 1945 by 51 countries after a world war killed over 60 million humans, the UN was an intellectual and emotional reaction by humans to a world in chaos...The obvious mission of the UN is to prevent another world war and to accomplish that through a variety of cooperative services and commitments among the nations of the world concerning social progress, living standards, progressive human rights, sustainable development, environmental management, disaster relief, disarmament, non-proliferation of nuclear weapons, gender equality, women rights, governance, economic development, international health, landmine removal, food production, etc, and this organization is currently recognized and accepted in almost every nation on Earth...If resources from every nation were focused and used efficiently the UN could change civilization and in a distant future an organization like this will likely evolve into the 1st World Government, but in this century the UN is a consortium of human diplomats representing world populations and within this organization multiple members of the General Assembly do not understand the reality of existence, the morality of a sentient species, essential human rights, gender equality or the needs of millions of humans across the planet...The UN does a lot of reactionary good in many parts of the world but the true potential of an organization that includes membership from almost every nation on Earth will not be realized until Earth's IQ is increased and a clear majority of global participants understand human reality...Until that time, territorial conflicts, monetary demands and a hierarchical structure that allows only five permanent members on the Security Council to essentially determine all reactionary procedure will continue to render the United Nations ineffective in dealing with growing violence and human rights abuses worldwide...

The Security Council's five permanent members were the five largest military powers on Earth at the end of World War II and are the largest military powers on the planet in 2016...United States, China, Russia, Britain and France...With ten countries as rotating members, the Security Council is the most powerful organization within the UN...There is pressure for the UN to give Africa a permanent seat on the Security Council given that Africa countries make up more than 25% of the General Assembly and is the largest continent on the planet while Germany, Japan and Brazil also lobby for permanent seats on the council...

On August 19, 2013 the UN 'celebrates' World Humanitarian Day while millions of children go hungry worldwide and continued fighting in Syria displaces over two million humans...In the excerpt below of just one of Secretary General Ban Ki-moon's Generational Imperatives and Opportunities listed under The Secretary-General's Five-Year Action Agenda, *the goals set forth by the UN are lofty and ambitious and expressed in very generalized language that has little specificity as to exactly what will be done and when...*With 193 nations focused on their own territories, populations and future resource allocation, each goal on this list creates it's own timeline and funding challenges with little cooperation in this organization to effectively address the needs of a crowded planet...

I. SUSTAINABLE DEVELOPMENT

1. Accelerate progress on the Millennium Development Goals:

•Keep the world solidly on track to meet poverty reduction targets focusing on inequalities, making particular efforts in countries with special needs and in those which have not achieved sufficient progress.

•Complete the final drive to eliminate by 2015 deaths from top killers: malaria; polio; new paediatric HIV infections; maternal and neonatal tetanus; and measles.

•Fully implement the global strategy on women and children's health to save tens of millions of lives, including through the provision of reproductive health services to meet unmet global needs.

•Unlock the potential of current and future generations by putting an end to the hidden tragedy of stunting of almost 200 million children by mobilizing financial, human and political resources commensurate with the challenge.

•Stimulate generational progress by catalyzing a global movement to achieve quality, relevant and universal education for the twenty-first century.

2. Address climate change:

•Facilitate mitigation and adaptation action on the ground:

  •Promote climate financing by operationalizing the Green Climate Fund and set public and private funds on a trajectory to reach the agreed amount of $100 billion by 2020. Ensure effective delivery of all fast-start financing. Deepen understanding of the economic costs of climate change, and the corresponding financing needs, including through mapping regional and sub-regional vulnerability hotspots.

  •Facilitate and execute agreements on reducing emissions from deforestation and forest degradation (REDD+) to protect forests and sustain the livelihoods of the people who depend on them.

•By 2015, secure a comprehensive climate change agreement applicable to all parties with legal force under the United Nations Framework Convention on Climate Change.

•Strengthen, defend and use climate science to make and promote evidence-based policy.

3. Forge consensus around a post-2015 sustainable development framework and implement it:

•Define a new generation of sustainable development goals building on the MDGs and outline a road map for consideration by Member States.

•Mobilize the UN system to support global, regional and national strategies to address the building blocks of sustainable development:

  •Energy: Mobilize a broad multi-stakeholder coalition under the Sustainable Energy for All initiative to achieve universal access to modern energy services, double the rate of improvement in energy efficiency and double the share of renewable energy in the global energy mix, all by 2030.

  •Food and nutrition: Adopt globally agreed goals for food and nutrition security, mobilize all key stakeholders to provide support to smallholder farmers and food processors and bolster the resilience of communities and nations experiencing periodic food crises.

•Water: Launch and execute a UN-wide initiative to provide universal access to safe drinking water and adequate sanitation globally.

•Oceans: Agree to a compact on oceans that will address overfishing and pollution by improving the governance of oceans and coastal habitats and by developing an institutional and legal framework for the protection of ocean biodiversity.

•Transport: Convene aviation, marine, ferry, rail, road and urban public transport providers, along with Governments and investors, to develop and take action on recommendations for more sustainable transport systems that can address rising congestion and pollution worldwide, particularly in urban areas.

•Work with Member States to make Antarctica a world nature preserve.

Deepen the UN campaign to end violence against women by enhancing support for countries to adopt legislation that criminalizes violence against women and provides reparations and remedies to victims, provide women with access to justice and pursue and prosecute perpetrators of violence against women...*(This item simply asks countries to be nicer to females and will change nothing in the many nations where women are treated like second class citizens with abusive males, no access to education and no standing of any kind)...*

The UN has 15 peacekeeping operations worldwide using over 116,000 troops in situations of violence or potential violence...Resources for these operations total $7.54 billion as of July 2013...They have maintained a presence in the Western Sahara since 1991, Darfur since 2007, Syria since 1974, Cyrus since 1964, Lebanon since 1978, Sudan since 2011, Republic of Côte d'Ivoire since 2004, Kosovo since 1999, Liberia since 2003, India & Pakistan since 1949, Middle East since 1948, Mali since 2013, Haiti since 2004, and the Democratic Republic of Congo since 2010...Total fatalities for UN peacekeepers is 3,136...Over half of these operations are in Africa...

UN WORLD FOOD PROGRAMME

In 2015 the WFP received $5,058,450,351 to feed over 100 million hungry humans in a variety of locations...Nations worldwide donate funds for this program with the U.S. contributing almost three times the amount of any other donor nation and the top 25 donors being some of the richest nations on the planet...China, one of the most prosperous and successful economies in this century, has pledged around $10 million in 2015 while the U.S. pledged over $2 billion...The food distribution attempted by this organization is essential in a world with a growing population and a changing climate despite the inefficiencies, corruption and crime that is inherent in any world program with global ambitions...

G8

The G-8 is a consortium of the eight richest countries on Earth: United States, Russia, Japan, Italy, Germany, France, Canada, United Kingdom and the entire European Union...This is supposed to be a private, informal meeting among the leaders of the most influential nations of the world to discuss problems and solutions...The intention is an examination of the world's most pressing problems without aides and advisors present in an effort to agree on effective policies and procedures to address these growing problems...Begun in 1973, each nation hosts the forum on a yearly rotating schedule and the group releases a report at the end of these discussions setting new policies that no nation is required to implement...In the June 2013 meeting Syria was discussed along with tax evasion, trade policies and government transparency...No timetables or declarations...Nice hotels and good food...

The G8 in 2015 was the G7 as Russia is uninvited due to actions in Ukraine...The Russian minister dismissed the decision as "insignificant" saying that almost all subjects to be discussed were addressed in other forums...

G20

The G-20 is a consortium of finance ministers and bankers from the 20 largest economies on Earth to discuss policy issues concerning the international financial system...Begun in 1999, the G-20 has no formal ability to set policy or make rules and faces continued criticism of the validity of it's declarations and conclusions...They meet bi-annually with the host being one of four nations within one of five groups...Members include South Africa, United States, Australia, Germany, France, Italy, United Kingdom, Saudi Arabia, China, Japan, South Korea, India, Russia, Indonesia, Turkey, Brazil, European Union, Argentina, Mexico and Canada...Also included in the meetings are the president of the World Bank, the chairman of the Development Assistance Committee, the International Monetary and Financial Committee and the managing director and Chairman of the IMF (International Monetary Fund)...The G-20 is a self appointed group without transparency, accountability or any kind of formal charter who call themselves "the premier forum for international cooperation on the most important issues of the global economic and financial agenda"...The summits always draw large crowds of protesters and criticism of it's stated goals to "promote policy coordination between its members in order to achieve global economic stability, sustainable balanced growth and to promote financial regulations that reduce risks and prevent future financial crises...*This has not happened...*

NYT's reported that 2016's G-20 wrapped up after 2 days with a rough consensus: "The markets worry too much"

> *The incident with Edward Snowden seeking asylum in Russia highlights the immaturity of a planet...In the posturing before a planned meeting between Obama of the U.S. and Putin of Russia prior to the 2013 G-20, many politicians encouraged Obama to cancel his meeting in protest with the American press reporting the two will have to "hold their noses in the same room" and there are "too many problems" so a meeting would be insignificant...They act like two small children in a schoolyard instead of two sentient creatures on a world with two large populations and enormous weapon stockpiles...They should meet precisely because of many problems and this immature posturing by leaders in the United States Congress is counterproductive, scary, silly and sad...Obama could have been so much more if he would have met Putin knowing the problems of the world are so beyond triviality and the actions of a single man (Snowden) making transparent what would already be transparent in an intelligent civilization...What these sad men in Congress were worried about is that "Putin would look big" and "Obama needy"...Very childish...*

NATO (The North Atlantic Treaty Organization)

Headquartered in Brussels, Belgian, NATO was created in 1949 after WW II to provide for a United European security against a growing Soviet threat and as a deterrent to the rise of nationalistic militarism across the European continent which allowed the rise of the Nazi party and a war involving most of the world..."NATO's essential purpose is to safeguard the freedom and security of its members through political and military means" (NATO)...Currently focused on a continuing expansion, NATO is cooperating to build a missile defense system throughout Europe as a countenance to Iranian threats guided by a Strategic Concept adopted in 2010 that embraces a policy of doing whatever they believe is necessary in any situation using diplomacy, military force and post-conflict stabilization...The alliance was a primary force in Afghanistan for protection and training and NATO was a major player in the fall of Libya and Muammar Qaddafi in 2011..Essentially a security alliance with a challenging political structure that has member states worried about future resource allocation, leadership responsibilities and mission uncertainty in a new global strategy demanded by a shrinking world...The process by which membership is offered to other nations can be problematic with regional loyalties and economic dependence taking precedent over multiple security concerns and a growing apprehension among Russia and Middle East nations concerning NATO's push into Eastern Europe...NATO is funded by 28 member states dependent on current mission priorities and administrative needs...2016's civil budget is $253,179,900 and 2016's military budget = $1,824,720,000...The alliance members collective defense budgets for 2015 totaled $892 billion with the summit talks dominated by Syria, ISIS (Islamic State of Iraq), territorial incursions, migration, aggression, weapons and troops...

THE WORLD BANK

A financial arm of the United Nations which provides money to developing countries to repair infrastructure, initiate capital projects and promote development...The World Bank loans approximately $50 billion per year with goals including financing the delivery of vaccines, preserving biodiversity by refusing to finance environmentally destructive projects, promoting education for males and females, implementing new strategies to reduce poverty and just in the last decade trying to find a way to make renewable energy cost-effective in developing nations where coal is the primary source of energy...An organization with over 10,000 employees and many offices in many regions of the world, the bank is undergoing a complete restructuring to deal with the new global reality of cost effectiveness and a shrinking budget...Another global structure with many voices influencing decisions and policies...

In 2013 the U.S. spent over $4 billion every two weeks to continue activities and a presence in Afghanistan...

MANAGING A PLANET

If you consider the enormous reach of the hundreds and thousands of organizations on Earth that seem to deal with every aspect in the management of a world it makes one wonder why or how the discord, suffering and fragmentation in nations and societies everywhere on Earth continues to grow...If you read stated goals and missions of these various agencies in concert with funding and organizational structure you soon realize how difficult it is to implement sensible policies and change in a civilization where over six billion humans cannot understand the consequence of their collective actions...The UN, an international organization with the membership of almost every nation on Earth, has a well known Children's Fund (UNICEF) yet thousands of children and infants die and suffer from chronic abuse and malnutrition...A Human Settlement Program and a Refugee Organization Program in a world with over 21.3 million international refugees at the beginning of 2016 and over 40 million refugees displaced in their own nations...There is a Joint Program on HIV/AIDS so why isn't everyone with HIV/AIDS given medication...Eradicate it...And there exists a UN Environmental Program which is "currently helping developing countries implement environmentally sound policies and practices" on a planet with more coal being burned in developing nations than ever before...Canada's determined to destroy half of their country bringing to market oil of the worst kind, recycling is ignored in almost every corner of Earth while toxic runoff and dumping in the oceans is destroying food chains that support all life on Earth...

There are multiple organizations to "protect the ocean" and some with environmentally correct names and lofty mission statements who lobby worldwide to prevent regulatory restrictions that would affect corporate profit...There is an International Organization for Migration (IOM) who are committed to the principle of "humane and orderly migration to benefit migrants and society" while the EU pleads for help as migrants and refugees tear down fences to get in...And it continues...How is it possible for a self-aware sentient species to manage a planet in a way that will provide basic securities and services for every human who requires it, design a finite world for maximum efficiency to provide a sustainable food allocation system, limitless pollution free energy, the eradication of diseases that are unnecessary and controllable with current technology and provide endless opportunities for a choice of existence in a limited lifetime...Some humans exist just to watch the sun set every day and some wish to fly to the moon...All these possibilities are real and all choices should be available...

This is a very small window into thousands of organizations on this planet that collect data on a daily basis and many offer workable solutions that get lost in the ignorant and competitive majority who are determining the future of Earth...IOM knows migrants are invading Spain and UNICEF knows of hungry and dying children on Earth, but what's important is all these organizations and technologies (GOSAT, a satellite measuring CO2 and methane from some 56,000 locations) are providing data on a daily basis modeling an accurate picture of global activity to inform Earth's population of what must be done to survive in a finite ecosystem...Humans who understand and who care are trying but territorial mindsets and ignorance of consequence continue to dominate global policy in every nation...

A non-profit organization, The World Government (1951), seeks a world where the laws and territorial boundaries of nations do not interfere with the 'right' of every human to explore this world without violence...

The problem with managing the Earth is the absence of a cooperative global coalition with the intelligence to understand the reality this data reflects or a consensual awareness to do what's necessary to operate a world in an effective, efficient way...In the future intelligent governments worldwide who understand existence and have an awareness of causality will get together and do what is needed and it won't be about money, profit, power or territorial posturing, it will be about the survival of a species and the health, safety and freedom of every human on Earth...It will be about solving real problems thousands of years in the making with technology, tools and knowledge acquired only in the last century, but not yet...In this century the majority of humans who inhabit political coalitions, influential corporations and many of the organizations whose primary focus is on accumulating wealth without an awareness or concern of the negative impacts on the ecosystem that supports their existence continue to compete in primitive ways without a connective neural density able to recognize causality, impact and consequence...Within these organizations are many who *do* recognize possibilities and realities but they are overwhelmed by the majority who don't and human history is filled with intelligent individuals and thousands of incidents where humans who insisted that "leaders" recognize obvious truths were killed or imprisoned for ideas that threatened existing power structures controlled by ignorance...

Wiring a world can be learned in a single century but improving the operational capability of a human brain, the most complex object known to exist, will take many centuries...Many of these world organizations are covers for self-serving people who want to do exactly the opposite of what their title and mission statements imply but many are genuine attempts to make Earth and human civilization a more efficient habitat, very difficult on a primitive planet of seven billion and the only way this species will survive into a distant future is a global focus on what is necessary to maximize initial brain growth in every baby...This is mentioned many times in this book...The first 1,000 days of life...Not possible in this century, the human species must make infants a priority to develop a future with a planetwide intelligence necessary to create a majority who will understand the nature of human existence and what is needed to survive...Every world organization with resources and access needs to be unflinchingly empathic to provide for babies worldwide with nutrition, stimulation, caregivers and knowledge in every culture and every society on Earth...It will take a thousand years but it's imperative we start now...Every time Gaza fires a missile at Israel, destroy the site with technology and precise targeting and then flood the Gaza strip with humanitarian aid to create new generations to change this primitive posturing...Do not punish populations for actions by those who have no ability to understand existence and don't traumatize their babies with reflexive acts of violence...Massive humanitarian aid for Syrian refugees and citizens with a inflexible stance against Al-Assad's illegitimate government...*And create a global economy that includes everyone on Earth so babies and children are not hungry and the billions who have nothing can have something so they will have a reason to take care of their world instead of destroying it in hopelessness and desperation,* changes that will not happen in this century with the political/corporate structures that dominate today...Earth is a planet of 'tribes' with the leader of each willing to destroy populations and species so their tribe can have more but this will change if there is a focus on the security, nutrition and nurturance of small humans in every culture on Earth...It must change if there is any hope of surviving this pivotal, critical stage of human development...

EARTH'S IQ
"a planetwide intelligence"
A "planetwide intelligence" will evolve when all babies on Earth are loved, cared for and stimulated during the first 1,000 days of life to create a consensual intelligence, an awareness of reality and a human technology capable of inhabiting a star system and advancing human progress for the next billion years...

OCEANS

96% of the world's oceans are affected by human activity...

The ocean covers over 70% of Earth's surface in four great basins labeled the Pacific Ocean, the Indian Ocean, the Atlantic Ocean and the Arctic Ocean, boundaries that are nonexistent conveniences for human navigation and mapping...There is only one ocean on Earth with an average depth of 4,200 meters (14,000 feet) and a continuous 40,000 mile chain of volcanoes that circle the planet identifying tectonic plate boundaries and hot spots where crust is forming at a rate of inches per year...Four billion years ago there were no oceans and liquid water was almost nonexistent on Earth, a young, hot, molten planet venting hot gases into a newly forming atmosphere of methane, carbon dioxide, water vapor and ammonia...When the gases cooled after millions of years, water vapor condensed and the rains began, dissolving minerals with salt on a geological time scale of millions of years to flood a world located in the sweet spot of the solar system, an orbital location that allows liquid water to exist...Fossil records prove life existed in the ocean around 3.7 billion years ago with the earliest known fossils of evolving land animals less than 600 million years old...Today life is everywhere on Earth and all reptiles, fish, birds, amphibians and mammals have ratios and proportions of sodium, potassium and calcium similar to those found in the ocean...Life on Earth came from the sea...

Salt is a staple of human history with records as far back as 6,000 BC proving salt an essential part of all societies...It's estimated the ocean contains millions of cubic miles of salt with a salinity of around 3.5%...

Geographically, the southern ocean circling the Antarctica continent is the only place on Earth where the ocean circles the globe with no interruption from land...It's the largest and strongest ocean current on the planet insulating Antarctica from warmer waters with strong circumpolar westerly winds that make exploration difficult and navigation treacherous but exploring Earth from the ocean is what humans do, what we've always done and in a history beyond record, transportation, fishing, commerce and adventure have made the ocean an essential part of human activity and a vital continuing source of food in a mostly unexplored environment that's being changed dramatically by human activity...

In this century our ocean is threatened as never before by a world that uses the sea with no binding global agreements regulating or balancing human activity to prevent the depletion of essential fish and no ability to control the pollution which threatens species who need the ocean to survive (It's possible that every species on Earth needs the ocean to survive)...In a modern age where humans have the science, technology and intelligence to leak proof all vessels transporting resources around the globe and the ability to monitor and control pollution from the shore, there is no global awareness of what is necessary...With technology unheard of just 50 years ago we have the data and knowledge to regulate fishing to restock protected marine areas and the ability to create a sustainable ecosystem that will prevent the collapse of a vital global food chain but there is no global consensual awareness that recognizes causality or recognizes a proven reality that clearly reflects the negative impact of current human activity...In the crowded unregulated rush for 'more' of everything, the motivation of large corporations and governments who control most of the world resources is profit, expansion and growth and the only hope is that Earth's ocean is more resilient than current science predicts and it will wait until primitive political structures evolve and the continuing negative impact of today's careless activities inform a global population that seems blind to the realities of human survival...

Civilizations and cultures have depended on the ocean for thousands of years for survival, expansion and defense...Even in this century many of the largest cities on Earth are coastal, integrated with the sea....

Theories that surfaced recently suggesting Earth's water was delivered to the planet via comet impacts is being challenged by recent measurements of hydrogen isotope ratio's in three comets that don't agree with the ratio of hydrogen in oceanic water and it remains to be seen if these three comets represent the chemical ratios of the majority of comets believed to originate in the Oort Cloud...This coupled with questions of how comets could have accounted for the enormous amount of water required to fill the oceans has demonstrated how little is known of the processes that created this water world whose tilt and orbit are perfect for the existence of carbon based life..

Ocean Acidification

The increasing acidity in the ocean is worldwide, well documented and is the cumulative effect of the amounts of CO2 humans now emit to supply energy to populations who ignore fossil records that prove extreme effects in the past from increased acidity...Scientists had long believed the absorption of CO2 from the atmosphere would be moderated and diluted by the natural mixing of shallow and deep water but recent studies have shown the majority of CO2 is concentrated in surface waters increasing concern for coral reefs, fish and many animals who depend on the ocean for food...Higher acidity slows the growth of calcium-carbonate shells many ocean creatures need to survive and high acidity damages coral reefs, huge infrastructures of life protecting coastlines and home to thousands of species, part of a food chain in existence for millions of years...Toxic algae blooms are becoming more numerous and increasingly toxic as the ocean absorbs more carbon and in this century levels of acidity will continue to increase in a single ocean on a single planet still powered by fossil fuels...

In 2012 over 500 global researchers of ocean acidification met in the U.S. to analyze data and write an impact statement (Ocean Acidification Summary for Policymakers) released in late 2013...The conclusion of this report cited human activity (adding 22 million tonnes of CO2 to the ocean every day with current energy policies and the continued burning of fossil fuels) as the catalyst for an increasing rate of acidification not seen for over 55 million years...30% of ocean species in 2014 will not survive an expected increase in acidity of more than 170% by the end of this century, a rate of increase many times faster than any event in the past...Scientists believe acidity has increased more than 25% since the industrial revolution began and many predict various sea life will adapt in ways unknown while many others will disappear, unable to survive a new reality with senses altered and habitat destroyed...Short-sighted unintended consequences of human progress that could affect all life on this planet...

Higher acidity in the ocean reducing the amount of dissolved oxygen and an increase in water temperature planetwide have been factors in every mass extinction event in Earth's history...

In September 2013, XPRIZE announced a $2 million competition to spur global innovators to develop accurate and affordable ocean pH sensors to transform human understanding of ocean acidification...The oceans absorb about one-quarter of the CO2 humans release into the atmosphere causing the chemistry of the water to change and the oceans to become more acidic...As a result of increased CO2 emissions, ocean acidity in this century are currently at unprecedented levels...With inadequate or unaffordable sensors currently available on the market, ocean acidification is well documented in only a few parts of the world...To fully understand and adapt to this threat, better pH sensing systems are needed to continually monitor and collect ocean pH data...The Wendy Schmidt Ocean Health XPRIZE is a 22-month competition with two $1,000,000 purses, one for accuracy and one for affordability...

March 2014...Several of the world's top fishing powers, the European Union, the United States, Japan, Columbia, Indonesia and the Philippines signed a declaration in Greece to promote the sustainability of world fish stocks...They promised to support effective measures to limit the destructive overfishing of the world's oceans including eliminating fisheries subsidies, limiting licensing, vessel tonnage, and developing an international record keeping system...

The journal Science published a study in 2006 that predicted the collapse of all the world's fisheries by 2050 if the depletion of marine animals continues at it's current rate...(Collapse defined as 90% depletion of species abundance)

Overfishing

The UN Food and Agriculture Organization has stated that more than 30% of fish species are over-fished and in this century no global organization controls fish conservation as efforts to control human activity on the ocean is determined by regional organizations who set limits through multilateral agreements developed from scientific research, historical evidence and global demand...By the end of the last century fishing fleets of every nation had harvested most of the available fish at shallow depths and now large-scale commercial fishing is concentrated at depths of more than a mile to harvest what is left...This is bringing new fish to market previously thought of as unappetizing with the United States importing more than 80% of the fish they consume and the European Union importing almost all of the fish they consume, mostly from the South Pacific and Africa...The depletion of large fish in the coastal waters on Earth has resulted in frequent algae blooms and a proliferation of jellyfish with no natural predators and in a civilization where profit determines policy, fishing fleets roaming the ocean in this century are huge floating industries with precision location devices and trawl nets, an inhumane and devastating way to fish with nets designed to drag the bottom and catch anything and everything with little concern for the species or it's value to the food chain...Much of the environmental damage to seafloor ecosystems (some visible from space) is due to these large trawl nets used globally without regard to consequence, some as large as multiple football fields creating havoc on the ocean floor with unwanted species thrown back dead into the water, an estimated 26 million tons every year in an industry with a singular goal of profit, ignorant of the balance of life on a single planet and the long term impact on the survival of all species on Earth...There are continuing efforts to create protected marine areas where a limited amount of fish can be caught to allow replenishment of stock and to balance the food chain (less than 1% of the ocean is protected) but any change is challenged by large-scale corporate operations who know that fisheries cannot be monitored without an effective real-time surveillance program to insure compliance with regulation and it's increasingly clear the consumption of fish and the continuing waste associated with human activity are exceeding the ocean's ecological limits...Another reason to determine a sustainable population on one world with finite resources...

Species of fish in danger include bluefin tuna, flounder, cod, marlin, halibut and more are added every year as fishing industries prowl the ends of the Earth looking for more profit...Whales were devastated by fishing in the 19th century in a world demanding whale oil and it's now estimated that industrial fishing, in just the last fifty years, has reduced the populations of all large ocean fish to less than 10% of what they were 200 years ago...The harvest peaked 26 years ago when 86 million metric tons of fish were caught (1996) and the yields have declined ever since...Scientists predict all fisheries on this planet will begin to collapse by 2050 if humans continue to fish as aggressively as they do in 2016 even though new techniques, strict compliance and global regulation could be used to manage fisheries in an intelligent way to increase and sustain multiple fish populations...In the 21st century, illegal fishing, unsustainable harvests, nations protecting industry, ongoing pollution of the ocean and no consensus among industries give the fish very little chance...The ultimate impact of this continuing activity on the future of human survival is unknown...

It's estimated over 100 million sharks are killed each year for their fins with species depleted by over 90% to satisfy ignorant traditions in Asian countries that continue to consume a tasteless shark fin soup for abstract reasons based in superstition...Only three shark species are protected globally under the Convention on International Trade in Endangered Species...Hong Kong imports more than 50% of sharks harvested each year...2012

A new study published in the journal Nature in 2016 says the amount of fish taken from the ocean over the last 60 years has been underestimated by more than 50%...The authors say the reason for this is statistics have not included small scale commercial fisheries, subsistence fisheries, illegal fishing and the amount discarded by catch...The report estimates around 32 million tonnes of caught fish are unreported every year...

The Law of the Sea

The Law of the Sea Convention is an international treaty written to give nations control over coastal waters out to 200 miles and set up guidelines and rules for all activities impacting the sea...It's been ratified by 162 countries with the European Union and came into effect in 1994...America has not ratified the treaty due to political pressure, an unexplainable fear of territorial incursion, a misguided sense of exceptionalism and pressure from corporate interests controlling American lawmakers...With multiple nations and differing interests this is a reasonable agreement to help coordinate security, pollution, disputes over waterways and drilling rights for the nations in the convention...

Exploration in the 21st Century

There is more land under water than not and humans are now developing the technology and science to explore the ocean floor in unprecedented detail... As is common during this era of history, it's about profit...Nations and private companies everywhere on Earth are claiming "rights" to large areas of seabed in a rush to strip profitable minerals and metals from underwater deposits that have been undisturbed for millions of years...China has claimed exclusive sulfide rights in almost 4,000 square miles of the Indian ocean, Russia and France have claimed territory deep undersea in the Pacific ocean and private companies are working out leases and contracts in the territorial waters of island states to mine minerals for many years...The environmental impact of the industrial activities needed to recover these resources is always negative with a history of destructive techniques used in the past without real concern about environmental damage that's always secondary in the rush for more of everything... A lot of these resources exist on the moons circling the planets and in the thousands of rocks in the Solar System to be recovered one day without a negative impact on Earth...Minerals and metals available in enormous quantities if humans survive their adolescence and develop technology to leave the planet...

In 2007, Russia deposited a flag on the seabed at the North Pole in a symbolic move to enhance the government's disputed claim to nearly half of the Arctic Ocean floor for oil or other resources located in this hostile environment...In one of the submarines that submerged over 2 miles to the seabed was a member of Russia's parliament to give validity to the claim...Eight nations, including Russia, have Arctic Ocean coastlines and under international convention have rights to economic zones within 200 miles of their shores...

In 2012, the U.S. Interior Department Bureau of Safety and Environmental Enforcement approved Shell Oil's plans to drill in the Beaufort and Chukchi Sea's over the next 2 years, an action that was challenged in court by a coalition of environmental groups claiming Shell had not demonstrated an ability to react to spills and leaks that would threaten endangered wildlife and the fragile ecosystem in this part of the world...Despite this challenge, in late 2012 Shell was allowed to proceed with the drilling of exploratory wells but suspended operations in 2013 due to unpredictable weather and questions of safety in current drilling procedures...In 2015 Shell renewed their push to drill in Arctic waters after America's Interior Department approved operations that could result in six wells operating in dangerous waters without contingencies for disaster...There is opposition from many directions and permits to be obtained before operations can begin but the money and influence of energy corporations has been very successful in obtaining the support of the largest governments on Earth to continue destructive activities without restraint...(Shell suspended drilling operations in late 2015 after costly operations found no large oil deposits)

The first maps of the ocean floor were not produced until the 1990's when declassified satellite sensor data was combined with data from European and NASA satellites to generate computer model maps in unprecedented detail...Today mapping the seafloor is ongoing worldwide with a growing interest in circulatory currents affecting weather and territorial posturing by many nations anticipating a new frontier of resources and profit...In 2010 the first Census of Marine Life Catalog was released with a database covering distribution, abundance and diversity of many marine species that cultures and civilizations have interacted with over thousands of years and with human technology so young it would seem sensible to explore and understand this underwater world before it is changed in ways that are not recoverable but this will not happen...Without a global awareness of causality the ocean is in great danger *(water is the first feature visible when approaching the Earth from space)*...All life originated in the seas that cover this planet and the ecological balance of Earth is anchored in the food chains and the diversity in the oceans essential for the survival of many species, maybe all species...

The International Seabed Authority (ISA) has issued a total of 26 exploration permits for deep-sea mining covering an area of 1.2 million square kilometers to government-owned companies in India, Germany, Brazil and Russia and to privates corporations including UK Seabed Resources (a subsidiary of Lockheed Martin)...Preliminary surveys have indicated some of the metal reserves could be very lucrative but scientists are urging extensive environmental assessments before mining in the ocean...Potentiality catastrophic for the marine life nearby, mining is expected to begin in a few years although no rules or guidelines have been established at this date...July 2014

In April 2014 the ISA applauded an agreement between Nautilus Minerals and Papua New Guinea to extract copper, gold and other minerals at a depth of 5,000 feet with Papua New Guinea receiving a 15% stake in exchange for a $120 million subsidy...The operation will destroy hydro-thermal vents where minerals form in higher concentrations with a fleet of robotic machines that will destroy the seafloor and recover the minerals as slurry...Mining is expected to start in five years...Nautilus insists the life destroyed in this operation will recover in "5 to 10 years"...

The International Seabed Authority is an autonomous international organization established under the 1982 United Nations Convention on the Laws of the Sea and a 1994 agreement relating to the Implementation of Part XI of the United Nations Convention on the Law of the Sea...The Authority is the organization through which State Parties to the Convention shall, in accordance with the regime for the seabed and ocean floor and subsoil thereof beyond the limits of national jurisdiction (the Area) established in Part XI and the Agreement, organize and control activities in the Area, particularly with a view to administering the resources of the Area...

Imagine the oceans drained...There would be thousands of artifacts, shipwrecks, airplanes, mysteries solved and mysteries created as one walked through a debris field representing thousands of years of human history...

Pollution

The Scripps Institution of Oceanography in 2012 reported the quantity of small plastic fragments floating in the north-east Pacific Ocean has increased a hundred fold in the last 40 years...During the same voyage 9% of the fish collected had plastic waste in their stomachs...An expedition in 2014 by the Algalita Marine Research Institute found over 35% of the fish they sampled had ingested plastic that fragments into smaller and smaller pieces...

The Earth has five large gyres, giant circular swirls of water, each one thousands of miles across with their location and size determined by the planet's rotation and the ocean currents that move and distribute water around the world...These five gyres are collection points for ocean debris and it's clear from examining this debris that Earth's ocean is the primary dumping ground for the human species, past and present...An activity which could be mostly ignored before the invention of plastics (less than 100 years ago) and before an explosion in population that has grown by over 4 billion in just 60 years...

It's been estimated the five gyres contain over a million tons of human trash...

For centuries the ocean has been the perfect aesthetic repository of all the used and unwanted stuff humans create because the vast majority of pollution was beneath the surface and invisible to our eyes...This is changing...Measurements and collections in these five trash accumulation areas verify the extent of human pollution, the composition of human waste and the potential impact of such irrational behavior on the creatures in the sea and the health of all living things on Earth...A recent survey by researchers at the Scripps Institution of Oceanography found the amount of plastic in the North Pacific has increased by 1,000% over the last four decades...Researchers skimming the surface of the North

Atlantic gyre for just one mile captured large amounts of polypropylene, the tiny plastic remains of fishing gear, clothing, plastic bags, plastic food containers and Styrofoam that's ingested by marine animals and sea birds, broken down into harmful chemicals and harmful persistent pollutants like PCB's and absorbed into the animal's fatty tissue to become part of the food chain...There is little data on exactly what all this plastic is doing to ocean life but almost anywhere you dip a net you will find the waste of our species, the leftover debris of human civilization...

An investigation published in Marine Ecology Progress Series estimated the fish at intermediate ocean depths in the North Pacific Ocean could be ingesting plastic at a rate of roughly 12,000 to 24,000 tonnes per year...

Plastics are just part of vast amounts of *toxic waste* continuing to contaminate the ocean from human landfills, mines, farming and industry in every part of the world with inadequate regulation, corrupt governments, lack of enforcement and a shared ignorance of consequence...A rapid increase in population (over 6 billion in just 300 years) has challenged a historical belief that the ocean is 'endless' and can absorb any impact without real change as human science and multiple negative effects growing increasingly obvious confirm the damage being done to a vital source of life...Every boat that uses an internal combustion engine leaves gasoline or diesel in the water, in fact every internal combustion engine on the planet leaves residue on streets everywhere that find a way to the ocean and these engines emit CO_2 which is partially absorbed by the ocean...Every piece of trash manufactured for human use breaks down when submersed in sea water and releases lead and other chemicals which end up in fish, end up in humans, and can cause severe neural damage in children and adults...Sewage systems all over the developed world share their drainage with storm drains that feed into the ocean (18 million tons of sewage were dumped in the ocean in 1980, legally) and industry globally discharge millions of gallons of contaminated water or chemical byproducts into the sea with little or no oversight, even more so in developing countries trying to catch up to the modern industrialized world...Agricultural operations using pesticides and fertilizers to grow and manufacture food contribute large amounts of pollution to oceans, rivers and lakes with careless irrigation and runoff without adequate filtration...Oil pollution from spills and leaking vessels have effected almost every continent on Earth and it grows worse as a world desperate for oil uses aging fleets of tankers and poor industry regulation to meet the demands of the many...Dumping nuclear, industrial and other waste in the ocean was legal until fifty years ago and despite regulations it continues today in many parts of the world...It's estimated 80% of the waste material dumped in the sea is from dredging, frantic human activity changing a world to fit our needs...

More than 30% of shellfish growing in waters around coastal America are adversely affected by pollution.

There are few real solutions to ocean pollution without a consensual awareness, world unity and cooperation among nations...In a "modern" world the volume and pace of human global activity uses more and more resources with diminishing efforts to recycle the recyclable material and insure the proper disposal of everything else...Today the demands of growing societies to have more 'stuff' is overwhelming...This is a solvable problem with current technology but money and profit determine all activity on Earth and it's cheaper to mass-produce products as inexpensively as possible and just throw it all away...Governments are weak in regulation and policy, industries dedicated to recycling the tons of trash accumulating daily don't exist and corporations in almost all nations bear no responsibility for the ultimate destination of their 'stuff'...

In July 2013, 91 dead dolphins washed up in New York, New Jersey, Maryland, Delaware and Virginia...In 2012, those same states had a combined dolphin death toll of nine...It is unclear why...Possible causes include disease, pollution, interaction with fishing gear, environmental issues or underwater military sound sources...

It's estimated there are more than 86,000 large ships operating commercially on the ocean surface with more than 6,000 intentionally polluting the sea ignoring regulations set forth by MARPOL (The International Convention for the Prevention of Pollution from Ships) established in 1973 to reduce pollution in the oceans...Most of the oil contamination is from illegal dumping believed responsible for the discharge of 100 to 200 million gallons of waste oil each year...Cruise lines are fined millions of dollars for illegal bypass lines, record falsification and open dumping of sewage, oil and waste without regard for species or environment and these are not accidents...These are focused attempts to bypass regulations necessary to protect a single, closed ecosystem and the ignorance of so many humans who live in this ecosystem to harm the very thing supporting their existence is indicative of the developmental stage of human intellectual awareness on 21st century Earth...It continues today...

Sea Level

The amount of CO2 currently in Earth's atmosphere has guaranteed sea level rise will be a part of human existence for centuries with global sea levels increasing about 16 centimeters in the 20th century and an estimated increase of over a meter in the 21st century, a number yet uncertain...Ongoing studies of the Arctic ice melt and Greenland's ice melt make accurate calculations a rapidly changing science as cascading effects must be combined with more rapid heating of the Arctic ocean as sunlight warms water unimpeded by ice releasing methane which warms the Earth even more...The impact of changing climate on the vast continent of Antarctica must also be considered long-term with a total melt of this continent increasing sea levels by over 70 meters...It would be a very different Earth...

A recent reassessment of 18 years of satellite data reaffirms a sea level rise of more than 3mm per year on average and even though there is just one ocean, sea level increases can have regional differences due to currents and circulatory patterns with the rate of increase in the Philippine Sea over the same time period more than 10mm per year suggesting long-term trends that fluctuate over decades instead of years and with the greenhouse gases already present scientists predict a sea level rise of 3 to 7 meters in the next 300 years which will destroy almost every coastal city on Earth without dramatic changes...Although the melting of ice and the rise of the oceans will be a long-term event that humans will adjust to in an ever-changing world, sea level change may also be cyclic responding to activity inside the planet or precession involving the spin axis, a rotational alignment of Earth over long time periods increasing direct sunlight to affect different geographical areas with increasingly predictable effects...It's clear this current period of warming is directly linked to human industrial activity with the burning of Earth for energy and it's critical to understand and accurately model possible and predictable effects of future activity to relocate populations and continue to grow food in a sustainable way...In this century human activity continues to emit greenhouse gases at ever increasing rates...

The ocean is the largest repository of food on Earth...

Marine Life

The goal of the annual meeting of the International Whaling Commission (IWC) in 2012 was to set quotas for indigenous groups in the Arctic, consider a proposal declaring the South Atlantic Ocean a whale sanctuary and a proposal for the UN to take charge of whale conservation...Another point of discussion was the time it takes for whales to die after the initial human attack with reports from thirty minutes to two hours depending on the method used to kill the whale...

The proposal for a whale sanctuary covering a large part of the Atlantic Ocean south of the Equator from the west coast of Africa to the east coast of South America failed with less than a three-quarter majority...The proposal to move whale conservation to the UN also failed and in a move that was highly controversial, South Korea announced it was preparing to allow fishermen to hunt whales under regulations that permit this activity for serious scientific research...*The major accomplishment of this conference was an agreement to hold meetings at two year intervals*...The anti-whaling nations are a majority over the whaling nations but it's clear to intelligence that none of these creatures should be hunted and killed without understanding their role in the ecological balance of ocean life, a study of their intelligence and an awareness of their habitat...

In 2014, the U.S. designated the largest protected area of critical habit in it's history covering 300,000 square miles of ocean and 685 miles of beaches primarily for sea turtles...This area covers 84% of all known nesting sites for the Northwest Atlantic loggerhead turtle, an endangered species...The military is excluded from the habitat restrictions...

Drift Gill Nets are used by industry fishing fleets to catch everything that gets in the way with no consideration for species, large or small, endangered or not, edible or waste, throwing anything they don't want back in the water (usually dead) with a singular focus of profit...In a single year these nets drowned 16 sperm whales off the coast of California and killed sharks, turtles, dolphins and countless open water species without reason and the nets are a long-lasting nylon that can remain intact for years continuing to trap species after being lost underwater...Governments insist they regulate the size, length and depth of net use in fisheries but thousands of unwanted species are killed by these devices every year in an ocean that is more and more challenged by ceaseless, careless human activity...Drift nets as long as 35 miles were routinely used by industry until 1991 when the UN General Assembly called for the cessation of large scale drift net fishing in international waters limiting the legal maximum length to 2.5 kilometers (1 mile)...Compliance is hard to monitor, there are many violations and individual nations set their own rules of use in their territorial waters...Even with clear indications of cause and effect there is little hope in today's civilization of stopping this cruel and wasteful activity...

Over 80% of the world's fisheries are over-fished or "fully exploited" with no way to safely increase the catch...

Noise

Another serious hazard to marine life of every kind, especially the largest marine animals, is the proliferation of *noise* in underwater environments globally...Every marine protection act legislated and passed by any government addressing the impacts and dangers of high frequency and loud underwater noise has contained exceptions for "military operations necessary to the defense of the country" and for "scientific research" which includes private corporations mapping potential oil deposits beneath the ocean floor using high frequency sonar, radar and explosives, ongoing activity proven to disrupt the biological navigational systems of many oceanic animals leading to frequent beachings and death...*It's glaringly obvious that humans are destroying the ocean habitat when the animals native to that environment beach themselves trying to escape human activities*...No industry or corporation on Earth who profit from this activity will agree with the science and the obvious effects of human intrusion into an environment that was pristine and quiet until the whaling ships, steam powered ships and the loud obtrusive ships of today changed a dynamic and balance that had existed for millions of years...Every government assert their right to create noise to test weapons, seek resources, train personnel and improve detection systems without regard for species...

Michael Jasny, a senior policy analyst at the Natural Resources Defense Council, said the Navy routinely underestimates the effect of it's military activities on marine mammals...He referenced a study by government and private sector scientists published in the journal Proceedings of the Royal Society just a month before the two environmental impact studies by the Navy (cited below) showing mid-frequency active sonar can disrupt blue whale feeding...The study says feeding disruptions and the movement of whales away from their prey could significantly affect the health of individual whales and the overall health of baleen whale populations adding that the Navy's ocean activities are "simply not sustainable"..."These smaller disruptions, short of death, are themselves accumulating into something like death for these species and death for populations," Jasny said...

In July 2014, the U.S. Bureau of Ocean Energy Management approved using seismic cannons to survey the ocean floor for energy deposits...These cannons are louder than commercial jet engines and are "fired" every 10 seconds for weeks at a time to locate deposits under the ocean floor...In an absurd statement, the director of BOEM said the "decision reflects a carefully analyzed and balanced approach" while acknowledging that this activity will harm or kill thousands of sea creatures...Estimates are of 37.5 trillion cubic feet of natural gas and 4.7 billion barrels of oil as humans with no regard for a "balanced approach" continue to harvest the world for fossil fuel energy...

August 2013...Two environmental impact studies released by the military reports Navy training and testing could inadvertently kill hundreds of whales and dolphins and injure thousands over the next five years...The studies focused on waters off the East Coast, the Gulf of Mexico, Southern California and Hawaii from 2014 through 2019 where the Navy tests equipment and trains sailors...The majority of deaths would come from explosives though some might result from testing sonar or animals being hit by ships...Computer models developed from estimates of the hours the Navy will spend *testing and practicing* with sonar, torpedoes, missiles, explosives and other equipment (over five years) predict the possible deaths of 186 whales and dolphins off the East Coast with 11,267 serious injuries, 1.89 million minor injuries (temporary hearing loss) and 20 million instances of behavioral changes...

Naval activities off Hawaii and California will kill an estimated 155 whales and dolphins with 2,039 serious injuries, 1.86 million minor injuries and 7.7 million instances of behavioral change...The studies were done ahead of the Navy's application to the National Marine Fisheries Service for permits to engage in these activities complying with federal environmental law...Rear Adm. Kevin Slates, the Navy's energy and environmental readiness division director, says the Navy uses simulators but sailors must train in real-life conditions..."Without this realistic testing and training our sailors can't develop or maintain the critical skills they need or ensure these new technologies can be operated effectively"

At the end of 2013 the National Marine Fisheries Service, an organization entrusted by the U.S. Congress with protecting marine life, gave the U.S. Navy permission to kill 155 marine mammals off the coast of California and Hawaii and injure thousands more by disrupting behavior patterns established for hundreds of thousands of years essential to these animal's survival...The Fisheries Service has authorized sonar testing and detonation impacts to exceed 10x's the previous limit with no requirements to consider any harm done to the animals when there is clear evidence of serious damage done by human intrusion into ocean habitat with a single purpose of testing weapons of war...Several environmental groups are attempting to challenge this decision in U.S. courts...

The U.S. Department of the Interior decided to postpone their decision on whether to allow the use of a seismic air-gun until March 2014 after pressure from environmental groups to review new data from the National Oceanic and Atmospheric Administration concerning levels of noise known to be harmful to sea life (*Approved in March 2014*)..The oil and gas industry routinely uses seismic airguns to find deposits beneath the ocean floor...Towed behind boats, these guns shoot blasts of compressed air every 10 to 15 seconds to create a geologic map from reflected sound waves...Government estimates predict seismic testing disrupts critical behaviors like feeding, calving and breeding for many marine creatures and disrupts threatened loggerhead sea turtles as they journey to nesting beaches to lay their eggs...It's estimated more than 138,500 dolphins and whales will be injured or killed by the deafening blasts including the highly endangered North Atlantic right whale...Commercial fisheries are at risk as airguns displace commercial species, kill fish eggs, larvae and lower catch rates between 40 and 80 percent impacting the economies of seven Atlantic states...The seismic testing zone would span more than 300,000 square miles of ocean (twice the size of California) and while current regulations ban drilling in the Atlantic Ocean until 2017, oil and gas industries could begin the mapping process while the drilling ban is still in place...*Oceana, Aug 23, 2013*

The IPSO (International Programme on the State of the Ocean) released a report in late 2013 warning of a rapidly deteriorating environment in Earth's ocean accelerated by the cumulative effects of CO_2 absorption, dead zones from pollution and bad fishing practices..."The public and policymakers are failing to recognize or choosing to ignore the severity of the situation"...The report urged world leaders to stop increasing CO_2 concentrations at 450ppm (CO_2 concentrations have been measured at over 400ppm in this century and the levels are rising), urged governments to find a way to monitor and enforce "sustainable fishing" and created a priority list of polluting chemical agents doing the most harm...The growing effects of inaction are clear but there is little hope in the near term that anything will change in a substantial way...

Aquarius, the size of a large bus, is the only underwater laboratory of its kind left in the world...Supported by less than $3 million a year, this single structure represents a failure to explore and understand an enormous unknown that is rich with diversity and is vitally important to human survival on a single world with a single ocean...

WAR

There are over 1.6 billion humans serving active military duty worldwide...

Wars and conflict killed an estimated 150 million to 220 million humans in the 20th century but the 21st century could surpass that number on a primitive and violent Earth with nearly every nation on this planet acquiring and distributing weapons of war, funding military organizations and ignoring the millions of children and poor who struggle to survive in a blind, uncaring environment of territorial priorities and military posturing to defend against perceived threats to every nation's sovereignty...To facilitate this destruction of life and resources humans have developed a formal/informal agreement among nations and military leaders worldwide referred to as the "laws of war"...This is absurd in any context...Why would the most intelligent species on a shared planet agree to a set of laws and rules that allow for the devastation of entire societies and the killing of millions of people?

Laws of War

An informal context of customs, practices, usages, conventions, protocols, treaties, laws and other "norms" governing the commencement, conduct, and termination of hostilities between warring nations...After a long history that has its roots in medieval times with influence from the Catholic Church concerning just wars, unjust wars and God's displeasure with certain types of weapons (?), the United States, Mexico, Japan, Persia (now Iran), Siam (now Thailand), and 19 other nations, including all the major European powers, signed the Convention with Respect to the Laws and Customs of War in 1899...By law, military forces are restricted to the type of weapons and explosives they may employ in war, weapons calculated to cause unnecessary suffering like cluster bombs and devices that can leave fragments of glass and plastic in the body...The use of poisons and poisonous gas is prohibited and the use of bacteriological agents to spread disease is prohibited...Thermal Nuclear Weapons are prohibited due to their inability to discriminate between combatants and noncombatants but there are exceptions with the U.S. deciding in Vietnam that incendiary weapons (napalm) and chemical herbicides (Agent Orange) were allowed because the enemy hid in the jungle...In this century no organization on Earth exercises sovereignty over any individual member state in any meaningful sense and no permanent or impartial international body has ever existed to administer the "laws of war"

Historical evidence shows these 'laws' mean little in real life...Israel and Syria use cluster bombs with thousands of children becoming the victims of conflicts decided by ignorant humans in safe rooms while millions of land mines continue to be designed and employed with a single purpose of destroying a human life...Why does an intelligent life-form capable of mapping the universe continue to develop devices to kill members of their own species and bury bombs in the ground to destroy any creature who steps on it? Humans have a long history of sending thousands of themselves to die fighting for territory and resources which could be shared while rationalizing killing and dying for imagined deities that do not exist...It's clear that any ideology or belief that demands killing and cruelty to maintain that belief system is by definition a system based on ignorance and primitive instinct existing in a construct blind to the possibilities of a shared human existence...In this century the dominant species on this planet is a dangerous animal unaware of the nature of it's own existence without the connective intelligence to understand it's own reality, an embarrassment to the possibilities of sentient life...

An enormous number of humans accept without question the horror of war and the killing and injuring of some of the youngest members of their species fighting over primitive ideals and territory that in the end are restricted to one planet...The morality of every human being is determined by a natural process of growth and development in the brain during the first three years of life and a human cannot become someone who kills without remorse if brain growth is maximized during these first years of existence...This connection between human morality, forced combat, PTSD and military suicides will be obvious when studied and understood in a more comprehensive way...

The Pentagon spends more on war than all 50 states spend on health, education, welfare, and safety...

The U.S. has 5% of the world's population and almost 50% of the world's total military expenditures...

War and the many boundaries wars are fought over emphasize everything the primitive human brain represents...Territorial instincts marking territory with natural boundaries, rivers, oceans, valleys, mountains, man-made walls and agreed upon lines in the sand...Even a casual look at global resources confirm almost every nation has a military force, no matter how small or large, militaries allocated far more resources than any other segment of society...Territorial justification for death and destruction in unimaginable ways as groups of modern humans unable to understand the reality of the planet they live on or the possibilities of a shared existence defend boundaries and actions based on fear, confusion and primitive survival instincts over a million years old...This sad, senseless posturing will only end when new generations of reality-aware humans become the majority of Earth's population and understand the absurdity and primitive ignorance of such behavior...Until that time millions of humans will continue to send their children to die in a patriotic effort to defend sovereign territories that are all part of the same planet...A single blue world that supports human existence...

In the 21st century, governments of the world continue to be dominated by humans who cannot understand why killing is not 'noble' and 'right' no matter how many dead bodies or dead children, the saddest victims of war, or the thousands of soldiers who come home damaged physically and mentally by doing the thing their country is most 'proud' of...Israel with cluster bombs and dead children in Gaza, Bashar al-Assad in Syria with cluster bombs and dead children, extreme violence worldwide and always dead children...What did the children do to deserve death?...Almost all wars fought throughout history have been about territory and boundaries including the religious wars, the infamous crusades that simply killed people using a non-existent God as a reason to slaughter, rape, steal, burn, torture, enslave and destroy populations under the protection of the church with a singular goal of conquest and expansion...European settlers slaughtered millions of native Americans to occupy their land, the Civil War in America killed almost a million brothers, sisters and neighbors over the right to buy and sell people and today the conflicts continue to escalate in multiple nations between differing communities with angry people killing other humans who are fundamentally the same as they are because they do 'something' in a different way...In the Arab world they are Shiia or they are Sunni and in the African world they are Hutu, Tutsi, Muslim or Christian...Millions of humans in Africa killed and uncounted people tortured and raped because of ignorance and a lack of intelligent awareness that would realize all humans are essentially the same...Wars reflect a very primitive stage of species development...

URANIUM-238 CONTAMINATION

Weapons

The most profitable weapon corporations in America all have former members of government or former pentagon officials on their corporate boards with a priority on maximum profit determining corporate policies which have increasing influence over the decisions of governmental bodies in areas of defense and conflict...When this influence and focus on money involve long, intractable wars that are increasingly profitable for the companies receiving defense contracts, a cycle of violence and profit exist that could one day have corporate representatives encouraging war as a pathway for corporate success...It's not hard to imagine powerful industries hiring mercenaries to enter a stressful region of the world to attempt to accelerate conflict and motivate foreign governments to purchase more weapons for defense and invasion...This growing influence and control over weapon spending that is wasteful and mostly unnecessary continues unabated between the military, defense corporate structures and the political leaders of the world as all three entities gain influence and wealth at the expense of millions, operating in a system of influence, corruption and greed that has been institutionalized and put into law by politicians who are bought and controlled in increasingly corrupt political systems...Conflict is now a growth industry, a legal paradigm for killing anyone anywhere on Earth...

Nuclear Weapons
(Nine nations on Earth have a combined total of approximately 16,300 nuclear weapons...SIPRI)
__China__ = 250+ nuclear warheads...__France__ = 300 nuclear warheads...__United Kingdom__ = 160 strategic nuclear warheads + 65 stockpiled warheads...__India__ = 100 nuclear warheads...__Pakistan__ = 110 nuclear warheads... __Israel__ = 80? nuclear warheads...__Russia__ = 1,600 strategic nuclear warheads, 6,400 stockpiled nuclear warheads...__United States__ = 1,900 strategic nuclear warheads and 5,300 stockpiled nuclear warheads...

In 2011, the world market for U.S. weapons hit an all time high with over $45 billion in military devices sold to foreign governments from American manufacturers...Weapons and equipment made by the human species to support, supply, transport and enable humans to kill humans...This singular reality defines the primitive evolutionary stage of human development and the absence of *reality intelligence* among the majority of the world's population...The irony of the global trade in weapons and death is a lack of any responsibility by the governments and corporations who control this market resulting in the indiscriminate use of force against any and all populations on the planet with many innocent creatures and sometimes the weapon makers themselves killed by these violent devices they continue to develop and sell...The deputy director of the Arms Control Association (a U.S. national nonpartisan membership organization dedicated to promoting public understanding and support for effective arms control policies?) emphasized the U.S. *loses control of a weapon system once the sale is made*...This ignorance and lack of awareness among the network of humans that create, support and justify this violent reality is stunning...Pentagon officials defend the sales in saying they are carefully regulated by the State Department but how is this possible if the U.S. *loses control* once a sale is made? Currently, Saudi Arabia, one of the most repressive governments on the planet, is buying fighter jets, 2000 lb bombs and attack helicopters in a $60 billion deal...This nation is ruled by a small-minded group of humans who treat females of all ages as children if they follow rules of no self-expression, no personal rights, no driving, no voting and they kill or imprison females who want anything more than what the males of this group will allow...A small group of wealthy monarchs with vast quantities of oil controlling a huge population with intimidation, weaponry and continual censorship will control an enormous amount of lethal weapons in exchange for their money...What will they do when the oil runs out? Egypt, Bahrain and Tunisia are among many nations who have used U.S. bought weapons on their own populations and Morocco, Iraq, United Arab Emirate, Britain, Australia are all seeking weapons which are designed and used only to kill humans...A promising, primitive species...

World military expenditures in 2015 is estimated to have reached $1.7 trillion dollars...
SIPRI (Stockholm International Peace Research Institute)

America is by far the largest supplier of military hardware globally selling weapons to defend and promote war all over Earth with hugely profitable military industrial structures who promote their products with no thought or concerns of the use or consequence of these sales...Four of the five biggest buyers of new military hardware from the U.S. are Middle Eastern countries...Russia is the 2nd largest supplier of military arms in the world exporting over $13 billion of weapons to more than 50 countries, including Syria (75% of all Syrian arms are from Russia), Iran (over $500 million over three years) and a proposed deal with Iraq for over $4 billion in military hardware (canceled in 2012)...

"The Arab countries are our traditional clients and there is no reason why Russia wouldn't sell weapons to Iraq...I wouldn't see any politics behind it...I think it's just money because Iraq is an Arab state like any other right now and Russia has purely commercial interests there..." Dimitry Babich, Voice of Russia

The Crusades
1095-1291

Seven Crusades (some say Nine) over nearly 200 years begun by Pope Urban II who called on the princes of Europe to recover the Holy Lands inhabited by Islam resulting in the deaths of three million to five million people with essentially no change in the permanent status of the Holy Lands and new levels of hate and intolerance between Muslims and Christians that still reverberate through world societies today...The only 'winner' was the Catholic church and the papacy which gained vast wealth, authority and influence over much of Europe that continues in this century...

The Thirty Year War
1618-1648

A series of wars begun as a religious war between Protestants and Catholics over the balance of power in Europe killed between 8 and 10 million humans and devastated entire countries with famine, disease and plagues...As is usual it was a war over territory with Spain, France, Sweden and Denmark wanting some control of German states and the independent states of the Holy Roman Empire...World populations grew and ignorant leaders both secular and religious postured for territory and control with little regard for the loss of life inflicted on their own subjects...With new weapons of war continuing to be developed at a rapid pace and communication limited, conflicts on Earth grew larger, deadlier and increasingly consequential with an escalating loss of life and destruction to the innocent people bound up in these conflicts...Another dominant human was always ready to take the place of the fallen, ready to sacrifice the soldiers at their command to maintain and advance territory driven by instinct, fear and an alliance with a leader or an imaginary God to justify their cause...

World War I
1914-1916

European armies faced each other along a series of trenches 500 miles long that stretched from Belgium to Switzerland and over 9 million soldiers died fighting for land and political ideology in an arena dominated by tanks, machine guns, poison gas and military aircraft...Mass slaughter and at the end of the conflict the opposing forces were essentially where they started...Only one hundred years ago...Almost seven million civilians died and the Ottoman Empire (arguably the greatest independent Islamic power in history) collapsed after Sultan Mehmed V declared a jihad against European powers Britain, France and Russia ending four centuries of relative stability and allowing victorious European powers to reshape a Middle East that is one of the most violent regions on Earth in the 21st century...

It's believed the Treaty of Versailles signed at the end of World War I was instrumental in creating a social and political environment which allowed the rise of extremist groups in a German nation that was ostracized, humiliated and devastated by the war, an environment responsible for creating the Nazi Party and leading to World War II...

In World War I manufacturers sold weapons to both sides in the pursuit of profit...It continues today...

World War II
1939-1945

Another territorial war with Germany invading Poland in a madman's quest to take over Europe and subsequently the entire world...With new technology in communications and alternative military possibilities, the leaders of the free world had many chances to stop the rise of Germany's military in the years before the war but looked the other way unable to comprehend or visualize a reality that was increasingly obvious...By the end of the war as many as 69 million humans had died worldwide with uncounted others injured and maimed for life...The destruction of life and property took place on four continents and ended with the first use of nuclear weapons by one nation against another nation...The two atomic bombs dropped on Japan killed over 130,000 innocent people and the extreme number of dead in this violent war included over seven million Jewish people systematically killed at the hands of Germany, an estimated seven million Chinese killed by the Empire of Japan and more than 18 million citizens of the Soviet Union...Over 40 million of the dead were civilians...

In 2009, the Pentagon awarded a contract to build the largest most powerful non-nuclear weapon in existence...The GBU-57, a 'bunker buster' carrying over 5,000 pounds of explosives that can burrow 200 feet into the ground before exploding...In an absurd statement that defines primitive thinking the Pentagon said "this has been a capability that we have long believed was missing from our quiver, our arsenal, and we wanted to make sure we filled that gap"...

The Future of War

Many human 'experts' since World War II have stated there will be no World War III as it would result in global devastation and the end of modern civilization...As a consequence we live with many wars as war in the 21st century has become very problematic for the conquering country...With the beginning of the 1st World Civilization and the use of technology allowing any action or behavior anywhere on Earth to be recorded and monitored by the rest of the world, a nation cannot simply 'exterminate' the population of another country and call it their own...The populations of the world have grown to a point where large numbers of civilians in all countries under invasion and occupation resist and condemn occupation, forcing conquering nations to find a protocol of believability and a reason to continue fighting on another nation's territory...In the modern age a nation cannot just kill everyone opposing them and take over the home of millions of innocents so the best that can be hoped for is an occupation of the invaded country in collusion with a puppet government, set up and controlled by the invaders to access wanted resources and establish a presence that is difficult to remove...An example of this is America's invasion and occupation of Iraq and Afghanistan in the first decade of this century...

Inventive and curious, humans have now created a new and better way to make war and destroy life...*Drones*...The United States has thousands of drones, more drones flying than manned aircraft and more personnel trained to operate these remote control machines than actual pilots being trained...There were no military drones just 20 years ago...With little publicity and minimal congressional oversight drones have killed over 2,000 people since 2001 (with less than 3% being "high value" targets) and it's only just beginning...Recent reports reveal the United States has carried out over 450 drone strikes in Pakistan and more than 60 strikes in Yemen and with advanced guidance systems, new advanced power systems, improved satellite communication, larger weapon payloads and improved stealth technology, unmanned drones are the military of the future for many nations who have equal technology to create and operate these lethal weapons...Silent death from the sky anywhere in the world...

A report released in 2013 by Human Right Watch detailed an excessive use of drones by the United States in Yemen which has killed many more civilians than targeted militants through indiscriminate use with no accountability or limits...Leaked documents reveal collusion between governments of both countries to "cover up" their involvement in specific attacks and a failure to compensate families or individuals for their losses...

It's estimated there is a landmine buried somewhere on Earth for every 52 people alive...It's been estimated that it would take more than a thousand years to find and clear all the land mines in the world...

189 nations are party to the Convention on the Prohibition of the Development, Production, Stockpiling and Use of Chemical Weapons and on their Destruction (CWC)...Israel and Myanmar signed the agreement in 1993 but have not ratified it...Angola, Egypt, North Korea, South Sudan and Syria are non participants...

An arsenal in America contains over 500 tons of deadly nerve agents VX and Sarin stored in more than 100,000 munitions that are old and increasingly unstable...Under international treaty these munitions were supposed to be destroyed in 2012 but the Pentagon estimates it will not be done until 2021 costing nearly $50 billion...VX nerve agent is extremely deadly with a drop on the skin causing death...*Why would a sentient species living on a defined biological world from which there is no escape create an extremely deadly product in such enormous quantities?* It's amazing how primitive structures in the evolving brain, instincts of fear and the inability to understand causality and reality can dominate a mindset and create policies and activities that are senseless and absurd...Were the military or civilian leaders intending to kill every human being on the planet? Was there any serious thought to the consequences of using this agent? And why so much? As an emerging intelligence humans are just coming out of the dark ages of religious nonsense and scientific ignorance and realizing the potential and possibilities of existence in this star system...Without reality intelligence, without dense neural connective networks throughout the entire brain structure to enable communication, rationality, logic and an understanding of existence by a majority population on this planet the human species is in great danger from it's own technological achievements...

Nuclear Strategy
In 2010, the United States and Russia signed the 10 year New START treaty in an agreement to reduce their deployed strategic nuclear warheads to 1,550 within seven years...Stockpiled nuclear warheads on both sides were not included in the treaty...The Russian and the U.S. arsenals contain over 90% of the world's nuclear weapons with no obvious targets for these explosives that would change the world if used in any quantity...

It's estimated over 40 tons of highly enriched bomb grade uranium remains stored at multiple research facilities around the world, the result of a cold war strategy of the past...

China has over 300 nuclear weapons and has stated they will only use them for self-defense...Pakistan has about 100 nuclear weapons and have not clarified their intentions...

Nuclear capable missiles are those which can deliver a 500-kilogram payload to the target and are limited to missiles with a range of at least 300 kilometers, considered to be the smallest possible distance between two nuclear combatants...(The distance and carrying standards are laid down by a 33-nation Missile Technology Control Regime, a consortium that restricts the export of flying rockets)...America's military has the technology to miniaturize nuclear arms virtually at will...The U.S. military has nuclear-tipped versions of both the Cruise and Tomahawk missiles...

The United States budget for nuclear weapons in 2010 was estimated at $25 billion...

In March 2012, 53 countries met to discuss nuclear terrorism across the planet in an ongoing effort to secure the world's fragmented stocks of bomb grade plutonium and highly enriched uranium...The summit asked only for further "voluntary" reductions by 2013 in a world where there are over 1,300 tonnes (almost 3 million lbs) of highly enriched uranium and over 450 tonnes (almost 1 million lbs) of separated plutonium scattered across more than 30 countries, most of it in Russia and the U.S. ...

Russian military officials stated with clarity they would use pre-emptive weapon strikes in Europe and NATO countries to target missile defense sites the US military and NATO are determined to build and deploy as part of a NATO shield if Russia determines these missile sites to be a deterrent to their ability to launch and target their missiles anywhere in the world...

U. S. Intelligence Agencies spent over $50 billion in 2009 to spy on America's citizens, allies and enemies...In 2013, it became obvious to the world these intelligence agencies were infiltrating and collecting information on multiple governments worldwide and monitoring the private communication of every citizen in America...The U.S. government defends this activity as "necessary" and anticipates no change in their policies going forward...

Cyberwarfare

The newest version of territorial warfare, *cyberwarfare*, is just beginning with an internet still in it's infancy, fiber optic networks revolutionizing internet speed and access, wireless intrusion systems containing increasingly complex algorithms and a world-wide web where everything is connected to everything...The danger is nations can develop cyberweapons outside any regulatory framework unlike nuclear and chemical weapons and every government globally has cybersecurity on the agenda...One of the unknown dangers of this uncontrolled 'arms' race in cyberspace is individuals who develop and release computer viruses for whatever purpose and then lose control of the virus as it seeks out targets which will invariably include victims other than the intended target...Flame, a large, invasive malware espionage program believed to be developed by U.S. security agencies to target Iran (the U.S and Israel deny any responsibility), is now in free circulation around the world as efforts to counteract it's effects continue...The Pentagon's Defense Advance Research Products Agency has launched 'Plan X' with its stated priority and purpose being "a need to dominate the cyber battlespace"...Using America's private defense corporations under intense secrecy, the Pentagon's goal is to engage in a robust cyberstrategy designed to infiltrate systems anywhere in the world to extract information, input false information and control operating systems essential to power grids, nuclear facilities and basic infrastructure in a way that disrupts or destroys their target...The U.S. does not deny it developed the "Stuxnet" digital worm responsible for disrupting Iran's uranium enrichment program and America's current administration has confirmed that their position in the global arena is the 'laws of war' are applicable to cyberattacks...

The U.S. government recorded over 77,000 "cyber incidents" in 2015...Newsweek

It's only a matter of time before a nation attacks another with military force after a cyberattack shuts down power grids and infrastructure...The internet is an invaluable tool for a global civilization in monitoring world events and informing every citizen and one day it will be an open transparent communication matrix but it will take a long time...Revealing information released by a U.S. National Security Agency (NSA) contractor revealed that America's spy agency intends to gather information in whatever way possible from every citizen in the country, from foreign embassies anywhere in the world (friendly or hostile), from allies, world leaders and world organizations without regard for privacy or protocol, collecting information from anyone this agency seeks to add to their database and they will defend this intrusion on the privacy on every single human affected as *"necessary and warranted actions to prevent a terrorist attack" "completely legal within the framework of congressional laws and oversight"* citing compliance and trust as the proper response to activities having no controls or limits in the global community...An organization who has interpreted "oversight laws" to mean what they wish as they continue to record every communication possible...What is true is future warfare and the threats of enemies real and imagined will spark innovation, new technology and a rapid growth in all areas of human/machine interaction leading to complex integrated systems for things unimaginable today...The human species will not revert to an agrarian society...Science and technology must be the bedrock of any civilization wishing to exist for millions of years...

No challenge to America's invasive collection of data from any human or organization on the planet has succeeded in stopping the ongoing compilation and storage of terabytes of data at remote sites around the world for analysis as this clandestine game of spy vs spy continues without any real oversight or checks and balances in an opaque system without limits...In real time there is not yet an efficient way to organize and search this vast amount of information in an effective way but that is rapidly changing as new algorithms improve...2016

Approximately 200,000 women serve in the U.S. military and one in three have been sexually assaulted, soldiers assaulting soldiers in a male dominated mindset that routinely ignores the crime and blames the woman...In the military command system a victim reports the rape to their commander and the commander is the sole judge of determining if any action is taken...Less than 10% of cases are prosecuted and 90% are by repeat offenders...

PTSD – Post Traumatic Stress Disorder

It's not yet clear why the obvious negative effects of repeated percussive explosions in the arena of war are not recognized for the damage they cause to a human brain in very real physical ways and demonstrated, debilitating mental ways...The Defense Department in the U.S. continues to downplay the impact of these events, continues to misdiagnose TBI (Traumatic Brain Injury) and still uses outdated diagnostic techniques to clear TBI victims for further military duty ignoring and dismissing a growing body of empirical and documented evidence, testimony by victims and behavior assessments of professionals in an outdated military tradition of 'toughness and silence' that permeates through an aging core of military leaders who put money, duty and pride above the recognition of reality and the welfare and health of soldiers...Treatment for PTSD is stigmatized by the military as an 'emotional problem' or 'a lack of the right stuff' and in many cases the diagnosis is made without testing for actual brain damage with diagnostic conclusions routinely changed and downgraded by doctors or clinicians who have never seen the patient to avoid paying benefits and compensation...

Mental health disorders are estimated to affect over one in five active duty military personnel with many service members not reporting symptoms as they try to 'be a man' and 'deal with it' coupled with the fear of being stigmatized by fellow soldiers and command in a way that restricts assignments and creates perceptions affecting their military career in a negative way...PTSD, called 'shell shock' in World Wars I & II, has been a part of every war in human history and still nobody can define what it is...A current definition is that PTSD might occur after someone has "experienced, witnessed or was confronted with an event or events that involved actual death, threatened death or serious injury"

There is a recorded history of soldiers returning from World War II with severe problems concerning PTSD and TBI and these soldiers were little discussed and they were not featured in the magazines of that time that profiled and applauded the 'real' warriors...Today there are many stories from families worldwide, familiar stories of men who returned from war and were silent and remote for the remainder of their lives, unable to understand the reality they experienced and unwilling to communicate their confusion and trauma to unsympathetic military structures...

Research will one day find PTSD is reactionary trauma within intellectual and logical circuits of the brain unable to rationalize and reconcile a deep-seated natural morality with the assigned murder of other human beings in any situation or circumstance...An ongoing and continuing mental conflict as these affected soldiers try to do what their human counterparts urge them to do, insist is the right thing to do and at some point becomes impossible to do...There is guilt, not always over killing another human, but guilt and a sense of failure over their inability to kill as expected...And there is confusion, nightmares, visions, depression and an inability to reconcile their purpose in life with past experiences and a continued existence that seems 'wrong'...War is an extremely primitive stage in the development of a higher and increasingly complex intellect and will not exist in ten thousand years...

Among all veterans in the U.S., a suicide occurs every 80 minutes...In 2012, 185 active duty military in the U.S. committed suicide...In 2013 the number was 150 active duty and 151 non-active personnel in the reserves...Suicide kills more military personnel than combat and 95% of military suicides are male...

$2.4 trillion, or 4.4% of the global economy "is dependent on violence" according to the Global Peace Index, referring to "industries that create or manage violence", the industrial military complex and the defense industry...

Approximately one of every 169 humans on Earth has been displaced by war, internally or cross-border...

THE MIDDLE EAST

There are no globally accepted boundaries defining the Middle East...An area of approximately 4 million square miles with Turkey in the north, Iran in the east, Yemen and the north of Africa in the south and the eastern coast of the Mediterranean Sea to the west...Current population is estimated at more than 360 million humans...

One of the oldest continually populated regions on Earth surrounded in controversy, unending violence, extreme oil wealth, a chronic mistreatment of females (misogyny) and a fear of women that is extensive in almost every Arab society...Science has recognized and proven in every way the equality and capability of men and women but ignorance, superstition, religion, tradition and an unusual fear of females influence the behavior of millions of males who are seemingly incapable of understanding this reality...Over thousands of years the Akkadian Dynasty, the Assyrian Empire, Persian Empire, Roman Empire, Arab Empire, Turks, Mongols, Christians and the Ottoman Empire influenced and changed Middle Eastern culture and territories determined by the religious significance of past events, natural geographical borders to provide defense, control of subservience populations and the ego and perceived righteousness of conquering civilizations and their leaders...The 20th century saw a growing dominance of rising European powers seeking wealth, control and stability create a Middle East of their choosing with arbitrary enforced borders and mandates of European rule...During World War I the allied powers of Europe, primarily Britain and France, subdivided the Middle East among themselves with the Sykes-Picot agreement (Asia-Minor agreement) a secret document giving them power to establish boundaries without regard to differing populations and promising a recognition of Arab states with protections and future independence...A subsequent document just 18 months later (Balfour Declaration) challenged the validity of Sykes-Picot with a promise to Zionist Jews to establish a homeland in Palestine (many Jews migrated to Palestine immediately after the war believing it was promised) creating the modern Jewish-Palestinian conundrum which has challenged the peace of this entire region for almost a century...

The Middle East today is among the most violent regions on the planet...The discovery of oil reserves has transformed the region into a consortium of nations ruled by dictators and monarchies who are dependent upon a single product, *oil*, to provide revenue necessary to placate restless populations and purchase weapons needed to exist in an area of continuous threats and unending conflict...The U.S. has long been the guarantor of stability in this region protecting nations and their enormous reserves of fossil fuels, but in the 21st century the world watches as violence crosses borders and frightened ruling coalitions kill their own people and implement draconian laws and restrictions to prevent the spread of unrest and dissent...Iran rules a modern population with intimidation and violence while continuing to develop nuclear weapons which could change the dynamics of the world...Israel threatens war with Iran while continuing to treat Palestinians as subjects, building on their land and stifling prosperity with movement restrictions and unending blockades...ISIS threatens Iraq beheading innocents and burning humans alive while the world watches their videos and does little...Syria is on the brink of a meltdown with Assad murdering his people (hundreds of thousands) in a desperate attempt to hold power...Saudi Arabia placates their population with money and spends billions ($) on weapons while treating females as slaves and second class citizens...Yemen is a chaotic enclave of terrorists located across the water from Somalia (a failed state) and multiple nations and members of the UN Security Council sit on their hands with ineffective interventions and a focus on sovereignty and profit being many times more important than the lives of millions of humans who live in this region...In the course of this century the ancient nations of Iran (Persia), Egypt and Turkey will likely keep their boundaries and sovereignty while Iraq, Saudi Arabia, Syria, Lebanon, Jordan, Palestine and the Yemen peninsula position for power and territory amid the unpredictable violence and politics that define this area of Earth...

Control over the production and distribution of oil is the decisive factor defining who rules in the Middle East...

In 2013, the U.S. approved the sale of more than $10 billion in weapons to the Middle East nations of Israel, Saudi Arabia and the United Arab Emirates..."This is more advanced weaponry than we've sold before," (Senior Defense Department Official)...For Israel, air-to-ground anti-radiation missiles, radar, KC-135 air tankers and the V-22 vertical-lift aircraft..."This year the U.S. provided $3.1 billion in foreign military financing to Israel, the highest the United States has ever provided," (Senior Defense Department Official)...United Arab Emirates got a $4.25 billion fighter-jet deal with an option to buy as many as 25 F-16 Block 60 multi-role fighter-jets, a variant sold to no other country, as well as long-range air-to-ground missiles...For Saudi Arabia, an advanced class of precision "standoff munitions" to fit the 84 F-15 fighter jets purchased under a previous arms deal...In a win-win deal for the U.S., record profit's for the world's largest industrial military complex and stability(?) in the Middle East...

BAHRAIN
(1.3 million humans)
Military Forces = 15,000 active duty / $700 million per year

A proxy of Saudi Arabia and ally of America because of it's strategic location and relationship with the Saudi's, America looked the other way while the leaders of this country and the military of Saudi Arabia crushed a Shiite led movement for transparency and power sharing in 2011...In September 2012, the government of Great Britain launched an inquiry into the relationship between Bahrain and the Saudi's concerning 'human rights abuses', an inquiry criticized by the Gulf Co-operation Council as 'foreign interference into our affairs'...As is always the case, the British confirmed Saudi Arabia and Bahrain are 'key strategic partners and close allies'...This island nation is home to the U.S. Navy's Fifth Fleet and is ruled by Sunni Muslim's over a Shia majority with a continuous ongoing tension in every segment of society between the Shiia and Sunni, humans who are fundamentally the same in every way without the awareness to understand this reality...In 2002, Bahrain became a constitutional monarchy allowing women as candidates for the 1st time but no females were elected (3 women were elected to parliament in 2014)...The king is supreme ruler with members of his family holding important positions in government while protests continue in the uncertain aftermath of the "Arab Spring"...Media freedom is limited with heavy censorship and arrests if the government deems it necessary for security...There is a high volume of foreign workers who are subject to abuse, forced labor and imprisonment...Extensive environmental degradation of multiple coastlines due to the high volume of ship traffic, oil spills and discharge from multiple refineries...Child labor force is over 5,000...Over half of Bahrain's economy is energy concerning transport and refining with 92 billion cubic meters of natural gas reserves and 124 million barrels of crude oil reserves...Part of a coalition active in air strikes on rebels in Yemen...

Over a million Palestinians live in Middle East refugee camps in a surreal existence of daily survival without destination or conclusion more than three generations after the 1948 war with Israel...

GAZA
(1.8 million humans)

In October 2012, the emir of Qatar became the first head of state to visit Gaza in thirteen years, a trip called "astounding" by Israel...Sheik Hamad bin Khalifa al-Thani crossed into Gaza from Egypt with a large international delegation and was given the VIP treatment by Hamas Prime Minister Ismail Haniyeh as Qatar's emir continued working to improve the diplomatic standing of his country and to mediate between Hamas and the Palestinian Fatah movement in the West Bank...In the short history of this territory vital to a future Palestinian state, violence, uncertainty and oppression have become a way of life with the uncertainty of the Hamas governing authority, the inconsistencies of a Fatah Palestinian government and stifling oppression and control by the Israeli government...A series of agreements over two decades resulted in Israel withdrawing settlers and soldiers from Gaza and most of the West Bank but Israel still has absolute control over airspace, borders and the sea...As a result, humanitarian aid to help children and the population of Gaza have been intercepted by Israeli's military and prevented from going where it's desperately needed, even when sent by non-violent groups with open inspections...

What is Israel's authority to close the Gaza coastline to all maritime traffic? Every legality cited by Israel continues to be murky without substance...How can they declare a maritime zone off the coast of Gaza and destroy an economy with restrictions and violence? Is Israel acting unilaterally? Can any nation do this? If terrorists in Montreal were shooting rockets at the United States would America have the authority to destroy the city suburbs and infrastructure killing innocents and children because they felt threatened by a small group of humans? By what law or decree can Israel control the existence of an entire human society restricting water, food and movement with devastating results? The Israeli government's response is always the same in claiming food, construction materials, toys for children and necessities for life will be used by the minority radicals in Gaza to build missiles and weapons to attack the Israeli state when Israel has a defense capability to prevent almost any attack from inflicting serious harm...Somehow they still don't understand that restrictive apartheid like measures will create a future with more violence and more determination from an oppressed population without hope...

If Israel would make a humanitarian effort to allow humans in Gaza and the West Bank to live a normal life with open trade, state of the art electronics for children and freedom of movement, Palestinians would act within their own population to stop aggression and the radicals that threaten a peaceful coexistence with the Jewish population...

In 1998, a treatment plant was built in Gaza to remove chromium 3, a product used by tanneries that becomes the carcinogen (cancer causing substance) chromium 6 if untreated...The plant's extraction process used sulfuric acid to remove and recycle the chemical...In 2005, Israel determined that sulfuric acid was a "dual use" chemical which might be used to make bombs and they banned the substance...The result of this action is water contaminated with chromium 6 allowed to flow into cities and out to sea...NYT's 2014

After the Israeli invasion of 2008, international organizations and governments pledged over $4 billion dollars to rebuild Gaza but little happened and in 2014 Israel destroyed even more of Gaza in reactionary ignorance...More billions were pledged to rebuild but Israel will not allow reconstruction to happen...Many of Gaza's families live in refugee camps, a quarter of dwellings with no sewage system, limited access to fresh water and few options for modern medical care...In a country just over 350 sq km this is beyond incompetent...For any country bordering a major water source, access to the water is vital for stability and commerce yet Israel has blockaded all maritime traffic for over three years while the rest of the world looks the other way...Israel maintains that the blockade has not contributed to the humanitarian crisis (?) a stunning ignorance of an obvious reality...This statement is a continuation of a transparent web of nonsense coming from a government frightened of reality without the strength or awareness to make the changes needed to promote and grow intelligence instead of destroying it...

Question...Can any nation act unilaterally to occupy and control a smaller nation by insisting they feel threatened by the population they are occupying and controlling to prevent progress and growth? Is a loaf of bread a weapon?

The UN Human Rights Watch and many influential international organizations agree Israel has successfully occupied Gaza by controlling every aspect of regional vitality and economic stability...It's not disputed that the Hamas organization has violated human rights, forcibly installed themselves in all institutions of control and lack the intelligence and insight to understand the needs of populations of poor and desperate people, but Israel has the power and the stability to change everything...They have little to fear from a Hamas military with no teeth and they should find a way to build a stable economy with international help for a majority of Gaza citizens who wish for a "normal" life...Give these people what they need to build a working society, allow imports, exports and freedom of movement *and* be intelligently vigilant, ready and capable of eliminating individuals or groups who commit unprovoked violence against anyone for any reason...Israel has the military, intelligence and security to target the right targets...Do not punish a million people, destroy homes and traumatize these children (the next generation) out of anger and desperation over a few thousand ignorant men who cannot understand the possibilities of peaceful co-existence...The population in Gaza is literate and young (median age is 18) with very high unemployment (45%) and widespread poverty (30% below poverty line) in a population increasingly dependent on humanitarian assistance...Exports and imports, the lifeline of any economy in a global market, are not permitted by Israel...How is this possible? It is far more challenging to solve problems with intellect, compassion, flexibility and determination than to secure an entire population with guns, walls and checkpoints...Until people in Gaza are given standing and opportunity equal to all the people in this region and until everyone understands the impact of this conflict on the developing brains of babies with continual violence, confusion, fear, stress, hunger and a desperation continued for decades, there will be no recognition that Israelis and Palestinians are the same people and this violence will continue generation after generation without end...

In 2015, the UN issued a report stating Gaza could become uninhabitable by 2020 with current border restrictions...

IRAN
(80 million humans)
Military = 538,000 active duty / $6.2 billion per year

In early 2012 elections in Iran resulted in a conservative parliament strengthening the power of supreme religious leader Ali Hoseini-Khamenei and diminishing the influence of the "green revolution" supporters who called for greater democracy and increased transparency in a country ruled by religious dictum's...A theocratic republic with a legal system based mostly on Islamic law, the supreme leader is installed for life and despite various councils, political parties and the Assembly of Experts, Iran is a religious state where clerics determine "behavior" and a military security structure controlled by a few maintain the status quo...A rigorous process to weaken opposition and dissent has been successful in muting a population suffering from world sanctions which has devalued the currency amid a reduction in oil exports, Iran's largest source of revenue (Energy wealth in Iran is 157 billion barrels of crude oil reserves and 34 trillion cubic meters of natural gas reserves)...In 2015, a coalition of nations agreed on a structured timeline with restrictions on processing and the storage of uranium while still allowing Iran enough capacity to fuel nuclear power plants...In return, sanctions would be lifted improving the economy and fulfilling a campaign promise made by the new president (Hasan Fereidun Ruhani) in 2013...However there is little doubt despite Iran's promises of cooperation on the international stage that some push for nuclear capability will continue with or without an inspection framework that can monitor all activities inside the country...Only change in the power structure by a secular majority will prevent the pursuit of nuclear weapons and military intervention would be disastrous no matter how it developed...Nuclear facilities are widely dispersed and it would be impossible to seriously damage their capability without widespread devastation across a nation of 80 million humans...

Strategic pinpoint strikes on any facilities by Israel or America would only entrench the Iranian government in negative ways with serious political and military implications for the entire region...The rational powers of the world must be vigilant and ready in strategic and cooperative ways to deal with the use of nuclear weapons anywhere on Earth but the isolation of any government is not helpful in the pursuit of cooperation and military action at any scale on a planet of seven billion is almost always destabilizing and counter-productive...

Iran is a country of mostly intelligent rational people who would not choose such a government if there were free and fair elections but the interests and direction of this country are in the control of the few with substantial internal security forces to maintain that control...*This will not last*...Khamenei is elderly and it's unknown what will happen when this leader is no longer a voice in the direction of a nation internally divided and conflicted in how to protect domestic interests and national resources while establishing foreign partnerships and becoming a force of unity and cooperation on the global stage...With their wealth and influence (Iran has large oil reserves and some of the largest natural gas reserves on the continent) Iran could be pivotal and instrumental in improving the volatile and unstable situations that exist in the countries that surround them promoting honest communication with regional populations about changes needed to compete and prosper in a growing global community...But Iran's economy is still suffering from sanctions and ineffective strategies with an unwavering focus on a pious agenda and war (Iraq & Syria) muting a population with threats, a moralistic agenda and controlling women with draconian laws...With Iran's location, history and resources, this nation could be the center of an Asian renaissance if there were aware and intelligent leaders...

The government controls all broadcasting and influences the print media in a large population of young humans (median age is 29) starved for information and inclusion in a vibrant, exploding global community...Before the Islamic revolution of 1979 females were a valuable sector of the work force with the freedom to vote, dress as they wished, attend college and marry by choice...In 21st century Iran women have very few rights, they are forced to observe an Islamic dress code, they are segregated from men in almost all activities, they are forbidden to become judges or attend school after marriage and are subject to imprisonment and lashes for exposing any part of their body except the hands and face...Girls are considered adults at age nine (in lunar years=354 days), can be tried in criminal court and even put to death...The legal age of marriage is just thirteen although the father can force his daughter to marry earlier...Women must have permission from a father or husband to travel and they cannot be financial guardians for their own children...Men can divorce at will and marry multiple woman...Women can be stoned to death...Despite this, Iranian women have a storied history in this nation (the great empire of Persia dates back to 6th century BC) and intelligent women will continue to resist and organize across many layers of this society to win back their rights and the freedoms all humans are entitled too...

In 2012, Khomeini's government was the only government in the world to execute juveniles determined by the age at which a child reaches puberty...Boys were eligible at age 15 and girls at age 9...In 2012, Iran introduced a new penal code which they said abolished the execution of anyone under age 18 but the new code still considers the "age of responsibility" to be puberty which casts doubt on the literal interpretation of these new "laws" by Iranian judges...

IRAQ
(37 million humans)
Military Forces = 272,000 active duty / $6 billion per year

The British governed this country with occupation until 1958 when a series of assassinations, coups, internal rivalries and multiple successions over two decades led to the rule of Saddam Hussein in 1979...The shortsighted policy by the U.S. to invade and occupy in 2003 to insure the flow of oil and expand their footprint in the Middle East ultimately left Iraq with very challenging security problems, a fractured government and a centuries old dynamic between Sunni and Shia destroying a society with enough energy resources to rebuild a modern infrastructure inclusive to everyone in this nation...Once ancient Mesopotamia, Iraq is old and recognized as one of the world's first civilizations, a region that flourished with human activity and commerce on the banks of the Tigris and Euphrates rivers for more than two thousand years until being conquered by the Persians in 539 BC and undergoing a turbulent history until the Abbasid Caliphate conquered the Islamic world and established Baghdad...Occupied by Britain during WW I, liberated by revolution in 1920, occupied again during World War II, Iraq declared itself a republic in 1958 and had an opportunity after the U.S. ended their occupation to be a powerful democracy in the Middle East but the handpicked leader Nouri Al Maliki proved to be both politically incompetent and obsessed with ancient Sunni and Shite rivalries that were more important than good governance...Humans who are the same but believe themselves to be different continuing to kill each other in an ongoing conflict over long-standing ethnic and political divisions based on heritage and religion, killings which escalated in 2013 and killings that continue today...National unity impossible with ethnic militias and the invading forces of ISIS...Today Iraq is a broken nation with conflicts surrounding their borders sitting on top of large oil reserves that could rebuild this nation destroyed by war and conflict if an all inclusive government could moderate this primitive ignorance of religious and ethnic discrimination...*This will not happen...*

A law is being considered in Iraq to allow girls as young as nine to get married and would require wives to submit to sex when commanded by their husband...The legal age for marriage in Iraq is 18 without parental approval and 15 with a guardian's approval...The new law is based on the principles of a Shiite school of religious law founded by the sixth Shiite Imam...The draft measure was approved by the Cabinet in February 2014...Without mentioning marriage, the law sets rules for the divorce of girls who have reached the age of 9 in the lunar Islamic calendar (It also identifies this as the age girls reach puberty, essentially 8 years 8 months)...The father is the only one who can refuse or accept the marriage proposal for the "child"...This law also allows a husband to have sex with his wife regardless of her consent, prevents women from leaving the house without their husband's permission, restricts a woman's rights determining parental custody after divorce and allows men to take multiple wives...

An official report by the Human Rights Ministry said over 85,000 Iraqis were killed by violence between 2004-2008 including over 1,200 children and more than 2,300 women although international observers agree these figures are a very conservative estimate and the actual casualties will never be known...In May 2013, more than 1,000 Iraqis died in bombings and violence and in July 2013 another 1,000 died as sectarian violence continued under the government of Prime Minister Nouri Al Maliki who promised unity and inclusion but is revealed as a simple-minded man unable to understand reality or possibilities...Energy wealth is 144 billion barrels of crude oil reserves (5th in the world) and three trillion cubic meters of natural gas reserves (12th in the world)...There are over 700,000 children in the labor force and 27% of the population exist under the poverty line with an unemployment rate of 20%, humans who witness ethnic violence daily on a scale not imagined by western populations...In a violent region of the world with a delicate balance of power between multiple nations and the uncertain impact of Russian and American weaponry arming Middle Eastern countries, any positive future is cloudy and fragile as Iraq fights the invasion of ISIS amid an uncertain Middle East hierarchical dominance based on religious ideology, centuries old ignorance and profit for the few who control resources...

In all of 2013, over 7,800 humans died violently in Iraq at the hands of the military, government security forces, militias with no restraints, Al-Qaeda and civilians...

Civilian deaths in Iraq in 2014 totaled 17,049...IBC (Iraq Body Count)...Iraq Government reported 15,000 dead...

America gives an estimated $3 billion in military aid to Israel every year and 75% comes back to America's military industrial complex to purchase weapons of war...A very cozy hugely profitable arrangement...

ISRAEL
(8 million humans)
Military Forces = 167,000 active duty / $15 billion per year

The only Jewish state in the world, Israel exists as the result of a United Nation's vote in 1948 to partition Palestine and create a Jewish nation in the 'heart of the holy land', a region of Earth occupied since antiquity by Muslims, Christians and Jews, soaked with the blood of thousands of humans who have died over thousands of years because of territorial instincts, ideology and an obsessive focus on various religious beliefs that have little basis in fact...A continuing source of conflict since 1948, Israel continues to annex and steal Palestinian land recognized as such under international law, occupy and settle in Palestinian territory claiming it as their own and treating occupied residents as a lower class of human without rights enforced by a military that is uncaring and unilateral in its actions toward anyone or anything that will not bow to Israel's demands for obedience, ownership, control and territory...The Palestinians living next door to Jewish settlers have no right to vote, limited access to the justice system and very limited rights as an individual while the Jewish settler just next door has voting rights, full citizenship and open access to the judicial system to resolve disputes...

In September 2016, the U.S. and Israel signed a $38 billion dollar deal to supply Israel's military with weapons over ten years ($3.8 billion per year)...Israel must spend the majority of the money on American military industries...

Israel's Declaration of Independence promises it will "ensure complete equality of social and political rights to all its inhabitants irrespective of religion, race or sex; it will guarantee freedom of religion, conscience, language, education and culture..."

It's clear to most of the world that Jerusalem should be the capital of both Palestine and Israel, clear that borders should be established guided by the boundaries that existed before the 1967 invasion and occupation, clear that negotiations should be flexible, transparent and fluid concerning Palestinian refugees as well as Jewish settlements and clear that military force and apartheid like policies enforced and demanded by the Jewish state and criticized even by their closest allies are unworkable and without reason...Extremists on each side of this conflict sustain each other...Israel's population of over 8 million live in a small geographical area just over 8,000 sq miles in an economy of high-tech service industries, agriculture, manufacturing and a continuing acceptance of foreign aid (mostly from the US) to supply a military with weapons to control an occupied population and defend itself from external threats...Given the current policies and political structure in both the U.S. and Israel, it's hard to see any kind of end to this conflict...Humans living below the poverty line is around 40% in the West Bank, over 50% in Gaza and less than 20% in Israel *and all these people exist in the same place*..In a 2008 invasion of the Gaza Strip, over 1,400 Palestinians killed (over 170 children) and 13 Israelis killed with the Israeli Defense Force firing phosphorous shells at targets in Gaza, including a UN Relief Agency, destroying food and medicine and accepting no criticism of their army's assault on what is essentially a civilian population...

More than 115 missiles were fired into southern Israel from Gaza and Israeli planes launched numerous strikes in retaliation...Seven Palestinians, three of them gunmen, were killed...Eight Israeli civilians were hurt by the rocket fire and four soldiers wounded by an anti-tank missile...Days later, Israel killed the military commander of the Islamist military group Hamas in a missile strike on the Gaza Strip and launched air raids across the enclave pushing the two sides to the brink of a new war which erupted for just over two weeks killing 160 Palestinians(40 children) and 4 Israelis until Egypt negotiated a tense cease fire...November 2012

Israel has the defenses, intelligence and the responsibility to counteract isolated attacks without killing innocents or destroying property...Israel has the right to defend themselves and they do with an advanced detection system which destroys many missiles before they can reach the ground but when Israel attacks the population of Gaza they always kill civilians and children while 'targeting' Hamas houses, offices and government buildings with a stated intent to 'disrupt communication'...Somehow civilian infrastructure is always destroyed and children always die as the elected government of Gaza is targeted to kill those Israel 'believe' are complicit in the acts of the few who are firing missiles...Israel should target missile launch sites every time it detects a launch and do it with pinpoint accuracy and little collateral damage but the reality that no one seems to understand is instead of destroying life and property, instead of blocking aid and existing in this continual state of paranoia, if Israel (clearly in charge of transport, resources and freedom) if Israel would flood the Palestinian population *and* the Jewish population with food and resources, electronic games and learning devices for children and babies while allowing aid workers to help impoverished people survive and nourish their children, this conflict would disappear in two generations and long before that the people would act on their own to stop militants from launching rockets at Israeli cities...Future generations of Jews and Arabs would have the consensual intelligence to recognize the reality of this situation, they would both understand that sharing the land and sharing Jerusalem as the capital of each nation would be a symbolic gesture of peace and unity to the entire world and they would recognize that a shared wealth and prosperity is possible for all genders and ethnicities in this region...

If the world were to offer another location on this large beautiful planet to establish a nation with running water, fertile soil and 10x's the territory, neither the Jews or the Palestinians would accept because of the historical religious nonsense associated with this small desolate location...

This region has beauty as do all places on Earth and there is an attachment through all cultures with the geographical location of one's birth and adolescence but the violent extremism centered on this focal point of ancient events is very clear evidence of the human species intellectual awareness and primitive priorities at this early stage of development...The history and possible future of the Jewish state are obvious and respected by billions of humans around the world who promised to support and defend a progressive nation who is not a threat to it's neighbors, Arabs and Palestinians, who also have serious and valid concerns about their future in a region of the world they have inhabited for centuries and the question is how these two populations can move past what has been unsolvable to create a real peace which could change the dynamics of this entire region, maybe the entire world...The Israeli government must quit believing the world wants to eliminate Jewish populations from the planet (the few who believe such nonsense are recognized as ignorant by almost every nation on Earth) and establish an effective state of the art defense with help from allied nations...The Israeli's need to stop living behind walls and remove these barriers to the rest of the world that are unnecessary hardships on innocent people who want to live life as equals with freedom of movement, healthy, happy children and prosperity...The violent and ignorant humans on this planet who are unable to recognize reality are mostly obvious and with the technology today these individuals and groups can be identified and located with precision and accuracy in what will be a continuing vigilance lasting for years but without the restrictions and barriers that restrict the growth of an entire society...Change this with empathy and generosity to grow a new generation who understand the real possibilities of a shared existence...The Jewish heritage, population and culture is a growing dynamic in a global society and has equal standing with any ethnicity and culture on this planet...Rational, intelligent humans understand that Israel has a right to exist but Israel must recognize all humans are the same species and the only path forward that will be both profitable and secure is an open society with equal rights for all humans and the sharing of a land with a history like no other, belonging to no one and belonging to everyone...

Benjamin Netanyahu will never sign a peace agreement with the Palestinian government...His policy for many years has been to talk of peace in any forum but do nothing behind a blanket of misdirected attacks, conveniences and demands that are impractical or impossible...He believes that Jerusalem belongs to Israel and all land occupied belong to the state...There will be no compromise with Netanyahu in office...

In the West Bank there are a 1,000 Jewish settlers living in Hebron, home to more than 150,000 Palestinians with a massive system of checkpoints, road closures and guard posts operated by the Israeli military to protect the settlers...B'Tselem, the Israeli human rights group, have said that the Israeli authorities "have expropriated the city center from its Palestinian residents and destroyed it economically"...Over 2,000 Palestinian shops have been closed, some welded shut, thousands of people driven from their homes to accommodate the settlers occupation of illegal land with Palestinian babies born at checkpoints where mothers are not allowed to travel to hospitals and water diverted to Israeli citizens and businesses in obvious unfair and subversive ways...

In late 2012, a UN resolution approved to upgrade Palestinian status to a non-member observer state was vehemently opposed by the U.S. and Israel even as both these nations stated the vote meant nothing...*then why the opposition?*...Israel has even threatened punitive action against the Palestinian people for seeking this vote...The United States and Israel are worried this will allow Palestine to join the International Criminal Court (*why is this a problem?*)...Worried that the Palestinians might use this status to seek membership in specialized agencies within the UN (why is *this* a problem?) and insist this will not help in negotiating for a two state solution...*Most of the world has left the U.S. and Israel alone on this stage as the ignorance and hostility existing in the pro-Israeli lobbies of America and the right-wing parties in Israel is staggering and obvious*...This highlights the inability of leaders in both nations to understand causality, understand an obvious reality and rise above ancient religious dogma to recognize fair resolutions creating a stable, secure and prosperous society...They cannot understand that a legitimate Palestinian state would allow both nations to end these ceaseless hostilities...

July 2012...An Israeli government appointed commission said Israel's presence in the West Bank is no occupation and recommended the state grant approval for unauthorized Jewish settlements...

August 2013...New peace talks brokered by U.S. Secretary of State John Kerry were agreed upon with a nine month timetable...The day after this announcement Israel approved building 1,000 new homes on internationally recognized Palestinian territory confiscating land by seizure and occupation...

"The Palestinians must be made to understand in the deepest recesses of their consciousness that they are a defeated people..." 2002 Moshe Yaalon, Israeli Defense Force Chief of Staff

October 2013...Israel announced plans to build 1,500 settlement homes on land designated to be part of the future Palestinian State...Almost all nations see these settlements (whose population has tripled in 20 years) as illegal...

April 2014...Benjamin Netanyahu ended peace negotiations with the Palestinian government after Mahmoud Abbas said Palestine would seek to join United Nations organizations...

In July 2014, Israel launched an invasion into Gaza to 'go after' missile sites and as an act of revenge after 3 Israeli's were killed by Palestinian individuals...At the end of seven weeks Gaza infrastructure was destroyed, 1,492 civilians were dead (UN) and over 500,000 Palestinians in Gaza 'displaced' by the destruction...It's almost 2 years since this invasion and almost nothing has been done...$5 billion has been pledged to help rebuild this society and provide for the homeless but Israel won't allow it...This country continues to destroy the lives of innocent people with no remorse, innocent humans trying to live some kind of normal life amid hostilities and restrictions that never end...In March 2015, Netanyahu is in the US eating good food in a 5 star hotel while thousands in Gaza sit in the rain and cold under tents with nothing to eat, women, children and babies...The Jewish people have a right to live as anyone else but not the right to prevent other people from controlling their lives...Insane paranoia and religiosity...

Somehow Israel has complete control over Palestinian life...By who's authority?...The UN? A consortium of nations? Is this international law? Who is the authority??? Can the U.S. occupy Mexico by saying they feel threatened by people who hate them firing isolated, ineffective ordnance over the border? If a Mexican kills an American can the U.S. government cross the border and destroy the family home and property?

JORDAN
(8 million humans)
Military Forces = 115,000 active duty / $1.5 billion per year

A constitutional monarchy bordered by Saudi Arabia to the east, Israel and the West Bank to the west and Syria and Iraq to the north...One of two Arab nations that maintain peace with Israel, Jordan's economy is dependent on foreign aid and tourism with the king having absolute power to determine government structure and functions...This remains a stable and conservative society based on Islamic law with an acceptance of western values and modern new technology...The state controls all radio and television media with restrictions and threat of imprisonment for critical reporting of the government or religion and in the summer of 2013 the government instituted a series of laws and regulations to control independent media on the internet and censor free speech and opinion...The Syrian refugees escaping violence by coming into Jordan are 630,000 registered with over 100,000 at the Zaatari refugee camp, more than 400,000 living in Jordanian society and a government estimate of a million unregistered in the nation creating a huge problem in this tiny country with few natural resources and a dependence on aid for survival...Just a million barrels of crude oil reserves and 6 billion cubic meters of natural gas reserves...Median age of around 22 with a growing youth unemployment rate of 30%...Over 15% of the population exists below the poverty line, a number that is growing with the influx of refugees from Syria...Tension and scattered protests in the wake of the Arab Spring, unending violence in Syria and a large Palestinian refugee population (over two million)...

Ongoing long standing problems in the Arab world are the lack of individual freedoms, a minimal investment in modern education and the ignorant abuse of females in a culture of dominating males who fear women...

KUWAIT
(2.8 million humans)
Military Forces = 15,500 active duty / $5 billion per year

Kuwait is a small country on the northwestern edge of the Persian Gulf surrounded by Iraq, Iran and Saudi Arabia...A major producer of oil for half a century, Kuwait is ruled by the Al-Sabah family with an elected parliament inclusive to women and subject to political tensions within the ruling family who have dissolved the parliament multiple times in less than a decade...In 2012, the Emir of Kuwait blocked a proposal to amend the constitution to make all legislation in the country comply with Islamic Shariah law and dissolved an Islamic majority parliament...This wealthy country has almost no poverty among citizens supported by subsidies and governmental programs but there is a large population of foreign workers living in substandard conditions, common in this region where trafficking of humans for labor is widespread and abusive...Over 100 billion barrels of crude oil reserves and 1.8 trillion cubic meters of natural gas reserves...Kuwait has allied with western powers in both Iraq and Afghanistan conflicts while facing unrest and protests from it's citizens over a monolithic system of government control and perceived corruption at all levels of governance...Media is restricted with penalties, fines and imprisonment for reporting critical of Mohammed, God or inquiry into the activities of the ruling family with the censorship of all films and books deemed "immoral" routine...Median age of 29 with a life expectancy of 78 years...A population of over 90,000 "stateless" humans with few rights (bidun – Bedouin tribal descendants) who have been designated "illegal residents" by Kuwait authorities...

Almost the entire population of Kuwait live in the capital city Kuwait (2.3 million) located on the west coast of the Persian Gulf...Urban living, a reliance of foreign workers for labor and a lifestyle reliant on government programs to sustain a population, Kuwait now ranks 9th in the world in obesity, almost 50% of it's citizens...

2015...Kuwait is part of a strike-force against Houthi rebels in Yemen to return President Hadi to power...

LEBANON
(6 million humans)
Military Forces = 130,000 active duty / $1.7 billion per year

Located on the Mediterranean Sea with Israel to the south and Syria on the east, Lebanon is a hodgepodge of populations and ethnic tensions in a country only half the size of Israel...More than 10% of the population are Palestinian refugees in substandard living conditions with almost no legal rights and over a million Syrian refugees fleeing the crisis across the border in a diaspora that threatens every sector of Lebanon's fragile society...Israel has invaded this country multiple times fighting Hezbollah in southern Lebanon with Syria occupying parts of the country for almost 30 years...The government is an amalgamation of Sunni, Shiia and Christians ruling in an unstable coalition that has seen assassinations, resignations, boycotts, parliaments delaying elections, a prime minister who has assumed the duties of president (position is empty since May 2014) and militant political group Hezbollah (Shiite Muslims), a group involved in every twist and turn of this nation's unpredictable future who have resisted all attempts to reduce their forces in southern Lebanon...A self-appointed police force in the continuing absence of a stable, structured unified government, Hezbollah is actively fighting in the Syrian conflict allied with al-Assad's regime...The organization is estimated to have tens of thousands of missiles (many aimed at Israel) with military hardware, weapons and money supplied by Iran using Syria as a conduit...A powder-keg awaiting a match...Median age around 30 years, 50,000 children in the labor force with 5% of children under age five malnourished...30% of the population live below the poverty line...There are around 450,000 Palestinian refugees in country, over 60% living in 12 overcrowded refugee camps with limited rights, poor housing and poverty...

Prime Minister Tamam Salam issued new restrictions in 2015 to stem the tide of refugees from Syria after parliament extended their term in office to 2017 citing security concerns that would disrupt planned elections...

OMAN
(3.1 million humans)
Military Forces = 72,000 active duty / $6.6 billion per year

Oman's Sultan Qaboos Bin Said Al Said has ruled for over 40 years as defense minister, foreign minister, finance minister and prime minister in a nation strategically located on the western coast of the Arabian Sea near the entrance to the Persian Gulf...An ally of the U.S. with oil as it's main export, Oman is a stable society with geographic variety, a large fishing industry and government controlled media with extensive censorship of the internet...Women are permitted to hold a position in Oman's Consultative Council and all citizens over age 21 are allowed to vote...In response to citizen protests following the Arab Spring in 2011, Oman has reformed legislative and regulatory structures to allow accelerated job creation and the election of municipal councils in 2012 with the power to advise the Royal Courts on situational needs of citizens in Oman's eleven districts...Malnourished children under age five at over 8%, unnecessary in a society where subsidies and government welfare support a large sector of the population...A singular reliance on energy reserves (crude oil reserves 5.1 billion barrels and natural gas reserves at 850 billion cubic meters) has Oman actively seeking foreign investment, promoting tourism and promoting job creation in public and private sectors to replace foreign workers and lessen the need for subsidies to it's citizens...In a volatile and unpredictable region of the world, Oman spends 10% of GDP on the military and maintains active relations with the west...Governmental authority to censor media publications as necessary...

With over 2,000 kilometers of coastline with waters of high visibility (15- 25 meters), tourism is a growing industry in Oman with whale-watching and diving popular, festivals according to season and tradition and many ruins and artifacts from a history measured in thousands of years..

QATAR
(2.1 million humans)
Military Forces = 12,000 active duty / $1.9 billion per year

Third largest natural gas reserves in the world (25 trillion cubic meters), a wealthy country with the world's highest per capita income and a rapidly growing economy with a ruling family focused on building a modern society in education and art with a vision of the future...An emirate and an advisory council of 15 appointed members (plans to add 30 elected members to the council were put on hold in 2013), Qatar has been a pivotal player in the dynamics and politics of the Middle East as mediators and support for foreign forces in the many conflicts and wars that dominate this region of the world...Emir Sheikh Tamim bin Hamad al-Thani is expected to continue the cultural expansion and modernization of this country which is a peninsula off the eastern border of Saudi Arabia in the Persian Gulf...The Qatar government owns and operates Al-Jazeera, a global voice of the Middle East whose reporting covers in depth the politics of the region...The government "oversees" radio, TV, internet and publications with censorship and legislation which prohibit any criticism of the regime or inciting dissent among Arab populations...Lowest unemployment rate of any nation, 25 billion barrels of crude oil reserves, Qatar has a high literacy rate and a life expectancy of 80 years...A report in 2013 by Amnesty International highlighted the situational reality for over 1.3 million foreign nationals working in Qatar who in many cases are treated as servants and slaves with restrictions on movements, poor health care, inadequate or no salary and no way to address abuses or leave the country... A tiny country in a very strategic location across the Persian Gulf from Iran, Qatar's survival is dependent on it's foreign relationships...

"If a woman drives a car, not out of pure necessity, it could have negative physiological impacts as functional and physiological medical studies show that it automatically affects the ovaries and pushes the pelvis upwards...That is why we find those who regularly drive have children with clinical problems of varying degrees." Saudi Arabia cleric Sheik Saleh bin Saad al-Lohaidan in 2013...(There is no medical evidence for this nonsensical statement)

SAUDI ARABIA
(28 million humans)
Military Forces = 233,000 active duty / $66 billion per year

One of the heavily weaponized nations in the Middle East with the United Kingdom supplying Typhoon Jets and the U.S. selling 84 F-15SA Fighter jets in 2010 with an upgrade for 70 F15S fighter jets along with 36 Apache helicopters, 72 Black Hawk helicopters, tons of bombs, missiles and support equipment in exchange for almost $100 billion in just 6 years...21st century military technology in a social structure from the 7th century...Saudi Arabia is a country where women have no voice and their every activity controlled by archaic tribal rule and patriarchal religious dictates...Currently the Ministry of Labor is compiling a list of permissible jobs for woman in this society as females slowly demand an active role but in opposition to any participation of woman in the workforce are the religious leaders and social conservatives who argue that Islamic law require the 'woman' to stay in the home, argue that any establishment where woman work would have to be segregated and heavily cloaked to prevent them from being seen, argue that women would need to be driven to work because females are not allowed to drive (Saudi Arabia has almost no public transportation in a culture where 50% of the humans are not allowed to drive?) and argue that any tiny permissible freedom for women will change attitudes in conflict with the dictates of Islam...A stunning ignorance of reality...

The males in Saudi Arabia (as in most countries with draconian, ignorant laws against female freedom and participation within society) are very afraid of women...No other explanation fits...Afraid of female sexuality, female intelligence, female potential and female power, these fearful, ignorant men create laws and restrictions based on claims of religious tradition to prevent half of their population from participating with equal standing...A deep seated primitive ignorance to 'control' their women without any ability to communicate or understand reality...

Mecca, Islam's most holy site, is a pilgrimage for millions of humans every year for a variety of ancient rituals and tributes...In response to expectations of 16 million visitors in a single week in 2025 (determined by the moon and the Muslim calendar) Mecca is being transformed into a vast modern human mobilization machine with new seven star hotels, thousands of public bathrooms to cope with crowds, underground parking, new metro lines in and out of the city, extensive restructuring of the Grand Mosque with 3,000 extra square meters of space for pilgrims, large walkways to accommodate thousands of wheelchairs, escalators to over 10,000 ablution sites (where one can wash or cleanse their body in a ritualistic manner) and a finished capacity to hold over 1.6 million humans with more than fifty gates, more than 100 elevators and four new minarets...Estimates are the entire project will cost billions ($) and take decades to complete...

Most of Saudi electricity is generated by power plants burning crude oil, up to a million barrels a day in the hottest months...Enormous subsidies to the population for fuel make this an inexpensive country to drive in (men only) and subsidies for all citizens to prevent unrest are an acceptable expense to this government who report unemployment statistics for males only, officially around 11% but as high as 30% in reality...Political parties are banned...Crude oil reserves of more than 268 billion barrels and 8.2 trillion cubic meters of natural gas reserves...Almost 300,000 refugees from the Palestinian Territories and 70,000 stateless Arabs (bi-duns), both groups having few rights...Saudi women cannot pass citizenship to their children, a right reserved only for men...Since this government survives in a situation that by any measure is temporary without enormous social and institutional reforms, the Saudi government is currently buying large amounts of military hardware and weapons to safeguard their future options in a part of the world that has few resources other than oil in mostly barren desert with isolated population centers and has very few choices for the humans who occupy this land who have no voice in the administration and control of commerce...New weapons to protect the rulers from unrest or invasion from within their own country...Median age of 27 years with nearly 6% of children under five malnourished in a wealthy nation...

Saudi Arabia is estimated to have more than 25% of the world's oil reserves...

When the government opened the woman's branch of it's Human Rights Commission, no women were on the board...

Public sports events for women are banned in Saudi Arabia while physical education is banned in girls' public schools and gyms have been closed to women since 2009...A widely held belief in Saudi Arabia is that "once women start to exercise they will shed modest clothing, spend unnecessary time out of the house and have increased possibilities of mingling with men..."

Every workplace in Saudi Arabia that has men and women working together is required to install five foot high partitions to separate the sexes...A country with a "religious vice and virtue squad" that patrol the streets monitoring the behavior of the population...How can one rationalize such a completely stagnant society ruled by ignorant old men who are afraid of females? How can they be educating and raising their children in such a constricted way to keep an entire population (almost 30 million humans) unaware of possibilities and realities?

In January 2015, King Abdullah of Saudi Arabia died at age 90 after almost 20 years in power...An opponent of the freedoms sought by citizens in many Middle-East nations, he backed Bahrain in sending troops to quell uprisings and supported the military rule in Egypt which has shut down a free media and jailed reporters...Eulogies have focused on how transformational he was (he let males and females share the classroom at University, gave women a seat on the advisory Shura Council and promised women a vote in municipal council elections) but he was as misogynistic as his predecessors in believing females were inferior with no voice in the affairs of state...A man with limited education who supported extreme clerics and practitioners of the Wahhabi interpetation of Islam, Abdullah's every action was determined by the ethnic impact of Sunni and Shiia on the security of a state which buys billions of dollars of deadly weapons from western powers to build walls and control citizens...He is succeeded by Crown Prince Salman (age 79) who assured the world no policies will change...Abdullah's personal wealth was over $20 billion...

7th century barbarism...In 2015 Saudi Arabia executed (beheaded) the most humans (157) since 1995 (192)...

The obsessive focus in this region of the world on the perceived difference between Sunni and Shia is so primitive it almost defies explanation...A majority of Muslims are either one or the other and there are many more Sunni's (over 85% of all Muslims) than Shia's (8% to 15% of Muslims)...An estimated 1.6 billion Muslims in the world and just 200 million are Shia with Saudi Arabia the most powerful Sunni stronghold and Iran the heart of Shia dominance...An ancient rivalry 1,400 years old that began over who should succeed the prophet Mohammed has become so much more than that in the 21st century despite hundreds of years of peaceful coexistence and cooperation...

SYRIA
(17 million humans)
Military Forces = 180,000 active duty / $1.8 billion per year

In a United Nation's report released in early 2014 documenting Syrian government activity between March 2011 and November 2013, thousand's of Syrian children held in custody had been beaten with whips, tortured with electric shock, cigarette burns, solitary confinement and sexual abuse while being used as human shields and forced into combat with almost all mass killing burial sites containing dead children among the adults...In 2013, the abuse by opposition forces on innocent children caught in the conflict rose dramatically with virtually no intervention from world powers to stop this senseless territorial slaughter of an entire society...Both sides violently ignorant...

Basher al-Assad murders thousands to retain control of government with Iran a huge advocate of the current power in Damascus having signed a mutual defense agreement between the two countries with both governments supporting Hezbollah, Hamas and Islamic Jihad and both moving weapons and resources through their respective countries to supply and support these groups...Located in a strategic and sensitive area that borders Iraq, Turkey, Lebanon, Jordan, Israel and the Mediterranean Sea, Syria is a serious player in Middle East politics under an authoritarian political structure that will kill their citizens rather than compromise on any issue that might give the people a stronger voice in the policies that determine their quality of life...Assad is another intellectually challenged dictator who was handed power by his father and maintains control with violence and intimidation...His time is limited but it's unknown when and it's unknown what kind of chaos and order will rise after the current government falls...Syria has allies in Russia (a primary supplier of weapons to Syria) and China who is a consistent negative vote on the UN Security Council concerning actions against other countries in this region...As everywhere in the Middle East, religion and ethnic identity play a huge role in Syrian politics where the ruling party are Alawites, a sect of Shiite Islam, with the majority government in Iraq also Shiite and the rulers in Iran Shiite surrounded by gulf states that are majority Sunni...Opposition organizations, the Syrian National Council and the Free Syrian Army based in Turkey continue to challenge Assad with most of the world afraid of direct intervention for fear of making the situation worse...

Between 2007 and 2010, Russian arms deals with Syria totaled over $4.5 billion...

Over 470,000 have died in Syria since this conflict began with many more injured...In 2013 new reports of Assad using chemical weapons and poison gas on his citizens motivated the United Nations to initiate another "investigation" with no teeth that resulted in a Russian brokered plan to have Syria become party to the Chemical Weapons Convention and destroy their chemical weapons under UN supervision...In 2014, peace talks in Geneva were unsuccessful and Syria launched renewed attacks against civilians and rebels...Now a new player (ISIS) in this death and violence is actively fighting for territory in Iraq and Syria and in 2016 western powers are supporting al-Assad's fight against ISIS...

April 2015: Over 50% of the population is homeless (11 million humans), 250,000 Alawites dead...

Median age is 24 years, population growth is negative and 85% of the population live below the poverty line...Over 10% of children under age five are malnourished, unemployment rate is almost 60% and the child labor force is around 190,000...(*These statistics are variable and dynamic in a nation that has been at war with itself for over five years*)...An authoritarian republic with seven year presidential terms and no term limits, Syria's government regulated economy is in a continuing state of collapse with sanctions, high unemployment, violence in every part of the country, water shortages, declining industrial production and it's increased reliance on foreign allies to supply weapons and aid to sustain a war with it's own people destroying the lives of millions with no predictable positive outcome...

In May 2014, the top diplomatic negotiator for the UN (Lakhdar Brahimi) resigned saying the scheduled elections in Syria expected to give Al-Assad another seven-year term make further negotiations pointless...

"For kids coming out of a war zone, it's not just blatant trauma they may have witnessed or experienced, it's the cascade of disruptions they experience as their community collapses, loss of home, food instability, family break-up, lack of education and the absence of health care"...Dr. Katherine Porterfield NYU Program for Survivors of Torture...

In a fall academic study conducted in a Syrian refugee camp in southeast Turkey, 75 percent of the children said someone close to them had been killed in the conflict and 45 percent of boys and girls scored above the clinical cutoff line for a PTSD diagnosis...This is the impact of war: trauma and survival as ignorant adults destroy one another...

The end of 2013 saw the release of hundreds of photos from Syrian government torture chambers with starvation and obvious, systematic and hideous torture by Al Assad with eyes gouged out, people flayed alive and skeletal humans reminiscent of the holocaust...The world continues to watch while al-Assad destroys this nation...

During a week of international peace talks with Syria in 2014 that attempted to find some way to convince the Syrian government to allow in humanitarian aid, al-Assad's forces killed 1,900 citizens...The world shrugged...In his attempt to "punish" his own citizens Al Assad has destroyed entire neighborhoods, hospitals and schools all around Damascus creating a surreal landscape of utter devastation...World leaders continue to support his behavior...

13 million humans in Syria are in desperate need of aid, 6 million of them are children...2016

TURKEY
(80 million humans)
Military Forces = 450,000 active duty / $18 billion per year

Turkey is a member of NATO and the G-20, a democratic republic and a Muslim majority with real democracy a work in progress amid a contentious debate on a new constitution as the military and the Islamist AKP government party of Prime Minister Erdogan continue to challenge one another over the direction of a secular society backed by military force and a conservative Islamic power structure in the government...An ongoing tension which has led to multiple protests that have all been dealt with harshly by the government...Located in a strategic region on the Black Sea and the Mediterranean, the country is cautious and deliberate in dealing with other nations continuing a long standing dispute with Greece over the division of Cyprus, an aggressive approach to Israel and conditional support for the rebels in the ongoing Syrian conflict accepting 2.7 million refugees fleeing Syria since 2011...

Turkey is one of the largest economies in the Islamic world with the 2nd largest military in NATO (United States is #1)

In 2016, there's movement toward a more conservative society based on religious entanglement with government continuing a culture where women are not equal to men and violence against women increasing despite efforts to introduce reforms...Eighty million humans with a median age of 30 and a child labor force over 300,000...A free market economy with a large service sector, industry and a labor force of 29 million with 20% of the population living below the poverty line and an unemployment rate around 10%...Energy wealth is 296 million barrels of crude oil reserves and 6.8 billion cubic meters of natural gas reserves...A long conflict with the Kurdish rebel group PPK thought to have ended in 2013 with the rebels withdrawing into Iraq and the Turkish military withdrawing from Kurdish regions in the south was re-engaged in 2015 as President Erdogan seeks more presidential power and grows more militaristic...Turkey has many private media outlets although it remains a crime to "insult the Turkish nation" and according to a report released by the 'Committee to Protect Journalists', Turkey has more journalists in prison than any country in the world...Egypt will soon be #1 if not already...

In November 2013 the Turkish parliament voted to allow women to wear "trousers" in the assembly...How is it that men in multiple nations around the world "control" what women wear?

The refugee camps are full and the conflict in Syria continues with no resolution...The posturing of world leaders on the UN Security Council and the poor planning that continues without any agreed upon contingency for dealing with violent, ignorant leaders who cross a clear threshold of cruelty and chaos demonstrate the lack of a cohesive awareness among a global civilization that manifests itself in this confusing inability of the most powerful, influential leaders on Earth to act together in any decisive way to remove al-Assad or any ignorant ideologue who murders innocent humans from an entrenched position of power...Primitive territorial instincts, a misguided confusion concerning sovereignty and a paralyzing fear of consequence blinds these world "leaders" to the reality of human singularity, human rights and human diversity...No human being who lives without violence should be killed by primitive monsters incapable of understanding the complexity of human existence but in 2016 the leaders of free nations everywhere who are expected to understand the value of every life and the necessity of human progress are increasingly unable to understand the complexity of human existence while hiding behind "regulations", "laws of war" and the "sovereignty of all nations" to ignore situations and actions that are untenable and inconceivable to any sentient, self-aware intelligence...

A failed coup in July 2016 will result in more authoritarian power for President Erdogan with little restraint...

Turkey remains obsessed with the denial of their own history concerning the targeted killing of Kurds and Armenians in the early part of the 20th century during the chaos of World War I and the disintegration of the Ottoman Empire...

UNITED ARAB EMIRATES
(5.8 million humans)
Military Forces = 63,000 active duty / $14 billion per year

A coalition of seven states after Britain ceded control in 1971, the UAE is an oil rich nation on the Persian Gulf bordered by Saudi Arabia and Oman...A federation of seven emirates with half of the legislature (FNC) elected and the other half appointed by these rulers with a Supreme Court whose members are appointed by the president...No political parties allowed...6 trillion cubic meters in natural gas reserves and almost 100 billion barrels of crude oil reserves...Recently began to restrict internet content to prevent the free use of social media for activities "unhelpful" to the state...Publications are licensed and there are laws prohibiting criticism of the state and an open contempt of religion...Home to the tallest human structure on Earth (2,722ft)...Varied landscape with steep mountains and coastal plains that contrast new urban centers with modern technology making UAE a rapidly growing tourist destination in a strategic, pivotal location on the Persian Gulf...Approximately 75% of the population is made up of foreign workers and expatriates...A strong ally of the United States...

WEST BANK
(2.7 million humans)

Just 5,600 square kilometers in area, the West Bank is the Palestine homeland that is somewhat recognized by Israel even as Israel annexes more land and builds more settlements (estimated Israeli population in West Bank settlements is around 400,000 with more than 340,000 Israelis living in East Jerusalem)...Landlocked with no access to the Mediterranean Sea and very limited access to the Dead Sea, West Bank Palestinians exist in a surreal conundrum without resolution with the state of Israel in charge of all sectors of their society controlling the air, the sea and the land without regard to suffering while slowly annexing and stealing Palestinian land despite international law and criticism...In 2004 the International Court of Justice (ICJ) ruled that all settlements in the West Bank were illegal but it did not change anything...Today there are more than 120 Israeli 'approved' settlements and over 100 'unofficial' settlements with 13% of Israel's population living on Palestinian land and calling it there own...It is stunning to watch Israel slowly occupy another country with every intention of taking all the land while disregarding human rights and international law...All of these settlements are protected by the Israeli military with barriers, checkpoints and buffer zones in an occupation without end, controlling disputes with military law for the Palestinians (over 90% of cases prosecuted) and Israeli civil law when Jewish settlers are accused (under 10% are prosecuted) while the world watches and does nothing...

The West Bank population of 2.7 million is challenged by a lack of fresh water (controlled by Israel although the aquifer lies mostly under the West Bank), an inability to travel freely within their borders and commerce controlled by Israel and a security enforced by Israel...The Israeli Defense Force occupy and control the West Bank under virtual siege with curfews and interference inhibiting activity in every segment of society while continuing to build walls and dig ditches around Palestinian cities to prevent people from leaving, destroying homes at their discretion, closing Palestinian businesses for almost any reason, separating families, uprooting entire olive groves and killing almost 4,000 humans in an effort to suppress and control a population...Israel dismisses all criticism of it's activities and even with the UN's frequent declarations of Israel in violation of International Law, Israel continues to build settlements on land they do not own...Even a cursory viewing of a map denoting the internal borders in Israel make little sense to any intelligence aware of the history and tensions existing here...How did the West Bank become landlocked and how did Gaza end up a strip on the coast? There are many people with many answers but the real answer is going forward...If humans wanted territorial security, trade, commerce and peace, they would redesign the boundaries to makes sense to the millions of humans who live a desperate life with little promise of solutions or resolutions...*Here is a rational solution that cannot possibly happen at this primitive stage of human habitation on Earth*...Israel keeps the western border (halfway between Nablus and the Mediterranean Sea)...Extend the West Bank north to Lebanon including coastline north of Halfa so a newly created Palestinian state can have access to the sea...Israel gets everything south of Jerusalem, including Gaza...Israel would have the west coast of the Dead Sea, Bethlehem and Hebron...The Palestine's would have the northern coast of the Dead Sea, Nazareth and access to the Mediterranean Sea...Jerusalem would be shared as a capitol for both nations with respect, cooperation and bilateral security...Generously compensate relocated people with land and money, allow refugees to relocate to the Palestinian state with guidelines and join the 21st century...

The legal age for marriage in Yemen is 15 although it is common for girls to be married at age 12...

YEMEN
(26.8 million humans)
Military Forces = 66,600 active duty / $1.4 billion per year

Yemen is a country on the southwestern corner of the Arabian Peninsula bordered by Saudi Arabia to the north, the Red Sea to the west and Oman to the east...Yemen was unified in 1990 and has a population of approximately 27 million people (65% Sunni and 35% Shiia) with more than half the population under the age of 19...The country is violent with clashes between northern rebel fighters and government forces and a recently growing presence of al-Qaeda that has brought drone attacks, missile strikes and 'trainers' from the west (U.S.) to fight terrorism and stabilize the country...Many believe the actions by the U.S. were having exactly the opposite effect and in 2014 this became apparent when Houthi rebels took control of the capital and forced the president to flee in 2015 resulting in chaos, a resurgence by al-Qaeda, an insurgence by ISIS and the beginning of Saudi Arabia airstrikes that have killed thousands of civilians...America's support for a corrupt leader with millions of dollars that does not go to the population leads only to resentment and anger and results in more terrorists and activities against western governments...In 2012, UNICEF reported 750,000 malnourished children in Yemen under age five (43%) and a very wealthy ruling family...A child labor force of 1.3 million children with an average school life of only nine years...Unemployment rate over 30% with 60% of the population living beneath the poverty line...The president of 33 years, Ali Abdullah Saleh, a corrupt leader and ally of the U.S. in return for money and weapons, stepped down after protests over the missile and drone strikes that targeted terrorists but often killed civilians...Saleh's vice president Mansur Hadi took over the country in an uncontested election (2012) with a promise of free elections in 2014 (delayed while seeking agreement on a new constitution) though it is known that Saleh's family and friends still control the intelligence and military operations in the country...The poorest nation in the Middle-East...

President Hadi resigned in January 2015 as Houthi Shiia rebels took control of the presidential palace in Sanaa even though the parliament is refusing to accept his resignation...The rebels are seeking to form a ruling council and with little support in south Yemen, many believe this will create a power vacuum with elements of al-Qaeda in the mix...

There has been a continuing challenge to the central government from separatists in the south and the Houthi Shiia insurrection in the north for many years and from a movement in southern Yemen called Ansar al-Sharia who promote Sharia Law and has been at odds with a government whose focus for the past decade has been to enrich themselves and ignore a growing population of poor...The Ansar al-Sharia movement is made up of tribesmen and elements of al-Qaeda who are moving into villages in the south and implementing their own form of government in an effort to give local people some kind of security and options in a country whose recent election was a choice of 'yes' or 'no' with only one candidate for president...Contributing to the violence and lack of direction in Yemen is the Saudi Arabia connection that uses money and influence to pay tribal leaders, political leaders and conducts its own intelligence operations with the Yemen security forces...In 2014, American soldiers who were training Yemeni forces and providing intelligence to fight a threat they could not identify were forced to leave a country in chaos with Saudi Arabia's stated goal to "return the legitimate government to power"...It is unknown when this will happen if it happens at all...All peace negotiations have failed...Economy is primarily energy with 3 billion barrels of crude oil reserves and 476 billion cubic meters of natural gas reserves...Over 200,000 refugees from Somalia...Many tribal customs, hierarchies, family loyalties, rivals and the chewing of khat still dominate a population of mostly young, mostly unemployed...It is difficult to see the way forward in a country that is by almost any measure directionless...

In March 2016, Saudi jets dropped two laser guided 'smart bombs' on a marketplace in Yemen killing over 97 civilians (25 children) to target an estimated 10 Houthi fighters...The bombs and weapon systems are mostly made in America who continues to arm one of the most repressive and ignorant societies on Earth...Human Rights Watch says the Saudi's use cluster munitions (banned by CMM in 2010) manufactured in America who also provide strategic support of Saudi operations...Over 3,000 civilians have been killed since the start of the conflict...

The Convention on Cluster Munitions (CMM) is not recognized by the United States, China, Israel, Iran or Russia...

The Houthi's replaced the parliament with a transitional revolutionary council but their rule is not recognized by the United States, the Gulf Co-operation Council or the United Nations...The violence continues...

Over 6,200 humans (half children) have been killed since Saudi Arabia began air strikes in 2015...In June 2016, the UN Secretary General Ban Ki-moon called the situation in Yemen "particularly worrisome" ?

October 2016...The United Nation's reported 10,000 people killed and over three million humans displaced in 18 months of civil war in Yemen with the majority killed by Saudi Arabia airstrikes targeting both civilians and rebels...

WORLD POLLUTION

World pollution and the increasing negative effects of the global waste culture on Earth's habitat and the species dependent on that habitat is no longer a 'moral' question or a question of choice, it is a question of survival...In an information age despite extensive and overwhelming evidence confirming scientific processes that are indisputable, billions of humans continue to ignore causality at every level of civilization unwilling or unable to recognize a balance in nature that has allowed their species to survive for thousands of years...There are many rationalizations on a planet with seven billion humans that attempt to understand this increasingly self-destructive behavior but many agree the rapid growth of Earth's industrial age, the large time scales involved in recognizing the long-term effects of pollution and the sheer volume of chemicals and pollutants humans release in the air and water just to live the way they do has created a convenient global blindness of the daily impact of billions of individuals on a finite Earth...Is it possible to find any balance where industry and society can exist in a sustainable, progressive lifestyle and not pollute and change the environment in a negative way? When will humans have enough 'stuff' to recycle everything and use raw materials only when essential? If recycling every product manufactured on Earth were a shared responsibility by everyone in an organized world effort it would create huge industries with a focus on preventing the release of pollutants into the environment in uncontrollable processes while continuing to supply the growing populations on Earth with products they seek to support a sustainable, innovative and creative lifestyle...A real possibility with cooperation, awareness, 21st century technology, creative solutions, proven science and a majority population who recognize causality, but very distant from the realities of today...

On a world with billions of people unable to understand the impact of wasteful consumption and billions who simply don't care what impact their activities have on the planet's ecosystem a global financial market exists that values consumption and profit above all else with a disregard for the impact of unchecked growth and unregulated pollution...Many humans are unaware of their rationalizations, dominated by instincts that prioritize individual survival, a competitive drive to have more and a belief that the perils of world pollution are overstated and ambiguous but when you examine the structure of every polluting society on Earth, it's always about money...Individuals who control governments and should have the ability to recognize and understand the negative impact of unchecked pollution are the same people who promote corporate policies and support the organizations who influence elections that determine political power...The majority of industry and governments in this century regard regulation of harmful chemical pollutants, waste discharge, recycling, environmental standards and guardianship of the planet to be a nuisance that interrupts the flow of business and costs money that could be used elsewhere...A representative structure of 200 years of unparalleled progress, technological wonders and discovery at all levels of human achievement with an exploding population represented in this century by a majority who are unable to understand their reality and the impact of a collective existence defined by innovation, experimentation and the very science that is challenged when profit is a priority...

The power to address world pollution and the problems and challenges glaring and obvious on a world with more than seven billion consumers is clearly in the hands of individuals, organizations and governments worldwide who control the enormous amounts of money that support world commerce and civilizations...But many of these 'leaders' do not have an awareness of necessity and are motivated by ancient tribal bonds that exist even in the most developed societies controlled by a self-defined 'free market society' where powerful individuals make the rules and pay the representatives of government to insure their rules are legal and written into law...Sadly, corporate/political leaders who do understand reality and work to alter policy and status quo are a small minority in the aggressive power structures that dominate in this century and large consolidated interests continue to grow and increasingly control activities that affect human life all over the planet in destructive ways...Food used to be produced and distributed by entities all over the world and now the majority of food and agriculture is controlled by a limited number of large corporations fighting regulation and controlling lawmakers to allow pollution and waste without limits, abuse of animals numbering in the millions, control of multiple seed products and control of many farmers required to use these products, diversion of natural resources for private profit and an increasing control over ingredients and hazardous chemicals without disclosure...

The world's energy resources continue to be controlled by a short list of large corporations and governments who dictate waste discharge, drilling regulations, shipping regulations, pipeline transport and environmental impact to maximize profit and restrict oversight...Coal mining is environmentally destructive in all circumstances but the burning of coal increases worldwide despite a clear threat to the future of all species on Earth...There are thousands of chemicals used in manufacturing that find a way to oceans, aquifers and lakes without oversight, without testing, without any clear idea of impact and this happens *even as humans have the technology to know exactly how these chemicals react in the environment and their impact on plant and animal life*...Money and ignorance control a world that could not turn on a light 150 years ago and little will change until people who understand reality are enough to alter political systems around the world...It's unknown if this will happen in time...

In December 2012 the International Energy Agency (IEA) issued a report projecting that coal will equal oil as the world's top energy source within 10 years...The report rationalizes that natural gas offers the best hope of reducing carbon emissions in the short term but the increased demand from India and China coupled with economic and population growth in developing countries are accelerating the drive for coal..."Coal's share of the global energy mix continues to grow each year," says IEA executive director Maria Van der Hoeven and by 2017 global coal consumption will stand at 4.32 billion tonnes of oil equivalence, versus 4.4 billion tonnes for oil itself...In 2014, China accounts for more than half the world's coal consumption with India overtaking the U.S. in second place...In 2012, the IEA reported that without a major move away from coal as an energy source, average global temperatures could rise by 6C above the planet's pre-industrial levels by 2100...

The U.S. Toxic Substances Control Act of 1976 listed over 80,000 chemicals currently used in commercial products but the EPA (Environmental Protection Agency) has required testing on about 200 and banned just five...2010

Mercury
Mercury is a naturally occurring substance released into the air as a byproduct of burning coal, a gas that can drift thousands of miles before settling back to Earth...Mercury is also released into the environment through a number of other industrial processes (mining, metal and cement production) and once emitted persists in the environment for a very long time circulating through the air, water, soil and living organisms...As a gas it can be absorbed by the leaves that fall to the ground where they are swarmed over by bacteria and other organisms capable of converting the substance to it's organic form, methyl-mercury, which is the most toxic form of this heavy metal...In this form mercury is a neurotoxin that enters the food chain contaminating multiple species with human contamination resulting primarily from the ingestion of seafood...After 10 years of a study starting in 1990 the EPA Administrator in the U.S. finally determined the regulation of hazardous air pollutants including mercury from coal and oil-fired power plants was appropriate and necessary...

A study published in August 2013 in Nature Geoscience confirmed that the mercury found in Pacific fish near Hawaii likely traveled through the air for thousands of miles before being deposited on the ocean surface in rainfall...The North Pacific fisheries are downwind from rapidly industrializing nations such as China and India that are increasingly reliant on coal-burning power plants, a major source of mercury pollution..."This reinforces the links between mercury emitted from Asian countries and the fish that we catch off Hawaii and consume in this country," said Joel Blum, the lead author of the study..."The implication is that if we're going to effectively reduce the mercury concentrations in open-ocean fish we're going to have to reduce global emissions of mercury, including emissions from places like China and India...This is a global atmospheric problem."...The developing brains of fetuses and young children are especially vulnerable to the effects of mercury ingestion in humans...

In December 2011, the Environmental Protection Agency (EPA) released new standards sharply limiting future emissions of mercury and other toxic pollutants from coal and oil burning power plants in the United States...In January 2013, after more than four years of negotiations the United Nation's Environment Programme (UNep) negotiated the Minamata Convention on Mercury, a legally binding international treaty with the participation of 140 nations aimed at curbing future mercury emissions by **regulating the supply of mercury and the use of mercury in products and industrial processes** *(light bulbs, batteries, dental fillings, switches, soaps and medical devices containing mercury, all banned in 2020),* **reducing emissions from small-scale gold mining** *(no dates, limits or targets have been set...Large mercury mines in operation can continue for 15 years),* **reducing emissions from power plants and metals production facilities** *(does not require new facilities to have mercury pollution controls for five years after the treaty comes into force and existing facilities have ten years before they are required to have mercury emission controls)*...The UNep assessment reports that concentrations of mercury in the top 100 meters of the world's oceans has doubled over the past century and estimates around 260 tonnes of the toxic metal has moved from the soil into rivers and lakes...

To ratify the Minamata Convention, nations must first enact domestic laws meeting the treaty requirements...The treaty comes into effect 90 days after at least 50 nations have ratified it, a process expected to take an unknown number of years...The pollution continues...

The first nation signed the Minamata Convention on October 9, 2013...Each nation has a year to sign the pact...

The World Coal Association (WCA) contends that burning coal for energy accounts for approximately 24% of global mercury emissions and estimates the implementation of new technology could reduce emissions of mercury from coal-fired power plants by up to 90%..."this treaty will ensure that countries are able to address the issue of mercury emissions from their coal-fired power plants via the application of technologies which are most appropriate in a given national context and for a given facility without having to restrict the use of coal as an energy fuel or to compromise their economic development goals." WCA 2013

Energy companies in America are still fighting the regulation of this toxic chemical through political lobbying to avoid doing what is obvious and needed *but the new rules set forth by the EPA do not apply to existing power plants*...Burning coal remains the way most of the world creates electricity to power civilization and due to rapid industrialization South-East Asia is now the largest regional emitter responsible for almost half of mercury's annual global emissions...Mercury becomes more concentrated as it moves up the food chain reaching its highest level's in the predator fish that are consumed by humans making it highly toxic to developing babies and small children ...

"Mercury is highly toxic to human health, posing a particular threat to the development of the (unborn) child " "The inhalation of mercury vapor can produce harmful effects on the nervous, digestive and immune systems, lungs and kidneys, and may be fatal" "The inorganic salts of mercury are corrosive to the skin, eyes and gastrointestinal tract, and may induce kidney toxicity if ingested." World Health Organization

Asbestos

Asbestos is a set of six naturally occurring silicate minerals used commercially for their physical properties...They all have in common their eponymous, asbestiform habit and long thin fibrous crystals...The prolonged inhalation of asbestos fibers can cause serious illness including malignant lung cancer, mesothelioma and asbestosis..

More than two million metric tons of asbestos were mined worldwide in 2009, mostly in Russia, China and Brazil with more than half exported to developing countries...Today the mining continues despite an increase in asbestos-related disease in Europe and North America, bans or restrictions in 52 countries, libraries of incriminating studies and estimates of up to 10 million asbestos-related cancer deaths worldwide by 2030...A global network of industry groups have spent nearly $100 million in public and private money since the 1980's to keep asbestos in commerce by promoting the "controlled use" of chrysotile or white asbestos, the only form of the fiber used today...

In 2009, a panel of 27 experts convened by the WHO's International Agency for Research on Cancer reported, "Epidemiological evidence has increasingly shown an association of all forms of asbestos with an increased risk of lung cancer and mesothelioma." They also found evidence that asbestos causes cancer of the larynx and the ovaries...

Despite mounting scientific evidence of the risks of white asbestos and calls from health experts for a worldwide ban, asbestos production is holding steady at about 2 million metric tons per year to support new markets in the developing world where there is an increased demand for cheap building materials...Asbestos is still legal in the United States with restrictions on use and a lengthy history of litigation that has cost the industry over $60 billion in damages...It is banned in the European Union...

Canada remains a major exporter of asbestos and a strong defender of it's mining operations (subsidized by federal and provincial governments) despite objections from health professionals and activists...In 2009, Canada sent nearly 153,000 metric tons of chrysotile abroad with more than half going to India and the rest going to Indonesia, Thailand, Mexico, Sri Lanka, Pakistan and the United Arab Emirates...Another clear example of an industry with political connections prioritizing profit over the health of Earth's native species and the long term viability of the planet's ecosystem...

The World Health Organization reports 125 million workers are still exposed to asbestos...

Spills and Mishaps
A few of the more than 100 significant oil spills in the last 40 years caused by human mismanagement, corruption, lack of regulation, poor decisions, market pressures and unpredictable chaotic weather events...

December 2011...a leak in a Royal Dutch Shell pipeline off the coast of Nigeria was reported by the oil company to leak over 40,000 barrels of crude oil into the ocean although some estimates are much higher...A resulting oil slick of almost 400 square miles...

July 1979...Two oil tankers (Atlantic Empress, Agean Captain) collided in the Caribbean and 90 million gallons of oil were spilled...The Atlantic Empress caught fire and subsequently sank...

January 1991...When defeated Iraqi soldiers left Kuwait at the end of the short-lived Gulf War, they released oil from offshore terminals and several tankers spilling over 400 million gallons of oil into the Persian Gulf...International researchers reported in 1993 there was little long term damage...

July 2003...oil tanker Tasman Spirit, carrying 67,535 tons of light crude oil ran aground off the coast of Pakistan in the Arabian Sea...Efforts to remove the oil from the ship were successful until the ship broke apart in August spilling more than 27,000 tons of oil into the sea...Over 10 kilometers of densely populated coastline were affected with health problems reported for years after the spill...

"Flag of Convenience" is the common practice of registering a merchant ship in a sovereign state different from the owners state in order to avoid the regulations of the owner's state and avoid responsibility in case of a mishap, maximizing profits by operating below the safety standards deemed necessary to avoid polluting the world's oceans...More than half of merchant vessels carrying cargo in 2016 have open registries...

July 2010...Canadian tar sand oil, a heavy crude with *bitumen* (similar to asphalt), contaminated the Kalamazoo River in Michigan with over a million gallons leaking from a broken pipe...Affecting only a 25 mile section of the river, the cleanup was estimated to cost less than $10 million and take less than one year...In 2014 sections of the river were still closed and cost of cleanup is over half a billion dollars...A stark reminder of the disastrous consequences possible in the spilling of this heavy crude oil into the environment...Approval is expected for a pipeline (Keystone) through the center of the United States to transport the Canadian crude to the Gulf Coast for refining and transport to distribution terminals...(*U.S. State Department rejected the pipeline in late 2015*)

The Keystone pipeline would be built by TransCanada Corp. from Hardisty, Alberta to Steele City, Nebraska (1,179 miles) connecting to existing pipelines to move the oil to refineries on the Gulf Coast...The pipeline capacity is over 34 million gallons a day and passes over aquifers that supply water to more than 2 million people...

TransCanada continues to remind the world that the Keystone XL Pipeline "is a critical infrastructure project for the energy security of the United States and for strengthening America's economy" Multiple studies by industry groups, environmental groups and governmental agencies reveal this project would do little to effect 'energy security' in America with the majority of the product being shipped overseas and would have a very limited impact on the U.S. economy with the majority of labor needed to build the pipeline coming from out of country and a very small contingent needed to monitor operations once the project is completed...After the Obama administration rejected the pipeline in 2015, industry groups in Canada and America are still determined to build it hoping a new administration in 2017 will support and reauthorize the project and higher oil prices will make it viable...

*March 1989...*oil tanker Exxon Valdez ran aground in Prince William Sound in Alaska spilling almost 11 million gallons of crude oil affecting over 1,300 miles of coastline...In 2001, researchers at NOAA Fisheries conducted a survey digging more than 9,000 pits at 91 sites and found over half of the sites still contaminated...Exxon was fined $150 million in criminal court with $125 million waived for Exxon's cooperation in the cleanup and settlement of private claims...Another $100 million was paid for "restitution" to state and federal governments and $900 million was controlled by trustees over a ten-year period to restore resources and cover injuries and claims after the fact...In 2013, researchers found oil contamination still present on Gulf of Alaska beaches...In 2012 Exxon-Mobil posted a profit of $44.8 billion on an annual revenue of $482.3 billion...

*November 2002...*oil tanker Prestige sank off the coast of southern Europe polluting thousands of miles of coastline and beaches in Spain, Portugal and France...The estimated 62,000 metric tons of fuel oil spilled impacted offshore fishing and continues to impact the marine life with hydrocarbons ingested and distributed throughout the food chain...The seaworthiness of the vessel was questionable as was the ownership due to 'flags of convenience' commonly used to defer responsibility and avoid regulations...Follow up medical studies found many volunteers suffering years later from the effects of cleanup and as of 2013, litigation and criminal responsibility were still in the courts...

*April 2010...*the Deepwater Horizon spill in the Gulf of Mexico continued for almost three months at a rate of more than 50,000 barrels a day resulting in the largest marine oil spill in the planet's history affecting marine life and human life in ways that will continue for an unknown number of years...An estimated total of five million barrels is still mostly in the Gulf in large underwater plumes, in marine habitats and in marine life...Reportedly worst than the oil is the huge release of the chemical Corexit by petroleum companies to suppress the oil, chemicals that are loosely regulated and toxic to ocean life causing mutations and deformities in the variety of sea life still being found and studied by marine scientists in 2016...This spill effected and threatened thousands of species including turtles on endangered lists and offshore Gulf species with a very uncertain future...BP was temporarily suspended from new U.S. government contracts in November 2012 which will have some effect on the business practices of the company but the enormous subsidies BP receives from the U.S. (billions in tax breaks) continue and profits continue to grow in a world that increasingly demands energy...

A report released in April 2014 by the National Wildlife Federation studying species in the Gulf of Mexico found most of the spilled oil still on the ocean floor and in the marshes with species dying in record numbers...Over 900 bottle-nose dolphins have died since 2010 and over 500 sea turtles have been found dead every year since 2011 with species of tuna, birds and whales all demonstrating adverse medical effects from the contaminants still present...

In 2015, BP agreed to a $20.8 billion settlement for Deepwater Horizon...Subsidies continue and profits in the final quarter of 2014 were just $2.2 billion...For all of 2014 profits were around $12.1 billion...

In the aftermath of the Deepwater spill in the Gulf, environmental scientists and the researchers who went to the area to identify the oil contamination found many leaks not related to the spill...Under U.S. law any oil spill must be reported to the Coast Guard's National Response Center (NRC) and it became increasingly clear this was not being done...In response, the Gulf Monitoring Consortium was formed to monitor and count the oil spills in this area along with the Gulf Restoration Network and a tech group, SkyTruth, who help locate and measure spills by satellite and use a formula to estimate the amount of the spill...They found that oil spills were on average 10x's larger than reported to the NRC which receives over 10,000 reports of spills on average *every year*...A single state, Louisiana, has more than 300,000 oil wells, thousands of pipelines and the state admits to over 300,000 barrels of oil spilled every year (self-reported by oil companies) giving every indication the actual amount is much higher...

Batteries

An analysis in 2011 by the New York Times found almost 20% of spent vehicle and industrial batteries from the U.S. are exported to Mexico for recycling, more than 20 million a year because of stricter laws in the U.S. concerning pollution and recycling techniques...There are no restrictions to prevent companies from exporting the batteries to other countries where no laws exist protecting the people or the environment from the toxic chemicals released when the batteries are broken down and melted...Lead is enormously toxic causing neurological damage in children and a variety of physical ailments including high blood pressure, kidney damage and neurological behavioral problems...19 out of 20 recycling plants in Mexico do not have proper authorization for importing dangerous waste and a review of manifests discovered more than 140 illegal shipments of car batteries not detected at the border...The top two battery exporters in the U.S. sent more than 250,000 tons of spent batteries to Mexico, a glaring example of a corporate mentality that works in every way to avoid legal structures and responsibility by ignoring the best solutions in dealing with extremely hazardous materials and a glaring example of the political corruption that allows it...Maximum profit with minimum investment and a disregard for the environment and human life if it affects the bottom line...Mexico shipped an estimated 150 million tons of lead to China in 2011 as the global trade in lead has increased sharply due the demand from China and other developing countries...

Pesticides

In 2013, the European Food Safety Authority (EFSA) asked that the recommended "safe' levels of exposure to pesticides containing neonicotinoid chemicals (systemic pesticides widely used in agriculture that make every part of the plant toxic to predators) be lowered after research showed that exposure to two specific types of neonicotinoids, imidacloprid and acetamiprid, could adversely affect the development of brain structure and neuron growth in young mammals...Recommendations were based on experimental research with rats that showed a clear developmental affect from exposure to these specific chemicals...Industry suggests any conclusions are premature and more study is needed....Pesticides containing these chemicals are implicated in the rapid decline in bee population's worldwide...

In 2011, the International Union for Conservation of Nature (ICUN) established an international scientific task-force to investigate the impact of "systemic pesticides" on Earth's environment and the many species affected by the widespread use of these chemicals...In July 2014, the group reported that the use of systemic pesticides, specifically neonicotinoids and fipronil, is a global threat poisoning the planet's atmosphere, arable land and water adding that these chemicals are up to 6,000 times as toxic as older pesticides and are currently around one-third of the world's insecticide market...*Manufacturers of these agents contend there is no evidence that their chemicals are harming species or the environment but the analysis of twenty years of impact studies suggest a direct causality...*The damage to species and the environment directly attributable to the increasingly widespread use of chemical pesticides in recent history is impossible to overestimate...In a continuing pattern of corporate irresponsibility coupled with the availability of a growing global market, large chemical companies who have a monopoly on crop production and "seed control' continue to develop and promote products destructive to biological life and whose long-term negative effects on the stability of Earth's ecosystem are still unknown...Because highly toxic pesticides (herbicides 2,4-D and Dicamba) kill crops as well as creatures that feed on the crops, Dow and Monsanto developed new genetically engineered crops (corn, cotton and soybeans) that can withstand repeated exposure to these toxic chemicals, allowing them to be used in greater quantities at higher concentrations...These deadly herbicides are being promoted as a "solution" to the chemical resistance weeds have developed to "glyphosate" an active chemical used so widely it became ineffective in an increasing spiral of more deadly pesticides to control a natural environment...Both 2,4-D and Dicamba are known to "drift" long distances from the application point and both have been implicated in toxicity to humans...2,4-D is linked to cancers, kidney damage, liver damage and birth defects...Dicamba is a neurotoxin linked to developmental and reproductive disorders, is water-soluble moving easily through soils and industry is combining these herbicides (2,4-D, Dicamba, Glyphosate) to create a 'super mixture' effective in any application (*The Environmental Protection Agency (EPA) does not test environmental effects or human health implications of chemical mixtures*)...Although the EPA has required Environmental Impact Statements on Dow's Enlist (genetically engineered corn) and Monsanto's Xtend (cotton & soybeans created in partnership with BASF, the world largest agrochemical company) these products are expected to be widely available for many years...

*Glyphosate, the active chemical ingredient in the popular weed killer Roundup, is a hormone-disrupting chemical used primarily on corn and soy crops genetically engineered to withstand heavy dousing of the chemical...In this decade over 275 million pounds of glyphosate are applied yearly in the United States...*Science News 3/2016

Glyphosate exposure is linked to obesity, learning disabilities, birth defects, infertility, and potentially irreversible metabolic damage...It's been detected in streams, groundwater, rain and the atmosphere...

Organophosphates, another class of pesticides used worldwide, are extremely toxic to humans with an active agent mimicking the nerve gas Sarin in inhibiting production of an enzyme essential in breaking down the buildup of neurotransmitters causing death by suffocation...Most studies of toxicity are concerned with high dose short term exposure, but chronic low level exposure has been implicated in neurological disorders in children and infants and the effects of chronic low level exposure on the environment and in species is unknown...Poor regulation in developing nations is very problematic with the United States setting an ineffective standard by asking manufacturers to *voluntarily* eliminate residential use...It remains on the market in a variety of products...

In 2010, a United Nations Environmental report estimated over 2 billion tons of wastewater including fertilizer runoff and industrial waste is discharged around the globe every day...

Chemical Pollution
Laboratory tests of the umbilical cord blood of ten newborns in 2005 found that the samples contained an average of 200 chemicals which can cause cancer, brain damage, birth defects and other health ailments...The lab tested for 413 chemicals and detected 287 in the blood samples with each sample containing between 159 and 234 chemicals...Of the chemicals detected, 180 can cause cancer in humans and animals, 217 are known to be dangerous to the nervous system and brain and 208 can cause birth defects in animals...Environmental Working Group (EWG)

*In 2016, the TSCA (Toxic Substances Control Act) reformed version was signed into law after forty years of inaction by a U.S. Congress controlled by special interests...One of the most important changes is requiring the EPA to test new chemicals for safety **before** they are allowed to be used repealing a provision that restricted the EPA's actions in regulating toxic substances and requiring corporations to generate toxicity data on demand...It also restricted testing toxic chemicals on animals but allowed individual states to retain existing chemical regulations...*

In response to concerns over flame retardants required in new furniture and multiple products by America's lawmakers, the Chicago Tribune published a series 'Playing with Fire' explaining how corporations that manufacture fire retardants manipulated the system with money and politics to enrich themselves and put thousand's of humans at risk in order to sell an unsafe product...Chlorinated Tris and Deca BDE (banned by the European Union), chemicals suspected of causing endocrine disruptions and other health problems with little oversight or regulation continue to be widely used...

Federal law governing industrial chemicals in the United States since 1976 required virtually no information on health or human exposure as a precondition for the manufacture or sale of these products...The result of this policy is thousands of chemicals used in industry today that remain untested for human safety and environmental toxicity...

In February 2014, the Harvard School of Public Health and Icahn School of Medicine published a study online identifying a growing number of "developmental neurotoxicants" that are indicated in the retardation of normal brain development in children, chemicals responsible for autism, attention-deficit hyperactivity disorder, dyslexia, cognitive impairment and diminished intellectual function...The source of these neurotoxicants are industrial chemicals widely used worldwide with little regulation or even a basic understanding of the effects of exposure on biological life...Researchers predict there are many more of these damaging chemicals in use than the ones they identified and suggest a restructuring of policy in nations around the world..."We have the methods in place to test industrial chemicals for harmful effects on brain development...now is the time to make that testing mandatory." "Untested chemicals should not be presumed safe to brain development, and existing chemicals in use and all new chemicals must therefore be tested for developmental neurotoxicity"...The chemicals identified in this study included lead, dichlorodiphenyltrichloroethane, arsenic, toluene, methylmercury, polychlorinated biphenyls, manganese, tetrachloroethylene, polybrominated diphenyl ethers, fluoride, chlorpyrifos and polybrominated diphenyl ethers...*Published in Lancet Neurology*

Nuclear Pollution
In 1995 the Department of Energy in the United States said it would take more than $30 billion and over 50 years to clean up the Rocky Flats Plutonium Plant in Colorado...After granting a $3.5 billion contract to Kaiser-Hill to clean this waste site, it's now called Rocky Flats National Wildlife Refuge...About 1,000 acres of the "refuge" has restricted access due to unsafe levels of radiation...

In 2013 around 120 tons of contaminated water containing 710 billion becquerels of radioactivity leaked from Fukushima's No. 1's underground storage tank...Water around the affected tank is highly radioactive and about 800 meters from the Pacific Ocean...Japan's government and Tokyo Electric (Tepco) claim the radioactive water is unlikely to reach the ocean...(Radioactive plumes in the Pacific drifting toward America have since been detected)

Great Britain is actively going forward with a plan to build a storage facility over a kilometer (3280ft) deep to hold toxic nuclear waste for over 100,000 years...Earth does not have a single deep disposal site to store nuclear waste...

It is estimated the U.S. has almost 70,000 metric tons of nuclear waste stored in over 38 states...

Nuclear power generation creates high-level waste (99% of nuclear waste), low-level waste, and a measurable environmental impact from the mining and transport of uranium to more than 400 nuclear power plants currently operating on Earth...There are thousands of uranium mines worldwide that are unregulated but continue to operate to produce the biggest profit without a real focus or concern of the byproducts of uranium recovery or the health of the humans who recover this mineral...Little is said in any world conversation of the impact of mining but fifty years of suggestions, policies, high-level talks and scientific studies have still not decided on how to deal with the highly radioactive waste produced by this process...Almost all nuclear waste is stored at the nuclear facility that created the waste, over 400 temporary storage sites that will have to be addressed at some point in time...The history of nuclear mishaps is a short list but the meltdown's at Chernobyl and Fukushima have left visible scars on this planet that will take a very long time to erase and it's clear there should be no compromise for safety and efficiency when building these power structures, no matter the "cost"...Humans have the skill, the technology and the intelligence to build every plant to the highest possible standards in a political and corporate world that will not let that happen...There will be more incidents and more accidents...

The primary byproduct of uranium enrichment is depleted uranium (DU) with more than 1.2 million tonnes stored...

In nations with nuclear power, radioactive wastes comprise less than 1% of total industrial toxic wastes...

A nuclear power plant produces approximately 300 cubic meters of low-intermediate level waste and about 27 tonnes of used fuel in one year of operation...A coal-fired power plant producing the same amount of energy will produce over 400,000 tonnes of coal ash, a highly toxic mix of heavy metals...

"One of the most ubiquitous and long-lasting recent changes to the surface of our planet is the accumulation and fragmentation of plastics," David Barnes, a lead author and researcher for the British Antarctic Survey...

Plastics

Plastics, one of the most durable and long lived popular modern conveniences in human history are rapidly becoming a huge problem in the environment used in inefficient ways for single-use items and then disposed of as waste that will last for centuries...Floating plastic can last for hundreds of years and when it breaks down it can disperse into thousands of tiny pieces ingested by fish and eventually humans causing health problems still not fully understood...The production of plastic products on this planet is currently more than 300 million tons annually, more plastic manufactured in the first 10 years of this century than in all of the last century...If these products were made in intelligent limited ways, used repeatedly for many years and recycled without exception, it could be useful in a myriad of ways with a limited environmental impact...But in a global world of ignorance without consistent regulation, corporate profit and consumer demand will drive industry to produce as much plastic as possible with little consideration to the long-term impact of a substance mass-produced for less than 75 years..

A study published in the scientific journal Earth's Future (May 2014) estimates that as much as seven trillion micro-plastic particles may be locked up in the frozen ice of the Arctic Ocean, particles that will be released into the ocean as sea ice continues to melt (micro-plastics are pieces of plastic less than five millimeters in diameter)...The study measured a extremely high density of 240 micro-plastic particles per cubic meter of ice...

Only recently have the effects of repeated exposure to the myriad of synthetic chemicals used to make plastics been implicated in the disruption of the endocrine system and it's been found that many of these chemicals used in manufacturing can leach out of plastics, including many containers used for food storage and liquids, plastic toys given to children (frequently chewed on) and the flexible plastics used in industries and applications worldwide (hospital intravenous bags)...These many *endocrine disruptors* that are found in thermal receipts, plastic wraps, linings of canned foods, cosmetics and the packaging of almost everything in our modern global society have been implicated in the hormonal disruption of developing fetuses and developing children causing problems many years after continued exposure including immune system suppression and cancers in reproductive organs...

George Bittner (a neurobiologist with the University of Texas) and a team of scientists tested 455 common plastic products and found that 70% tested positive for estrogenic activity...When they subjected these products to real-world conditions such as microwaving and dish-washing, the proportion rose to 95%...They concluded that "Almost all commercially available plastic products we sampled, independent of the type of resin, product, or retail source, leached chemicals having reliably-detectable endocrine activity including those advertised as BPA-free. In some cases, BPA-free products released chemicals having more endocrine activity than BPA-containing products."

A new study released in early 2015 at a yearly meeting of AAAS (American Association for the Advancement of Science) estimated the amount of plastics entering Earth's ocean every year at eight million tons with a prediction of 17.5 million tons every year by 2025 if this exploding waste culture continues...

The majority of plastics used on Earth today are disposed of as waste ending up in landfills or burned in incinerators releasing toxic chemicals and heavy metals into the environment...Dioxins, a group of toxic chemical compounds that share characteristics, are produced when waste containing chlorine is burned and when multiple products containing chlorine are manufactured...Dioxin is one of the most harmful synthetic chemicals ever used damaging immune systems and reproductive organs at very low concentrations...After burning the toxic ash from the incinerators is deposited in multiple landfills polluting the groundwater and leaking into surface water with unknown consequences...Also problematical is the recycling of plastics which involves a melting process that also releases harmful chemicals into the environment...Almost all of the chemicals used to manufacture these products are unregulated in the U.S. although many are being studied for cause and effect at the normal glacially slow pace of regulatory bodies...More promising is a recent accidental discovery of a new durable "thermoset" plastic that can be recycled instead of being discarded after use...These plastics are widely used in industry and this could be revolutionary on a world where the careless use and disposal of plastics is having an increasingly negative effect everywhere on Earth...

One of the most toxic plastics from start to finish is polyvinyl chloride (PVC)...Chlorine based with many additives (lead, cadmium and phthalates) PVC is routinely used to make children toys and a variety of products (shower curtains) used by consumers worldwide...The use, production and disposal of PVC has been implicated in damage to neurological systems, cancer, infertility and hormone disruption...Humans have the technology to make durable, non-toxic, recyclable plastic products but this is a world where profit is the motive for change and regulation is an unnecessary burden...Industries that make this product argue that PVC's are made from "common salt" and complaints of it's impact on health are "exaggerated"...

Atmospheric Pollution

Studies over two years (*2011,2012*) on atmospheric chemistry and the interaction of pollutants and natural chemical compounds in the formation and longevity of fine atmospheric particles have shown a higher density and a longer life than previously believed by atmospheric scientists...These particles, smaller than 1/13th the diameter of a human hair, are estimated to kill over 50,000 people in the United States every year and millions more around the globe...Aerosols, chemical pollution and gasoline emissions are all involved in the formation of these secondary compounds which are more numerous and longer lasting than the current E.P.A. models used to formulate pollution policies and emission regulations...This type of pollution is lethal to all biological life...

In 2013, the World Health Organization (WHO) finally classed air pollutants from car exhaust, power plants, agricultural emissions and industry as a leading environmental cause of lung cancer stating that "the evidence is clear"...In reality the evidence was clear and obvious many years ago to any intelligence who recognizes the physiology of biological life, but on a planet with competing cultures and nations with self-serving priorities, global organizations are unacceptably slow in confirming and embracing obvious realities...The International Agency for Research on Cancer (IARC), a part of the WHO, said air pollution had been known to cause heart and lung disease but evidence has now emerged that it was also causing cancer classing air pollution in the same category as tobacco smoke, UV radiation and plutonium...The most recent data analyzed suggested more than 223,000 deaths from lung cancer around the world were caused by air pollution with the majority believed to be in China and other East Asian countries due to rapid industrialization and a lack of emission controls...The WHO said the classification should act as a strong message to governments to take effective action but government and industry will continue to downplay it's significance citing studies by their "experts' that find the evidence "inconclusive" and "in need of more research" while continuing to pollute the Earth in whatever way necessary to increase profit...

In March 2014, the World Health Organization (WHO) released statistics reporting the deaths of seven million humans from air pollution in 2012 making it the largest environmental health risk on the planet...Over 33% of the deaths occurred in Asia, mostly China and India, with indoor air pollutants killing 4.3 million people (mostly cooking stoves and fires for warmth) and toxic atmospheric pollution outdoors responsible for some 3.7 million deaths...Unchecked urbanization with poor planning and the expected migration of more than 100 million humans to urban environments in the next two decades will make the problem worse in a global civilization determined to burn more fossil fuels in this century, not less...The greatest threat to human health is the burning of coal, wood and animal waste with over 700 million humans in India burning waste to cook indoors and a heavy reliance in China and India on burning coal for power with poor regulation and an absence of available, effective pollution controls...Current and future health-care costs to China from this activity are estimated at $300 billion...

In April 2014, the WHO released a survey of 1,600 cities in 91 countries revealing the extent of atmospheric pollution in urban centers worldwide...Nearly 90% of urban dwellers are exposed daily to air that exceeds safe levels and half of urban dwellers are exposed to air 2.5 times higher than recommended safe levels...

Arsenic

Arsenic is a naturally occurring element in Earth's crust, constituent in more than 200 minerals, a natural presence in bedrock and in this century it's predicted millions of people will suffer some level of arsenic poisoning from groundwater contamination, mining and industry...There are known arsenic containing bedrock formations in China, India and Bangladesh where the World Health Organization (WHO) reports an estimated 50 million people are at risk of drinking arsenic contaminated water and a report published in the journal Science in 2013 estimated more than 20 million humans are at risk in China from drinking wells that have not been screened for contaminants, toxicity or other known health risks...(China has over 10 million wells)...Groundwater contamination from this element is also found in Europe, Mexico, Asia, South America and the United States while mining operations release arsenic when smelting ores, zinc, gold and copper, the largest contributor of arsenic to the environment...Toxic levels of ingested arsenic and repeated exposure can cause cancer in multiple organs, heart disease, hair loss, convulsions and premature death...In a modern world there are technologies that can capture and prevent this element from entering the environment with filters, multiple scrubbers and treatments with iron compounds which react with arsenic to remove it from water but these technologies are expensive and in many areas of the world there is little regulation or no regulation at all...There is a long history of using arsenic in wood preservatives and pesticides which continues today, processes that have left many industrial facilities and mines contaminated with cleanup being too costly and in far too many cases ignored by the individuals or companies responsible...

In 2010, the world production of arsenic was more than 50,000 tonnes with 50% used to make arsenic based insecticides and herbicides, 30% to make wood preservatives and the remainder used in industry for semiconductors, glass production and as a catalyst to harden metal alloys in ammunition and mechanical bearings...

A study released in 2016 testing 2,103 underground drinking wells in China found over 80% of the wells contained water badly contaminated by industry and agricultural runoff unsafe for drinking or bathing...China's consumption was estimated at 110 billion cubic meters of underground water in 2009...

Burning coal is a source of arsenic pollution in many countries while the use of pesticides containing arsenic continues in Australia, New Zealand, and the United States...There are reports of arsenic contaminated food in China and exposure to wood preserving arsenicals (arsenic containing compounds) in Europe and North America...

Canada and the United States entered into a voluntary agreement in 2003 to ban the use of CCA in residential applications but CCA (chromated copper arsenic) is still used to pressure treat wood for industrial applications such as marine timbers and utility poles...In much of the world there is no regulation controlling the use of arsenic...

In the United States it remains legal to purchase and use organic arsenical herbicides according to label directions and precautions...EPA has initiated action to study the potential of these chemicals to transform over time and contaminate sources of drinking water but it will take an unknown period of time to do the study...

Of Earth's ten most polluted cities, one is in Iran, three are in Pakistan and six are in India...WHO

China's recent turnaround in the use of coal (less coal burned in 2014 than 2013 and a three year ban on any new coal mines) has much more to do with the populations in the urban centers of the nation than any concern for the environment...Over 1.6 million humans are estimated to die in China each year from air pollution...China is expected to invest billions in solar, wind and nuclear power generation over the next five years...

India's use of coal to generate power continues to increase as the government looks at how they can add over 300 million humans with no electricity to the power grid...Coal India Limited (CIL), a national coal producer responsible for over 80% of India's current production, plans to produce around 900 million metric tons of coal by 2020...

Pollution in America is Just Another Expense

The Kerr-McGee Corporation and its parent company Anadarko Petroleum Corporation have agreed to pay $5.15 billion for environmental contamination at sites it formerly operated in settlement of a fraudulent conveyance case brought by the U.S. and co-plaintiff Anadarko Litigation Trust in the bankruptcy of Tronox Inc. and its subsidiaries...The bankruptcy court had previously found in 2013 that the historic Kerr-McGee Corporation ('Old Kerr-McGee') fraudulently conveyed assets to 'New Kerr-McGee' to evade debts including it's liability for environmental clean-up at contaminated sites around the country...The defendants agreed to pay $5.15 billion to settle the case of which approximately $4.4 billion will be paid to fund environmental clean-up and for environmental claims...The Department of Justice said this case is one the largest environmental enforcement settlements in history...

$1.1 billion will be paid to a trust charged with cleaning up two dozen contaminated sites around the country including the Kerr-McGee Superfund Site in Columbus, Mississippi...

$1.1 billion will be paid to a trust responsible for cleaning up a former chemical manufacturing site in Nevada that contributed to contamination in Lake Mead which feeds into the Colorado River, a major source of drinking water...

Approximately $985 million will be paid to the EPA to fund the clean-up of approximately fifty abandoned uranium mines in and around the Navajo Nation where radioactive waste remains from Kerr-McGee mining operations...Additionally, the Navajo Nation will receive more than $43 million to address radioactive waste left at the former Kerr-McGee uranium mill in Shiprock, New Mexico. ..

Approximately $224 million will be paid to the EPA for clean-up of thorium contamination at the Welsbach Superfund Site in Gloucester, New Jersey …

Approximately $217 million will be paid to the federal Superfund to repay the costs previously incurred by EPA's clean-up of the Federal Creosote Superfund Site in Manville, New Jersey...

News of the settlement sent Anadarko shares soaring 14.5%...It's America...

A corporate structure who knowingly, wantonly and carelessly contaminated areas with high populations and there are no criminal charges for any of this destructive activity...The company stock is soaring and the irresponsible humans who caused this damage are free to do what they wish...In America, injustice can be bought...

Stunning are reports in 2015 that the recycling industries are losing business as oil prices drop making it cheaper to create products (plastic bottles) out of raw materials (oil) instead of recycled materials...An enormous window into the collective ignorance of a species whose singular motive for all activity is profit and greed...

ENERGY

18% of the world has no access to electricity...(1.2 billion humans)

In 2016 there is an optimism among energy companies and governments worldwide over the discovery of oil and gas in shale all over the globe, a growing excitement over new deposits of coal in Mongolia and other countries that will supply China with coal for a 100 years, eager anticipation in the melting of Arctic ice allowing the drilling for oil in remote, previously inaccessible places, a continuing investment in the tar sands of Canada with huge reserves of bitumen oil that is dirty, heavy and more carbon intensive than the light crude recovered in the last century, a smug awareness of the exploding populations planetwide who demand dirty energy to power their cities and their internal combustion engines and a growing confidence in the political systems of the world who are compliant and beholden to the energy money that control governments to allow and encourage multiple interests worldwide to burn every drop of oil, every lump of coal, and every cubic meter of gas on Earth...The equation is simple...Humans exist on a single world of life traveling through an endless emptiness in an indifferent universe...One atmosphere, one ocean, one planet...There is no where else to go...

Energy is motion and one of the most fascinating and interesting realizations in this universe is that *everything is in motion*...Humans exist on a complex collection of matter orbiting a black hole at 500,000 miles per hour, the Earth circles the Sun at 67,108 mph and the speed of the Milky Way galaxy relative to CMB (cosmic background radiation) is measured at 1.3 million mph...From the very largest structures known to the smallest particle ever discovered, everything is moving in a constant, consistent and mostly predictable motion that creates variable dynamic magnetic fields, speed of light particles and speed of light waves that represent the energy of the Universe...A vast 'emptiness' saturated with sub-atomic particles, energy fields and waves in endless motion begun some 13.7 billion years ago that will continue for a thousand billion years...*Electromagnetism* was "discovered" less than two centuries ago (1861) when James Maxwell combined more than 100 years of experimentation into mathematical equations and essentially changed the world...Today electricity is what powers the Earth and the manipulation and control of magnetic fields and particles moving at the speed of light will be how humans distribute energy for millions of years...The challenge is how we generate, store and distribute this energy...In a sustainable future, solar energy, hydroelectric, wind, geothermal and fusion energy could effectively (and efficiently) generate all of the energy needed on Earth and with nuclear energy controllable and "clean" if we can deal with the waste (new technologies are making progress) a path exists to scale back carbon pollution until the large-scale control and storage of renewable energy is available and capable of providing Earth with a stable energy source for millions of years...

In the 21ˢᵗ century, humans have the intelligence and an understanding of existing physical laws and limitations to engineer new technologies needed to use the energy of the Sun, the knowledge to control the power of the atom and are creating innovative materials and new designs to develop solar collectors with maximum efficiency, quiet wind turbines, increased battery storage capacities (very important), electric vehicles, global recycle technologies, carbon capture technologies and effective desalination facilities that could give humans the resources to manage a world...The first powered, controlled and sustained flight took place just over one hundred years ago and we have already developed the science and technology to interact with multiple planets, comets, asteroids and many of the moons in this solar system...The only thing missing is a global awareness of the danger of burning this planet for energy...It sounds simple, but today the extraction and burning of fossil fuels continues without restraint and the risk of this seemingly unstoppable process is clear if it continues to dominate energy policies...Short-term energy with unpredictable, unknown consequences...

All matter contains a measurable amount of energy that can be released and controlled given the necessary conditions and our Sun (this star system's primary energy generation device) will continue to produce enough energy to power any conceivable future for over a billion years...One day humans will build solar collectors the size of small moons to deliver solar energy to Earth, distributing this energy to globally strategic locations with large storage capacities using efficient networks with materials now being developed to transport energy with minimal loss and minimal heat...Clean energy with unlimited potential...Or we will find a way to install efficient solar collection devices on every structure on Earth eliminating the need for physical transmission networks...Real possibilities, but in the 21st century the future of energy is unknown...It's early in the growth of energy technologies and although there is no consensus on the energy fuel(s) of the future, there is a growing global awareness of the need for change and the necessity of a clear strategy with innovation and unwavering commitment to move from unsustainable 'dirty' energy to a world of limitless, clean energy that will power civilization for not just hundreds or thousands of years, but for millions of years...But not yet...

In this century billions of humans lead desperate lives due to the unequal distribution of energy amid crumbling infrastructure and ancient, unstable power grids that are still part of most of the world...The inefficient transmission, distribution and storage of today's energy wastes more than 50% of the power generated and the electricity generating devices still use mostly fossil fuels...New energy reserves and new technology have accelerated the extraction of gas and oil everywhere on Earth and despite the "climate panels", multilateral treaties and the commitment of nations, humans are burning fossil fuels at record levels and polluting the atmosphere in dangerous ways with a lack of awareness and no consensus among the people controlling Earth's resources to recognize a balance of the benefits and dangers in producing the energy needed to sustain human existence...

Electromagnetism...the interaction of speed of light particles and fields in motion that will power humanity for millions of years have been generated and controlled by the human species for less than 200 years...

If humans have the resources, knowledge and technology to burn fossil fuels at high-efficiency with multiple scrubbers, filters and capture devices with zero pollution, why is this not done with every power plant on Earth?...Why do large corporate structures, elected governments, political coalitions and individuals, lie, manipulate data and rationalize ignorance to avoid doing what's needed to produce the cleanest energy possible for all humans...Cost should never be a factor in this critical balance of life and energy...*One planet, one atmosphere, one ocean...*Money, individual wealth and greed should not determine the growth and survival of the human species but in the 21st century the control of money is essentially everything...Money is an old, ancient manipulative median used to barter all goods and services around the world and at this time in human history most of the money that controls most of the resources on Earth is in the possession of a small percentage of human beings...In the future there will be no money, but that is not now...*Today, money is what drives human activity across the entire planet regardless of consequence...*In this money driven global society there are millions of people who do not understand the impact of unequal resource distribution on billions of humans and many corporations and governments controlling energy resources will not or cannot understand the long-term or the short-term impact of their actions...With an inability to recognize the deep interactive connection all life has to the geological and biological processes on Earth, damaging the planet and vital species to *succeed* as a group or to enrich oneself is a "normal" and a very primitive instinct...

History is filled with individuals or groups who justified extreme behavior to gain an advantage with enslavement and extermination of entire populations for territory and the persecution of many millions for ideological or religious beliefs...Driven by fear and instinct with an inability to understand existence, this ignorance and lack of empathy has a direct connection with the world as it is today at a critical time when our species has the intelligence, the knowledge and the resources (and enough time) to develop innovative technology to power a world civilization for millions of years with a manageable environmental impact...One day every device will run on electricity (even airplanes) and the challenge is how to generate, capture and store this energy in a continuous, effective way with a minimal negative impact on Earth and the species inhabiting the Earth...

In a distant future, the sunlight striking anywhere on Earth will be absorbed into an energy network with storage capacities unimagined today...An entire civilization powered by a single star for millions of years...

One of the most challenging aspects of power generation is the distribution of energy in an efficient way...There are current discussions and plans going forward to build a 'super-grid' across Europe from Ireland to north Africa to transmit energy from multiple sources of solar, wind and backup power plants but it will take decades to move forward with the fragmented politics and economics that exist today...

The Future of World Energy

In a world where dirty energy is widespread, inexpensive and available, the biggest challenge is a transition from fossil fuels to clean energy, something that needs to happen in less than 100 years but in reality will take centuries...It would be possible to do this on a large-scale in this century if there were an intelligent consensus on Earth acknowledging the negative impact of fossil fuels but there is enormous resistance from powerful entrenched interests who continue to harvest fossil fuels for energy and challenging these interests is difficult...Serious change would require the safe use of nuclear energy combined with hydro, wind and solar energy and it would need to be a coordinated global effort, almost impossible given the political reality in this century...A balance heavily weighted toward fossil fuels in the beginning (now) but it would slowly change in the direction of clean energy and with a focused global effort humans could do this in an efficient way burning less and less fossil fuels until all energy needs have clean energy possibilities...An incredible challenge even with the multiple technologies and new innovations available and it would take over a century to make this transition even if there were agreement on the process and the cost of this critical historic transition essential to the survival of our species...*In 2016, it's clear this transition will not happen this century (the use of fossil fuels grows more not less) and it's unpredictable at this time when it will ever happen...*

Scientists and experts on global warming mostly agree that about 70% of discovered fossil fuel reserves known to exist *must stay unused in the ground* if humans are to have any chance at preventing changes in the climate patterns affecting long-term ecological stability...*This will not happen...*Energy generation across the globe is controlled by governments and corporations with a focus on short-term profit and status quo who exist without the intelligence or awareness to understand what is proven to be happening now while ignoring long-term effects demonstrated by past events, more than 150 years of measurements, current events, computer modeling and a worldwide consortium of scientists and energy experts who agree on the known and unknown impacts of a continued reliance on fossil fuels...

A report released by the Overseas Development Institute (ODI) in 2013 shows the wealthy nations on Earth spending 7x's as much money on subsides for coal, oil and gas than is spent fighting climate change and supporting clean renewable energy...The figures available are challenged in a global "normal" of secrecy and misdirection but the report findings clearly demonstrate patterns of collusion, opportunity, and a connected web of politics and corporate interests that profit at the expense of the poor while maintaining the "status quo"...In 2011, the U.S. gave $6 billion to a growing ethanol industry and $2.5 billion to oil, gas and coal for research and development...In 2011, Germany gave 1.9 billion euros to the coal sector and the U.K. gave almost half a million in U.S. dollars to oil and gas production...The report reveals the governments of Pakistan, Indonesia and Venezuela spend twice as much on fossil fuel subsidies as they spend on public health, poor nations with desperately poor humans suffering disease and malnutrition...And in a stunning revelation that clearly shows a profound lack of awareness among the powerful, wealthy humans who control the world's resources, ODI said 75% of energy project support from international banks were designated for fossil fuel projects in the 12 developing nations with the highest CO2 emissions...

58,394 tons of uranium were produced in 2012 (World Nuclear Association)...Kazakhstan produces the largest share of uranium from mines (36.5% of world supply in 2012), followed by Canada (15%) and Australia (12%)...

Nuclear Energy

In 2009, there were 436 commercial nuclear power plants in the world generating 15% of the world's power with almost no greenhouse emissions...(It would require almost 20,000 nuclear power plants to supply energy for the entire world by 2050)...Every nuclear power plant prevents the yearly release of over three million tons of carbon dioxide, *green energy* from the only large-scale continuous power generation source that does not contribute to global warming...With nine billion people on Earth by 2050 increasing demand for energy three to four times what it is today, nuclear energy is possible now and with regulation, design and new technology it can be as safe as the solar, geothermal and wind energy that must produce the majority of Earth's energy in the 22nd century and beyond...*The key to success is safety without regard for cost*...Again, if the human species has the resources and technology to build these devices to the highest standards possible, why isn't it done with all nuclear power plants on the planet...How can big money and corporate greed result in substandard construction, political manipulation, enormous waste and cost cutting measures that affect safety and reduce efficiency?

Nuclear power generation is not a panacea...Current reactors and nuclear power facilities globally release small amounts of gaseous and liquid radiological effluents into the environment...The total amount of radioactivity released depends on the power plant, regulatory requirements and operation...Does the technology exist today to capture these radioactive gases and reduce dangerous emissions to zero? If not, the world should rethink their dependence on a technology that could be an effective bridge to solar, wind, geothermal and eventually, fusion energy...Radioactivity does not "go away" for many thousands of years and even the slight emissions from thousands of nuclear power plants will have a cumulative effect on populations for centuries to come...In addition, the process of mining for uranium, the nuclear fuel necessary for generating energy, is environmentally destructive and haphazard in the fragmented and under-regulated political coalitions who too often are willing to ignore the destructive and deadly mining practices of corporations who use influence and donations to profit at the expense of the workers and the environment...Hundred's of abandoned mines have not been cleaned up and are environmental and health risks in many communities with EPA estimates of more than 4,000 mines with documented uranium production and 15,000 locations with uranium occurrences in the western states of America...In the U.S., environmental groups have alleged that uranium mining companies circumvent cleanup regulations that apply to abandoned mine sites by reactivating their mine sites briefly from time-to-time...These groups have filed legal objections to prevent mining companies from avoiding compulsory cleanups but collusion between big business and political bodies have a long history and any challenge will be delayed for years if not dismissed...

In a global society, a global cooperation is vital to implement new advanced nuclear technology and design with intelligence, awareness, partnerships and a rigorous universal safety protocol necessary to prevent accidents and incidents caused by poor design, substandard construction and an ignorance of the science needed to safely deal with nuclear power distribution and the byproduct of nuclear power generation, radioactive waste...Recent events in Japan with an earthquake and tsunami impacting a nuclear power plant with multiple reactors is a perfect illustration of a lack of awareness among corporations and politicians who control the money and resources needed to build these technological wonders...All intelligent humans would agree that no power plant of any kind should be built on a known fault line or in an active earthquake zone and enough is known of the structure and geology of Earth to accurately identify high risk areas that should be avoided...There are many places on Earth where nuclear power plants could be built with a minimal risk of impact from a naturally occurring geological or weather event...New technology and new power grids could route the energy through high-efficiency systems to the populations needing the power with an understanding that the world needs a cooperative blueprint to maximize energy resources with the smallest possible impact on species and natural habitat...

In August 2013, the Tokyo Electric Power Company (Tepco) admitted that around 300 tonnes of highly radioactive water had leaked from a storage tank at Fukushima...Some 1,000 tanks have been built to hold the highly radioactive water being used to cool the reactors but these tanks are believed to be at 85% of capacity and every day 400 tonnes of water is being added...Many experts in the nuclear field believe the problem is worse than what is being reported with highly contaminated groundwater and leakage into the sea...This event is a result of multiple short-sighted humans who compromised safety in the construction and operation of the plant and have subsequently misled the world about the aftermath of the 2011 accident...Universal safety protocols would make possible the unlimited potential of nuclear power to produce clean energy that could effectively power this civilization for centuries...

We have the technology and intelligence to make these complex power generating devices safe, efficient and sustainable but there is not yet a consensual awareness of cause and effect or a cooperative intelligence to do what's needed with every nuclear power plant on Earth to make the transition to renewable energy sources...

Japan initially shutdown all 54 nuclear power plants...They are slowly restarting their nuclear power plants in a future that will focus on multiple sources of energy with a priority on developing green energy...

An independent study published in March 2012 reported radiation from the Fukushima Daiichi disaster was detected in the water supply of more than a dozen U.S. cities and in multiple food products in Japan...

Nuclear energy's potential is as a transitional power source to a future of renewable energy...The two major concerns that have kept this from happening is storage of highly radioactive nuclear waste and the possibility of major "accidents" with very long-term consequences...The potential for accidents would be low if these facilities were built to the highest technological and safety standards possible but in this century money determines quality and safety and money is often the most integral and essential part of every energy project on Earth...As for the radioactive waste produced by this process, several nations are actively exploring deep underground repositories where they can store waste but in the next 100 years the problem of nuclear waste might be solved...New ideas using nanotechnology to remove radioactive ions from the environment and a fission-fusion hybrid to create energy while reducing the radioactivity of nuclear waste are being studied and it's unknown at this time what is or what will be possible with new technology and innovative ideas to redesign and build nuclear power plants with higher safety standards while finding ways to re-use the nuclear waste or render it inert...Generation IV reactor designs are changing the way fuel and coolant are used with a *molten salt reactor* that dissolves the nuclear fuel in a salt mixture for safety with the ability to use the remaining nuclear fuel found in the waste and a *traveling wave reactor* designed to eliminate the need for uranium enrichment and use depleted uranium found in the waste to generate power for decades without repetitive refueling...These ideas and designs are real and are happening now with funding from private capital and various governments who are slowly starting to understand the stark realities of a changing 21st century Earth...Instead of seeking "low-cost" ways to build the nuclear power plants that provide pollution free energy for years, the priority should be to design the safest and most technologically efficient reactors possible with 21st century science to help our species transition to renewable energy sources that will power this world for millions of years...

India has 22 nuclear reactors operating in six plants providing about 3% of their energy and 44 reactors slated for construction or already being built even as a recent auditor-general's report said India's Atomic Energy Regulatory Board is ineffective, mired in bureaucracy, and negligent in monitoring safety...Over 60% of scheduled inspections in operating or under-construction nuclear plants are late or never undertaken at all while an estimated 400 million Indians have no access to electricity in any predictable way...By 2050, India wants nuclear power stations to provide 25% of the nation's power with Russia, France and the U.S. building multiple reactors to achieve this...

The Energy Department in America appointed a 'blue ribbon' panel to look at options to deal with nuclear waste and the first suggestion was developing temporary storage sites to hold waste up to 100 years??? This is no time at all...With the current political gridlock in America, it would take 100 years just to build the storage sites...

Canada, Finland, Germany, South Korea, Spain, Sweden and the United States directly dispose of spent nuclear fuel in storage containers...Switzerland, United Kingdom, Russia, India, Japan, China, Belgian and France reprocess high level nuclear waste...Finland's Onkalo facility will be the 1st geologic repository for spent nuclear fuel in 2020...

The U.S. Energy Department is financing the development of mini-reactors, small nuclear power generation devices contained within a single structure (a rocket-shaped steel cylinder) that generate approximately one-tenth the power of a normal sized reactor and can be deployed anywhere in the world where there is a need...Safety, security and environmental concerns aside, these devices could be a reality by 2030...

France generates 75% of their electricity from 58 nuclear reactors while the United States supplies less than 20% of their electricity from 99 reactors...

Almost all of Earth's robotic planetary explorers are powered by nuclear fuel and this energy will be critical in maintaining life support systems when manned missions are sent to the planets and moons...

More humans die in one year mining coal and drilling for oil than have died in all the nuclear accidents to date...

Fossil Fuels

This is the energy powering most of the planet in 2016 and it grows more every day...Primitive in every way, fossil fuels are combustible substances that humans dig out of the ground and burn...Tiny creatures living on a huge rock who started with shovels and picks and now use large machines to dig and harvest stuff that burns...Coal, petroleum and natural gas provide over 75% of the world's energy and *these substances can only provide energy when burned*...The byproduct of this type of energy production is the emission of carbon dioxide (CO2) into the planet's atmosphere and current human activity emits more than 39 billion tons of carbon dioxide into the atmosphere every year...106 million tons a day, increasing global carbon emissions by almost 6% in 2010...The U.S. spent $1 trillion in Iraq fighting for oil but spends less than 10% of that annually on alternative energy and now *fracking* is driving an oil and gas industry determined to do anything and everything to extract large amounts of gas and oil embedded in shale rock without regard for the environment or nearby humans...

Why would a species of intelligent beings carelessly and dramatically change the balance of the atmospheric gases that have allowed them to evolve and survive for thousands of years?

Zooming in from space with a global picture of human activities on planet Earth, one would see millions of small creatures who exist on an enormous rock of abundant diversity, a rock that is the only possible home for these creatures, drilling holes everywhere in a desperate effort to supply dirty energy for a lifestyle that cannot be maintained long-term and is proven to be tipping their world toward a climatological disaster...Humans are now so consumed by the need for their small mobile vehicles and their various labor-saving devices that energy harvesting activities are causing earthquakes by the thousands and polluting freshwater supplies with dangerous, unknown chemicals whose long-term effects on human health have not been studied and can only be guessed...America's 2005 Energy Act **exempted** companies who inject fluid underground for gas or oil extraction from the Clean Water Act, fluids that include synthetic chemicals, glycols and alcohols, benzene (found in concentrations well above Safe Water Drinking Act standards) and excessive methane levels...Ignorance, collusion, bribery and corruption...Over 40,000 wells have been drilled in the bottom of the Gulf of Mexico and over 27,000 of those wells are now abandoned with an estimated 17% of the wells slowly leaking oil into the water...If this is the case in a small gulf in the western hemisphere, how many abandoned oil and gas wells are there across the planet?

There are an estimated 20 million to 30 million abandoned oil and gas wells on Earth, over 2.5 million in the United States...150 years of exploration, little regulation and an inability to seal abandoned wells in any permanent way...

The United States uses 19 million barrels of oil every day, 25% of all the oil consumed in the world in a country with just 4% of the world's population...

Oil

World demand for oil is predicted to be over 94 million barrels per day in 2016 and demand is growing...Reserves of more than 500 million barrels of oil have been found deep below the Atlantic ocean off the coast of South America and will be recovered at a rate of 80,000 to 100,000 barrels a day by 2020 in difficult and risky operations challenging new technology with unknowable consequences if something goes wrong...The Canadian oil sands are estimated to hold the world's 3rd largest oil reserves (more than 150 billion barrels), but extracting it is dirty, expensive and this oil releases more greenhouse gases when refined than more conventional oil...Strip mining the tar sands uses two to three gallons of water for every gallon of oil extracted and the wastewater pits are highly toxic and hazardous to wildlife and humans...There is interesting new technology being developed in the extraction process that would make it cleaner and less destructive but industry is slow to change if there's more profit doing it the old way...The Keystone Pipeline being built to transport the oil is controversial and risky as tar sand oil is not like conventional crude and accidental spills can be very destructive...This was made obvious by a spill of more than a million gallons in Michigan in 2010, the first major spill of tar sand oil, a spill that is proving enormously difficult and costly to clean up with numerous health problems reported by nearby residents and environmental destruction from the cleanup efforts...It's predicted the oil from the tar sands will reach a production level of almost five million barrels a day in less than 25 years, large-scale environmental destruction...*(Keystone was rejected by the U.S. State Dept in late 2015)*

The tar sands in Canada cover more than 50,000 square miles with production increasing daily...Hundreds of billions of dollars have been invested in destroying the environment and polluting the planet by corporations focused on profit without regard to a changing Earth turning more than 12 billion cubic feet of clean water into poison every year while destroying vast stretches of virgin forest to create a product proven to negatively affect this world's environment and kill the biological creatures who inhabit it...From any point of view this makes no sense...In a future civilization with consensual intelligence, an awareness of reality and an understanding of causality, resources and the manpower used for this project would be focused on building enormous, effective, renewable energy facilities across Canada's unpopulated landscape that merge with the forests essential to the replenishment of Earth's atmosphere and the protection of wildlife and water (Earth's most precious resource)...A promising, primitive species...

The extraction of oil from *shale* in the U.S. can produce up to *half a trillion* barrels of oil but it's destructive environmentally and produces more greenhouse gases during the refining process than light crude...This process is destroying communities and environments in various American states as oil companies rush into drilling situations without oversight in an effort to increase production and profit without a clear understanding of the impact on the people, the land or the water...Recent studies have shown there is measurable "leakage" of methane (a highly effective greenhouse gas) at almost all fracking sites and numerous wastewater pits holding toxic water...Regulatory agencies vary from state to state and there is obvious collusion in many operations between the regulators and the operators that allow dangerous practices to continue and "accidents" to be minimized in the reporting structure...A report on the Utica Shale formation under Ohio estimates there are almost a billion barrels of crude oil to be recovered from the shale and the only way to do it is 'fracking' with still unknown consequences and untested procedures necessary to harvest more stuff to burn...

In the United States there are tens of thousands of wastewater disposal wells containing fracking fluid that comes to the surface with the oil or gas and is then injected with an unknown mix of chemicals deep underground...

There are estimates of over a billion barrels of oil to be found in the Arctic regions of the world that will be accessible with the warming of the planet and the melting of the sea ice that has prevented drilling in the past...There is clearly enough accessible oil to run this world in the conventional way for more than a 100 years but even if our species is enormously creative and successful at harvesting these resources, burning it will be catastrophic to life on Earth and will negatively affect the delicate balance of clean air and water that has allowed life to evolve and spread across a world...

This product has been an energy source for just over 150 years...In that short time oil and its increasing value to an exploding population has begun to affect the environment in profound negative ways and disrupt a balance in nature that has nurtured life and allowed the development of human intelligence...

There is no argument among intelligence that governments and individuals everywhere on Earth need to start using fossil fuels as intelligently and efficiently as science can determine while expanding research and development into alternative energy sources that do not threaten the continuation of life on this planet...It's difficult to overstate how essential it is to quit burning the resources found everywhere on Earth...If this process increases for the next 100 years and every indication is that it will, there will be negative consequences for thousands of years...Today that future seems almost a certainty...

"When taking into consideration the environmental destruction that is required to produce tar sand oil to begin with, an energy intensive process that emits three times more greenhouse gases than conventional oil, requires two barrels of water for every barrel of oil and the clearing of boreal forest for wastelands of tailing ponds, it becomes clear that tar sands is the wrong energy choice for Americans," Jenny Pelej, National Wildlife Federation...

*Coal mining reclamation can give the surface landowner many more options for developing his land...In the mountainous terrain, a mining process call mountaintop removal can create very valuable and usable level land for the surface owner...*Kentucky Coal Education

Coal

There's evidence in early cave dwellings and in early civilizations confirming that the properties of coal were recognized thousands of years ago and in 2016 coal is still the fuel used to generate the majority of Earth's electricity...World coal consumption totaled over seven billion tonnes in 2010 and it's estimated to be over nine billion tonnes by 2030...Carbon dioxide emissions (CO_2) are the leading cause of global warming and China is the world's largest emitter of carbon dioxide pollution (3x's more than 2nd place America) primarily through the burning of coal...Cause and effect...Technologies to deal with CO_2 emissions by capturing the emissions and storing them in deep underground rock formations are uncertain and problematic and are being challenged as too costly and too controversial...In 2011, two carbon capture and storage projects in Europe were canceled and more planned projects will be canceled in the future because of public concern, high cost, uncertainty in climate change strategies and regulatory challenges...By 2020, Europe will have just half of the plants promised to be running by 2015, plants designed to reduce greenhouse gas emissions to 80% of 1990 levels by 2050...The U.S. has four operating plants doing carbon capture and have plans for twenty more in the future...This is not the solution to climate change and burying carbon emissions to avoid and delay the development of renewable energy is very temporary...Humans need to stop burning everything on their planet for energy and after an industrial age of 200+ years, they don't know how...

At an environmental meeting in Brussels in 2012, the EU was unable to find a suitable carbon capture and storage project to fund despite having set aside 275 million euros (£224m) to kick-start the technology...Projects from different European countries including the UK had been in the running for a share of the money but all were forced to withdraw after failing to secure necessary financial support from their own governments...The International Energy Agency concluded that CCS technologies cannot impact climate change for at least five years and has awarded 1.2bn euros (£975m) to 23 innovative renewable energy demonstrations projects across the European member states...

To comply with a global agreement to keep Earth's temperature from increasing more than 2 degrees Celsius, IEA estimates there should be around 1,500 full-scale Carbon Capture and Storage plants in operation by 2035...In 2012 there were just eight...In 2013, Norway abandoned the continuation of one of the largest CCS facilities in the world at the Mongstad oil refinery...Political, technical and financial concerns were cited...

*Sensors in Greenland have recently recorded a new and highly symbolic level of carbon dioxide in the atmosphere at 400 parts per million...Many researchers believe the **average** global level of CO_2 will soon pass 400ppm...Scientists believe the uppermost level that will prevent serious and permanent climate disruption is 350 parts per million...*
(In 2016, levels were measured at 404ppm with a possibility of 1,000ppm in about 300 years...NASA)

World Resource Institute reported 1,200 new coal plants were being planned, mostly in India & China...2012

American coal companies are increasing exports to Asia to maximize profits without any long-term awareness or concern of the consequences to the world environment as their product is burned on one Earth in nations without pollution devices or any rational government controls over emissions that affect everyone on the planet...Carbon taxes are unpopular in almost every country that proposes them and with the enormous amount of money flowing from energy lobbyists to the lawmakers, carbon taxes seem very distant...Despite this, almost every "expert" queried about the increase of greenhouse gases worldwide opines that "carbon taxes are the solution", an entrenched belief in almost all the nations on Earth that money is the solution to any problem...

*Coal is nothing more than ancient wood which has been under pressure for millions of years and can be mined and burned with little environmental impact...*Kentucky Coal Education

Ironically, coal's biggest enemy may be the vast quantities of shale gas being found all around the world and the new technology that can recover this natural gas in a profitable way...New limits on carbon dioxide emissions in the U.S. can easily be met by efficient new gas power plants and big coal interests are on the attack sensing the eventual end of a profitable, environmentally destructive energy generation process that cannot continue if humans are to survive their industrial age...Nations with few restrictions on pollution from coal will be forced to consider new technology as international pressures increase (there's only one atmosphere and air pollution knows no boundaries) and populations demand change as urban areas grow dangerous to all the biological creatures who breathe the air...

The Environmental Protection Agency issued a proposal in September 2013 to limit the carbon emissions on new power plants built in the U.S., the first federal attempt to impose limits on pollution blamed for global warming and a proposal that was vehemently opposed by the coal industry and right-wing politicians...New coal-powered plants would be limited to 1,100lbs of carbon dioxide emissions per megawatt-hour (the average coal plant emits about 1,800lb of CO_2 per hour) and new gas-fired power plants would be limited to 1,000lbs of carbon dioxide emissions per megawatt-hour...To meet the new standard coal plants would need to install new carbon-capture technology which the industry contends is too expensive...After a period of public comment and more pressure from industry, the EPA stated in 2014 it will scrap this proposal and put forth another proposal that creates a separate standard of performance for fossil fuel fired electric utility steam generating units and integrated gasification cycle units that burn coal, petroleum coke and other fossil fuels based on partial implementation of carbon capture and storage as the best system of emission reduction....

In August 2015, the EPA finalized the Clean Power Plan to control CO_2 emissions from existing power plants with a stated goal to reduce carbon pollution by 30% from 2005 levels...States have until September 2016 to submit a plan (extensions of up to 2 years are available) with flexibility on how to meet the goal and the ability to delay implementation as needed...The first deadline for starting the reduction of CO_2 is in 2022 and full compliance is not required until 2030...The plan has been called a job killer by representatives of the coal industry and the politicians representing coal states...

After 29 states challenged the new regulations in court, the Supreme Court issued a hold (02/2016) on the application of the new rules that could last up to two years as the fight continues to pollute Earth without regulation...

The EPA regulates mercury and sulfur emissions but in 2016 there are no limits on CO_2 emissions...

Shale Gas
An energy source mostly unknown only ten years ago is now a major player in the world energy mix with the IEA (International Energy Agency) predicting humans are moving into a 'golden age of gas' with enormous resources of natural gas trapped in shale and new technology to recover it...The United States has large proven reserves of shale gas and it's estimated China and Europe together have more than twice what America has...China reported it has 1,115 trillion cubic feet of recoverable shale

gas reserves (2nd in the world to the U.S.) and Europe is moving in the same direction with concerns over energy prices and demand outweighing concerns over the environmental damage caused by the extraction process...Shale gas is cleaner than coal when burned and any energy source replacing coal is always a positive development, but it's still a fossil fuel that is burned for energy emitting dangerous greenhouse gases with increasingly serious environmental consequences and a finite supply...Also, a growing concern is the environmental impact of recovering shale gas with chemicals and the leakage of methane gas at a volume recently shown to be much higher than projected by the energy companies...

Italy, Germany and Japan which combined make up less than 4% of the world population imported over 30% of global natural gas imports in 2010...

In the end there may be no advantage in comparison to other fossil fuels making the extraction of shale oil and gas just another diversion by energy companies looking for profit...If the world were committed to transitioning to green energy, shale gas could help with natural gas producing 30% less greenhouse gas per unit of energy compared to oil and 50% less greenhouse gas per unit of energy than coal...Unfortunately, with a lack of intelligent consensus among a majority of Earth's population, many energy corporations and governments are embracing the abundance of natural gas with a false sense of security that will divert resources and exploration away from green energy sustainability increasing the focus on fossil fuel extraction with little concern for the proven negative impacts on environments and populations...Increased production of another substance that must be burned for energy...Because of groundwater pollution and the abundance of undisclosed chemicals used in the fracking process, some European countries have already banned the activity until environmental studies can be done to insure safety and responsibility but in most of the world and certainly in America, the priority is to drill now and deal with the negative consequences some other time....

Under the Barents Sea is a natural gas field (Shtokman Field) that holds an estimated 4 trillion cubic feet of gas but is located far above the Arctic Circle over 300 miles from the Russian coast...Gazprom, a state controlled energy company in Russia, has been trying to develop the field for more than 5 years without success...

Between 1967 and 2000 the central and eastern United States experienced an average of 21 earthquakes per year of magnitude 3.0 or greater...From 2010 to 2012 the same region experienced over 300 earthquakes magnitude 3.0 or greater...William Ellsworth, Science July 2013

"Our analysis showed that shortly after hydraulic fracturing began, small earthquakes started occurring and more than 50 were identified of which 43 were large enough to be located. Most of these earthquakes occurred within a 24 hour period after hydraulic fracturing operations had ceased." "A naturally-occurring rate change of this magnitude is unprecedented outside of volcanic settings or in the absence of a main shock of which there were neither in this region." US Geological Survey (USGS)

A lawsuit in Pennsylvania in late 2012 revealed state testing agencies did not report toxic metals found in the testing of groundwater ordered by the state while the lawsuit's plaintiffs (diagnosed with benzene, arsenic and toluene contamination) continued to have medical conditions including difficulty in breathing and severe headaches...The lab stated in court the toxic metals were not deemed 'relevant' to the case...The companies in the lawsuit insist fracking chemicals 'cannot enter public water sources'...collusion, corruption and deception

Bio-Fuels

A growing industry in America, bio-fuels encourage extensive deforestation that can make their carbon footprint larger than fossil fuels and bio-fuels can raise food prices as agriculture is focused on crop production for fuel, not food...Ethanol from the bio-fuel is only slightly cleaner than gasoline, decreases miles per gallon, takes energy to produce and transport and is not the solution to global warming...Citing cost and a continuing effort to reduce dependence of foreign oil supplies, the U.S. produced more than 13 billion gallons of ethanol in 2011 using 40% of the annual corn crop...Again, it's a product that must be burned...Brazil is the largest consumer of ethanol with the U.S. actively pursuing plans to use bio-fuels for a variety of applications including jet fuel...With mandates for use around the world and a growing industry seeking enormous profits, the push to increase bio-fuel production to provide more than 25% of the world's transportation fuels seems unstoppable...

A study published in the Nature Climate Change journal reported bio-fuels made with corn residue released 7% more greenhouse gases than gasoline...This is the first report of this type and created controversy in the U.S. where bio-fuel is subsidized by the federal government...2014

These products do not address issues involving climate change or environmental security and considering energy costs needed to produce bio-fuel using fertilizers, pesticides, harvesting techniques, irrigation and processing with growing populations and rising food costs everywhere, the emphasis on bio-fuel production is a wasteful, dangerous detour around what is really needed...More promising is the use of waste material converted into bio-fuels which would eliminate the food equation but it's still a pollutant when burned as all fuels must be...Promising are new facilities to turn waste into 'cellulosic bio-fuels' using wood waste and garbage to create ethanol with a Spanish corporation Abengoa saying they can produce up to 25 million gallons of ethanol annually from agricultural waste, non-food crops and wood waste...It could help transition to clean energy but it is not a solution...

Solar Energy

Solar energy is this century's fastest growing source of renewable energy but remains less than 5% of the world's energy...China is the largest manufacturer of solar panels but new technology in solar cell design, capture and storage will have to improve (and it will) for solar energy to become a large part of the global energy equation...On a smaller scale there are many current applications for solar power on individual structures and on boats to power necessary devices with solar power being used to recharge technical hand-held devices that exist in all societies and this will be a powerful technology in isolated regions of the world without power grids, but in a future without fusion technology solar power must play a larger role and be a critical part of Earth's energy solution...Humans are master tool builders and one can imagine large-scale solar collectors built in space with the capacity to provide unlimited energy for this planet, continuous uninterrupted energy for millions of years...It's been calculated that the solar energy striking the Earth every year is more than 20,000 times the energy used by all human activity in 2014, over 12 trillion watt-hours per square mile per year...And it will continue for a billion years...If solar-collectors could be built with technology to provide protection from solar flares, coronal ejections and space debris there would be no limit to the collection of energy...Not possible now, but one day we will harvest the energy of a star, an incredible technological challenge...

The largest solar thermal power plant in the world started operations in February 2014 in the Mojave Desert in California with a capacity to produce 392 megawatts of energy powering 140,000 homes...The Ivanpah Solar Power Plant cost an estimated $2.3 billion and uses hundreds of thousands of mirrors to turn water into steam to generate the power...While large solar power facilities will be instrumental and necessary in a transition from fossil fuels, research continues into the development of technology that will one day allow almost every building on Earth to have solar collectors on the roof with long-term high-capacity storage devices...No electrical grid needed...

To make solar power work there is an increased focus on greater storage capacities, collector efficiency and new transmission networks designed to transport this energy with maximum efficiency, minimal loss and progress is being made in all these areas by innovators and scientists who understand the urgency and necessity of transitioning from fossil fuels...QBotix has developed intelligent robots to dynamically operate solar power plants to maximize energy output with a measured increase of up to 40% over fixed mount systems...Researchers from UCLA and California NanoSystems Institute have developed transparent solar cells to give windows in any structure an ability to generate electricity with polymer solar cells produced in high volume at low cost...Rensselaer Polytechnic is researching ways to achieve four-layer anti-reflection transparent thin-film materials to capture more of the Sun's energy by achieving a low refractive index...Bandgap Engineering is developing nanowire-based solar cells that could generate twice as much power as conventional solar cells with the potential to convert 20% of sunlight energy into electricity and if advanced solar cells convert *only* 20% of sunlight energy, imagine the potential when we can build cells that convert 90%...Finally, NREL has demonstrated the first solar cell with an external quantum efficiency (EQE) exceeding 100% for photons with energies in the solar range (EQE is the % of photons converted into electrons to power a device)...

The limits of human technology are still unknown, but invention, growth and innovation are an integral part of human intellect...Building new power grids is difficult amid territorial boundaries isolating nations with outdated, fragmented technologies and a global political structure that will take decades to stabilize...This will change...New storage capacity devices are being envisioned, invented, tested and improved for applications from the very small to large facilities still being realized and solar cells will continue to improve until efficiency under cloudy skies exceed by multiples today's efficiency under clear skies...A future with solar devices on every structure on Earth and a new paradigm for energy generation and energy consumption with the potential to provide power anywhere across the entire planet without connecting to a grid...Limitless energy that's available, clean and necessary for survival on a world that depends on the Sun's energy for almost everything...

German firms Siemens and Bosch announced they are leaving the Desertec Industrial Initiative, a private industry consortium planning to install a total of 125GW of solar-power capacity throughout the Middle East and North Africa by 2050...By 2014, almost all initial shareholders had left the project leaving in it's wake a much smaller project as a globally connected Earth grows more fragmented and territorial...

Siemens announced it will completely pull the plug on it's solar-energy business to focus on developing wind and hydroelectric power units...The firm said it sees renewable energy accounting for 28% of global energy use by 2030 but solar power will be only 9% of renewable output compared with 54% for hydro and 27% for wind power...

In multiple states in America, energy corporations and conservative organizations spend millions of dollars to lobby against renewable energy...ALEC (American Legislative Exchange Council supported by Exxon Mobil and Koch Industries) are actively influencing legislatures to introduce bills to prevent the trading of renewable energy credits and to prevent business and consumers from breaking away from the grid to generate their own power...In the state of Oklahoma a bill was passed allowing utilities to charge consumers a fee if they install a wind turbine or solar energy in their home and Duke Energy is active in North Carolina in trying to change the net-metering policies that could affect the use of fossil fuel energy (coal)...At the announcement of new CO_2 emission regulations, 29 states sued the U.S. government so they can be allowed to continue damaging the planet...A stunning ignorance of causality...

Wind Energy

The potential of wind energy generation on a world that rotates and is enveloped by a moving, dynamic atmosphere is still unrealized with the U.S. getting less than 4% of their energy from wind and challenges worldwide of aesthetics and noise...Atmospheric movement is a continuous process and when technologies develop better storage devices that are environmentally safe with electrical grids able to power entire communities, wind power will prove to be an efficient and effective source of power generation with zero pollution...The U.S. has designated over 2000 square miles of the Atlantic Ocean for developing wind farms to provide energy for millions of human structures and there will one day be regions designated for large wind farms as human innovation in concert with consumer demand is creating smaller and quieter wind turbines, highly efficient capture and storage devices and an environmental footprint that considers the local effect of power generation on indigenous wildlife and humans nearby...Combined with solar energy, wind technology could generate and store electrical power without connecting to a grid and without dependence on power plant production...

The Netherlands, a country with big plans for wind power dating from the 1990's as an alternative energy source (Netherlands only get about 5% of their energy from renewable energy) have scaled back many projects because of market inconsistencies and resistance from their citizens concerned about about noise, transport and aesthetics...Over half of the planned onshore wind projects are currently in dispute by local residents making it difficult for Netherlands to meet its goal of 14% renewable energy by 2020...

The world's largest offshore wind farm located in Thames Estuary in England came online in 2013 and is currently producing over 350,000MWh a month...175 turbines will produce enough power to supply nearly half a million homes with electricity...Called the London Array, the farm occupies 245 square kilometers about 20 km off the coast...A Phase Two designed to bring the output capacity to 1000MV with a reduction of CO2 emissions totaling 1.3 million tonnes a year was canceled in 2014...

In 2013, the first competitive offshore wind lease sale in the U.S. was held covering more than 160,000 acres about nine nautical miles off the coasts of Rhode Island and Massachusetts with the potential to produce enough energy to power over one million homes...Current estimates put the offshore wind potential off the coastlines of the United States at more than 4,000GW, four times the energy needed to power the entire country... A second offshore wind lease sale of nearly 112,800 acres of Virginia's coastal waters is scheduled in the near future...

In late 2013, RWE Innogy discontinued the development of the 240-turbine Atlantic Array project saying it was "not the right time" for the project...Concerns were financing, anti-green opposition, lack of government support, and questions of how the project would impact marine wildlife in Bristol Channel...The Atlantic Array was planned in an area of 200 sq km (77 sq miles) about 16.5km (10 miles) from the north Devon coast, 22.5km (14 miles) from south Wales coast and 13.5km (8 miles) from Lundy Island nature reserve...The turbines would have been 220m (720ft) tall and capable of producing 1,200 megawatts of electricity, enough for 900,000 homes...

Geothermal

Geothermal is an growing field of energy production using the heat inside this planet to produce electricity with *flash steam* technology to tap into hot water reservoirs and generate steam to drive generators while injecting leftover water and steam back into the hot water reservoir...Using this technology to drive heat pumps with an exchanger and air delivery system to heat and cool buildings and houses with zero pollution could be very effective in cluster housing, individual structures and as a direct source of heat for agricultural applications...And there is research ongoing into EGS (Enhanced Geothermal Systems) that can go very deep in the Earth to generate power at a very large scale...It's estimated more than 100,000 megawatts, 10% of the power generating capacity in the U.S., could be available if the infrastructure were developed to use this energy source...

There is currently a large International project in France exploring the possibility of Fusion Energy, a process using two forms of hydrogen (deuterium & tritium), the most abundant fuel in the Universe, fused together in a plasma contained within a magnetic field heated to more than 200 million degrees Celsius to produce energy with no greenhouse emissions and little radioactivity...It's difficult to even imagine when fusion could be a realistic energy source on Earth as new technologies and techniques have to be invented along the way to begin to design a facility that could use the power of the Sun to produce a limitless supply of energy...In the global fragmentation of this century, the project is behind schedule and over budget with nations having differing expectations and manufacturing challenges but this is what humans do...Build incredible complex tools to implement ideas that initially seem beyond the realm of possibility, but eventually become a reality...

A fusion experiment at Lawrence Livermore National Ignition Facility (NIF) in 2013 was the first where the fuel released more energy than it absorbed...The energy released lasted 20 trillionths of a second, reached a temperature of approximately 50 million degrees Celsius, created pressures equaling 150 billion Earth atmospheres and was initiated by a laser burst of 500 trillion joules to yield 17,000 joules of energy...This is the challenge of fusion energy...

ASIA

Asia is the largest continent on Earth bordered on the east by the Pacific Ocean, the south by the Indian Ocean and the north by the Arctic Ocean...Almost 9% of Earth's total surface, Asia is home to more than four billion humans (over 2 billion in the two most populous nations on the planet), home to the highest point on Earth (Mt Everest at 29,028 feet), the lowest point on Earth (Dead Sea at 1,296 feet below sea level) with a topography varied and beautiful amid vast plateaus, basins, alluvial lowlands and some of the highest mountain ranges in the world...Extensive lowlands in the Western Siberian Plain (east of the Ural mountains) and the largest alluvial valley on the Indian subcontinent with rivers Indus, Ganges and Brahmaputra diverted for irrigation to feed the billions living in south Asia and a thousand year history written around and about the great rivers Huang He (Yellow River) in China, the Mekong flowing through China winding down to Cambodia, the Yangtze and the two great rivers in the Fertile Crescent, the Tigris and Euphrates...In Russia (the entire northern region of Asia) is the largest freshwater lake in the world (Lake Baikal) which contains more water than the five great lakes in North America, a single lake with around 20% of the freshwater on Earth and thought to be the oldest lake on the planet...An ancient continent with a history of human migration over thousands of years creating multiple cultures and multiple civilizations with archaeological artifacts that predate written history...In the 1st millennium Islam spread throughout Asia with teachings of the Prophet Muhammad, war against Persia (Iran) and multiple wars with China that established borders and boundaries which still exist today...A 1,000 years of territorial and religious wars, the rise of the Caliphate, the emergence of a Shiite and Sunni rivalry which continues to kill thousands in this century and the rise of an Islamic dynasty (The Abbasid Dynasty) which survived for hundreds of years...The second millennium began with the crusades of Christians against Muslims, a campaign of death embraced by the Catholic Church that killed thousands with Jerusalem a continuous focal point of violence and worship (much as it is today) and the destruction of Baghdad by Mongols as another thousand years of territorial and religious wars continued with the rise of the Ming Dynasty in China, the beginnings of the Ottoman Empire dominating much of central Asia for over 500 years (the "Holy Land" changing hands multiple times) and the rise of Russia...Russia gained power and territory during the 17th and 18th centuries leading to revolution in the early years of the 20th century, the rise of Lenin and the creation of the Soviet Union which became a major world power under brutal dictator Joseph Stalin...Russia in the 21st century is worrisome, unpredictable and still a major player in world politics with thousands of nuclear weapons, vast energy reserves, substantial mineral deposits and borders stretching 6.6 million square miles across nine time zones...In the far east of Asia is Japan who became a world power in the 20th century signing a pact with Germany and Italy (Axis) and joining a territorial war for world domination that peaked and ended with the nuclear destruction of two major Japanese cities...Despite this focus on war, religion and territory, Asia has produced some of humanity's greatest insights, inventions and intellectual advances in the development of civilization...Mesopotamia (located in a now violent region of the Middle East) was a center of human discovery with Sumerian writings, mathematics, Babylonian astronomers, the beginnings of planetary science, eclipses and the questioning of many natural events...Asia was the birthplace of Buddhism (a spirituality practiced by millions) and the birthplace of Islam (a religion practiced by over a billion humans), origin of new inventive agricultural techniques in China, the beginning of advanced medicine, multiple scientific discoveries and many works of art and culture which is now obscured by the violent reality of 21st century civilization with artifacts being destroyed by fundamentalists in an unparalleled era of ignorance and violence...

The Asian continent covers a total land area of 44,579,000 square kilometers...

Today Asia is dominated geographically by the Russian Federation and the Republic of China with India in the south and Japan in the east bordering a central Asia that is a geopolitical storm of uncertainty with recent wars in Iraq and Afghanistan, a revolt in Syria with Al-Assad killing the civilian population in his desperate effort to stay in power, rising tensions between Pakistan and India (both nations with nuclear weapons) and a growing awareness in China that threatens to expose the corrupt hierarchical structure rooted in all of China's governmental institutions that may challenge the power of the ruling party when over a billion Chinese citizens find their voice and seek to choose their leaders in a democratic way...Instability in the Middle East with the rise of ISIS, a self-proclaimed Islamic state who murder innocents and threaten war while the world watches and new disputes south over Asia's surrounding waters with newly discovered resources and rising sea levels threatening island nations southeast of the continent which will challenge multiple governments and affect the lives of millions of humans before the end of this century...In a future when humans understand diversity, equality and no one is restricted in the enjoyment of all the wonders on Earth, much of Asia will be a playground as well as a land of deep exploration with unknown species, ancient fossils and multiple archaeological treasures still to be found...Not in this century...Today this continent is in turmoil with multiple wars, terrorism, widespread famine, a continuous discrimination of woman unchecked by the powers in the region, religious ignorance threatening the structure of nations and a growing desperation in southern Asia with increasingly careless pollution of the water, land and atmosphere to supply a growing energy demand fueled by overpopulation and a rising global insistence of inclusion at any cost...Military posturing will continue over sovereignty of disputed ocean basins in both the South Sea and the Arctic Ocean while Russia rebuilds a military reminiscent of Cold War paranoia to control resources and the territorial disputes that have dictated human aggression for thousands of years...

AFGHANISTAN
(32.3 million humans)
Military Forces = 200,000 active duty / $11 billion per year
"Because I am a girl, no one knows my birthday"

34 provinces in the nation, one of the world's largest opium producers, one of the most unstable and violent environments in the world, Afghanistan is a stunningly beautiful part of planet Earth, a place of high mountains and deep valleys bordered by six different countries, Turkmenistan, Tajikistan and Uzbekistan to the north, Iran on the west, Pakistan on the eastern and southern border and China in the northeast corner...The country is smaller than Texas and home to some of the highest mountains in the world...When the planet is open to our species with a sense of shared ownership and responsibility Afghanistan will be an incredible destination but that possibility is many years away...Nearly 3,000 NATO troops including 2,000 U.S. soldiers have died and $500 billion has been spent since 2002 to stabilize the nation and combat terrorism and it grows still more violent with a majority of the money disappearing, wasted, stolen and grossly misappropriated while over 10 million humans exist below the poverty line...The World Bank estimates that 90% of Afghanistan's gross domestic product (around $20 billion) comes from military, developmental aid and spending by foreign interests in country...

An audit released by the Special Inspector General for Afghan Reconstruction in 2014 reported that not one of the 16 Afghan ministries is capable of preventing waste, fraud and theft within their own agencies...Current aid from the U.S. budgeted for Afghanistan is around $1.2 billion with most delivered directly to the corrupt ministries...

Afghanistan produces just 900 kilowatts of electricity and imports over three billion kilowatts but it is still not enough to run industries, hospitals and small business...This is after billions of dollars were spent on electric grids and generators with continuing violence and little oversight...There is a heavy use of diesel generators to supplement what electricity is needed creating pollution and demand for fuel and it's believed in ten years only 60% of the nation will have power in a reliable way...Solar energy could change that equation if such a thing is allowed to happen but the Taliban have regained strength and numbers since the end of the NATO operation (2014) which lessens the chance of real change in this culture...A culture that is pervasive in it's discrimination against women with almost 15 million females in country and less than 5% graduating from high school if allowed to go to school at all with most forced into marriage before the age of 16...They are captives in a world of men who have no awareness of reality, men of little education, men who fear woman and men who live in a world of ancient rituals, tribal customs and a strong belief in an angry God...In 2010, the U.S. reported that there were 8,000 schools in the country with several hundred just for girls insisting that Afghan women now had a voice in parliament and that women now have access to health care but the reality on the ground is very different from these statistics...Literacy rate is less than 40% for the total population...

There are an estimated 10 million unexploded land mines buried throughout Afghanistan...

Women have no standing in this society without a man or children and many commit suicide because of continuing cruelty and discrimination...Tribal law allows the exchange of women and the giving of a daughter to one's enemy to settle disputes (baad), an ignorant and obvious abuse of the individual rights of the 'girl' being given away who is often tortured and sexually abused for years...In many societal situations men are not allowed to look at women or interact with them for fear the men will "lose control" of their desires...There are women and men who understand this reality and work in the urban areas to bring an awareness of the necessity for serious change in the customs and laws of society but they are by far a minority of the population and the road ahead is long and dangerous...Over 2 million children in the labor force, a life expectancy of just 50 years and 50% of the children under age five are malnourished...Infant mortality rate is 1st in the world, birth rate is 11th and the death rate is ranked 9th...More than 35% of the population live below the poverty line in a nation with estimates of 50 billion cubic meters of natural gas...The government is a presidential Islamic republic currently in a power sharing agreement with President Ashraf Ghani and CEO Abduilah Abdullah in an attempt to unite a fragmented population...Entrenched corruption makes this very unlikely...

Insurgents beheaded 17 civilians in a Taliban-controlled area of southern Afghanistan because they attended a dance party (2012)...Under Taliban rule all music and film is banned and women are barred from leaving their homes without a male family member escort...Many Afghans and international observers have expressed worries that the Taliban's brutal interpretation of Islamic justice will return as international forces withdraw and it has...In 2016, the NYT's reported that the Taliban leadership is increasingly being integrated by the Haqqani militant group who are more organized with international support making the Taliban as brutal and as deadly as they've ever been...

BANGLADESH
(160 million humans)
Military Forces = 380,000 active duty / $1.6 billion per year

Terrorist Group is Hark-UL-Jihad-al-Islami, a Pakistan based terrorist group that has operated since the 1990's to establish Islamic rule by any means possible...Bangladesh has existed only since 1971 after Pakistan tried to bring order and control to East Pakistan in a bloody conflict that saw more than nine million East Pakistanis fleeing to India to escape a brutal Pakistan army who murdered over a million Bengali's, Hindus and dissidents who were systematically buried in mass graves...Bangladesh is a crowded, densely populated country in extreme poverty with as much as 50% of the 160 million humans living on less than a dollar a day, a fractured government with intense rivalries between the two parties affecting every sector of society, corruption at every level of existence and a growing religious extremism...Located on the Bay of Bengal with frequent flooding, monsoons and cyclones, Bangladesh is one of the most vulnerable nations in the world as sea levels continue to rise with more than 50 trans-boundary rivers and little or no control over the flow of waters that drain into the Bay of Bengal...The government estimates more than 60 million people are 'food insecure' meaning they consume less than a daily minimum to sustain health and chronic malnutrition among children under age 2 affects almost half of the nation's 20 million children resulting in growth retardation and mental deficiencies...66% of girls marry by age 18 resulting in a high percentage of newborns with low birth weight...These are factors that determine the density of neural growth in developing brains and consequently, almost 50% of these children never complete the five-year cycle of primary education...Literacy is just 61% with over 4 million children in the labor force...There is economic hope with newly discovered gas and oil reserves offshore (28 million barrels of crude oil and 264 billion cubic meters of natural gas) but rising sea levels in the 21st century and a fractured government without direction will contribute to the growth of problems without solutions...

In April 2013, the worst disaster in the history of the garment industry happened in Bangladesh when a factory collapsed killing over 1,000 humans with many injured...Minimum wage at these factories is $37 per month to create products for some of the most profitable corporations in the world in an industry that generates almost $20 billion dollars a year...Very little of this money goes to the workers...The government has promised $12,000 to the most grievously injured and Bangladesh's parliament is considering measures to allow workers to unionize and laws to strengthen building codes...The Bangladesh apparel industry, 80% of the nation's exports, is governed by the Bangladesh Garment Manufacturers and Exporters Association, a powerful lobby with influence and the ability to avoid regulation to maximize profit...Walmart and others have signed a five-year initiative for third body inspections and worker rights, but it's voluntary and the obligations of companies under this initiative are unclear...European retailers agreed to finance upgrades and fire inspections in an effort to continue a very profitable enterprise ...

BHUTAN
(750,000 humans)
Military Forces = 6,000 active duty / $13.7 million per year

25% of this country is designated as national parks for the protection of the Bengal tiger...An ancient Buddhist culture located between China and India in the Himalaya mountains, Bhutan is a parliamentary two party constitutional monarchy with a prime minister and a king...Life expectancy is around 70 years with a median age of 26...Literacy rate is less than 70% with 25,000 children in the labor force and 12% of children under five malnourished...Economy is hydro-power (to India), services and agriculture...China and Bhutan continue to dispute borders in the north...Isolated from the world until relatively recently, Bhutan continues to grow in popularity among tourists and mountaineers with its high vistas, stunningly beautiful geography and a government that is striving to provide equal rights and prosperity to it's citizens in a challenging part of the world...

BRUNEI
(420,000 humans)
Military Forces = 5,000 active duty / $300 million per year

A small nation that gained independence in 1984 on the island of Borneo (shared with Indonesia and Malaysia) Brunei is ruled by one of the world's longest ruling monarchs, Sultan Hassanal Bolkiah, an immensely wealthy leader with two wives and absolute power over his citizens who enjoy a high standard of living (mostly subsidized by the government) due to oil and gas exports (90% of GDP) that have brought stability and wealth to the ruling party...After decades of ruling by decree aligned with the tradition of the "Malay Islamic Monarchy", the sultan recently revived the parliament by appointing 21 members and then amended the constitution to allow elections though no elections have been held to date...The population is long-lived and is more than 60% native Malaysian, 15% Chinese and various indigenous groups with ethnic diversities...The royal family controls all media with threats of arrest and imprisonment for reporting news unfavorable to the monarchy...Brunei has been criticized for harsh corporal punishment laws and for human trafficking in forced labor and prostitution and adopted Sharia Law in 2014, a harsh Islamic criminal code that allows dismemberment and death by stoning (mostly for women)...Islam is the official religion and sale and consumption of alcohol is banned...

CAMBODIA
(15.3 million humans)
Military Forces = 125,000 active duty / $193 million per year

One of the poorest countries in the world with a history of almost two million humans tortured, murdered and starved by the Khmer Rouge in the 1970's while the world watched...A country littered with unexploded land mines that continue to kill and maim citizens and ruled by a corrupt government led by Prime Minister Hun Sen, a former member of the Khmer rouge whose ruling party has profited from illegal logging that has destroyed vast forests contributing to soil erosion and flooding...Almost three quarters of a population of 15 million humans exist on subsistence farming and depend heavily on foreign aid to meet basic needs...In 2012, a mystery illness surfaced in this country killing scores of children, all under age seven, a rapid onset and conclusion with most victims dying within 24 hours...It was later identified as a combination of pathogens (including enterovirus 71 and streptococcus suis) coupled with inappropriate medical treatment using steroids that worsened the condition...Cambodia is still plagued with dengue fever that has been a problem in Cambodia's past...The education system was destroyed by Khmer Rouge killing over 80% of educators and destroying all institutional infrastructure including more than half of all the written material in the country...Students of higher education were forced to work in labor camps or killed and education was seen as frivolous and unnecessary as most of the population was banished to the rural areas of the country to farm where over 80% of the country's poor live...Every election in this century has met with protest and criticism...Almost 30% of children under age five are malnourished with 1.3 million children in the labor force...Life expectancy is around 64 years...Corruption and poor regulation challenge the future of Cambodia and only in the first decade of the 21st century did trials begin to deal with the crimes of the Khmer Rouge...

CHINA
(1.4 billion humans)
Military Forces = 2.3 million active duty / $213 billion per year

More than fifteen million new babies born in 2012...The world's largest exporter and one of the world's most polluted countries, residents of Beijing have a life expectancy 5 years less than residents in rural communities...This country is ruled by the "Standing Committee", a group of seven individuals who are appointed, not elected, and control almost everything in the country with a focus on the central government's effort to control the population with censorship, reforms and dictates of solidarity that focus not on access to content, but rather on disrupting social networking in any form that could lead to the aggregation of large groups with a purpose or organizations that can communicate and coordinate activities that threaten or challenge the status quo...The greatest fear of this government is not that dissent and opposing views should ever exist, the fear is that dissent will grow and coalesce becoming organized and threatening to the stability of the nation...Still, it's hard to cover eyes once the people have been allowed to see and there is an awareness growing in this country of over a billion humans that will be difficult for a single authoritarian ruling party to manage and control in a way to safeguard the party's existence and command over the people of China...By removing capitalistic restrictions and allowing the country to become a commercial and economical powerhouse in the world, a cascade of cause and effect will enable a rapid and continuing growing awareness among the population who will in time demand new rules and new government for the people...This cannot be stopped without extreme negative actions by the Chinese government and military in the global spotlight of this century...Very effective at stifling unrest, the current system for dealing with dissidents and troublemakers is to allow local governments to break the law leaving them empowered and free to rule at the local level through corruption, threat and intimidation that will be less effective in a global world of instant communication and real-time awareness...The central government provide incentives and support for regional layers of government to control their local populations at any cost only stepping in when these abuses become a national or international story...If China is to become part of a global cooperation with stability into the next century they will have to begin a modification of central policies at some point with inclusion of the popular voice in a viable, functional political system...New president Xi Jinping (2013) is seriously challenging the status quo by illuminating activities that damage the "credibility" of the government in the eyes of the people and many observers say there has been more change in China in that regard over the last three years than in the previous twenty years...But many believe this new transparency to be a calculated attempt at dampening political unrest before the fact and many Chinese believe this new confessional tone to be superficial and government will return to the old tactics of control at any cost...

With oil company representatives embedded on committees at all levels of governance, China has successfully resisted upgrades to power plants and delayed improvements in fuel standards that continue the use of diesel fuel that emits over 23 times the levels of sulfur emitted by the diesel fuel in the U.S...Fines for the power plants who are found guilty of pollution control violations are capped at $16,000...

A report published in 2012 showed the leader of China to be as corrupted by money and power as any government in the world with his family members gaining control in key industries and a net worth of more than $2 billion during his time in power, corruption that is unchecked at every level of Chinese society...In 2012, the People's Bank of China estimated over 18,000 officials left the country rapidly and secretly taking around $2.5 billion out of the country and China has recently asserted it's sovereignty over disputed waters and islands in the South China Sea claiming territorial waters out to 200 miles instead of the 12 mile limit set by the United Nations Law of the Sea treaty in 1982...This treaty's been ratified and signed by 162 nations (the U.S. refuses to be a part of this) as an international agreement governing the open seas and waters that border nations to resolve disputes over minerals and resources still unknown in an ocean humans have only begun to explore...

A toxic heavy metal spill that released tons of cadmium in the Lonjiang River in 2012 killed an estimated 90,000lbs of fish and over a million fry...Residents along the contaminated river were not told of the spill for more than 2 weeks...

Estimates of energy reserves available for harvest are 120 billion barrels of crude oil and 800 trillion cubic feet of natural gas under the seabed of the South China Sea and this will be a point of tension and conflict going forward and maybe cooperation between China, Philippines, Brunei, Malaysia, Taiwan and Vietnam...China's National Offshore Oil Corporation has begun its first deep sea drilling project in undisputed waters and the U.S. has increased their military presence in the area to insure that vital shipping lanes remain free and open...In 2013, China announced an increase in military spending of over 10% to modernize forces and "ensure it's peaceful development"...China has a active space program that sent a probe to the moon in 2013 and plans a manned mission by 2020...

By 2030 China is expected to have 400 million cars...In 2014 there are just 90 million vehicles in the country...

China policy of actively and forcibly moving rural residents to cities will result in the migration of more than 250 million humans in less than two decades...

In December 2013, China successfully landed an exploratory rover on the moon with multiple cameras, ground penetrating radar(100 meters), spectrometer and multiple sensors to investigate the moons composition ...The most progressive part of the mission is a demonstrated ability to engineer a power descent and soft landing on the surface of another world, the first moon mission designed to land on the surface in more than 37 years...

In March 2016, China's new two child policy replaced it's one child policy in effect since 1979...

In 2016, China began a three year moratorium on the approval of new coal recovery projects with plans to close over a thousand coal mines and at the Paris climate talks in December 2015, China pledged to start reducing carbon emissions into the atmosphere beginning in 2030...China produced over 3.5 billion tons of coal in 2015...

INDIA
(1.3 billion humans)
Military Forces = 1.3 million active duty / $48 billion per year

1.2 billion humans live in India, 2nd most populated nation on Earth (China is #1) and the largest democracy in the world in a landscape changed dramatically in the last few decades...Rural lifestyles are giving way to mega-cities filled with smog and congestion in an explosion of mobility that has only just begun...Around 6% of Indian households own cars but yearly car sales are expected to exceed ten million over the next ten years with unknown emission control standards in an environment where cars and motorbikes today contribute a large majority of the CO2 polluting the cities...The 1990's saw India embrace an economic liberalism that has resulted in a profusion of wealth and investment that drive a uncertain economy in an uncertain direction...This rush for wealth and opportunity and a desperation "not to be poor" (in 2013 the UN said one-third of the world's poor exist in India) has accelerated the migration of rural populations to cities and left a centuries old agricultural society in a continuing state of decline...Many farmers in rural India have committed suicide (almost 300,000 since 1996) unable to deal with a globalized economy, unable to pay royalties on the "seeds" that Monsanto claimed were "intellectual property" (Monsanto controls over 95% of India's cotton crop) and unable to cope with new government policies that removed subsidies to promote a free market capitalism...In reality, the attention focused on the success of India's 'free' democratic society hide a reality of networks of wealth and patronage that keep the poor 'poor' and increase the wealth of the privileged, mirroring what is happening in many 'democratic' societies across the world...In this hierarchical class society, members of the merchant class (Vaishyas) learned over many generations how to manage and accumulate cash and they dominate any list of the wealthy in India...There are many levels of 'poor' in this country and despite electoral promises of reform and improvement, many of India's ancient problems of poverty, child labor, illiteracy, disease and a continuing caste system (ignorant, inhuman and 'illegal' for over sixty years) still prevail and limit inclusion throughout the nation...

All people in India can trace their heritage to two genetic groups...An ancestral North Indian group originally from the Near East and the Caucasus region and a South Indian group that was more closely related to people on the Andaman Islands...Today, almost everyone in India has DNA from both groups...

More than 40% of children under the age of 5 are malnourished, almost 27 million children are in the labor force, literacy rate for the entire population is around 70% and the nation is ranked #3 in the world both for people living with HIV/AIDS (2 million) and deaths from HIV/AIDS...India's gross domestic product per capita quadrupled since the 1990's but the percentage of underweight children under age three declined just 6% to a current level of 45%...Malnutrition is devastating in infancy, retarding the normal growth of neural connections in the first three years of life that almost guarantee the underdeveloped brains of these children will remain limited for their lifetime and result in large populations in the middle of the 21st century without the ability to understand human reality and what is necessary for change...A Right to Food Act under discussion for years that would guarantee a basic food supply to poor families was finally passed by a government that has spent billions of dollars on weapons and defense...India is the world's largest importer of weapons...In 2009, India symbolically launched the first nuclear powered submarine built in India becoming the sixth world power with this capability...The new submarine (activated in 2013) is named "Arihant" or "Destroyer of Enemies" with the capability to launch nuclear weapons with a range of 3,500 kilometers...

In 2016, the INS Arihant nuclear submarine (the first to be designed and built in India) is fully operational and ready for service...It's the first ballistic missile submarine known to have been built outside the five recognized nuclear powers (U.S., U.K., France, Russia and China)...This nuclear submarine will add a third dimension to India's defense capability as it has previously only been able to launch ballistic missiles from the air and from the land..."a giant stride in our indigenous technological capabilities" Prime Minister Manmohan Singh

The prime minister of India is Narendra Modi of the Indian People's Party (Bharratiya Janata), a Hindu who won elections in 2014 against the Congress Party with a campaign of good governance, economic growth and anti-corruption in an overwhelming victory that gave the People's Party a chance to enact real change with a majority in parliament and a mandate from the people...His priorities are to repair infrastructure and build with little talk of environmental responsibility but this is an opportunity for a new government to improve the lives of millions and the world is watching...Energy availability is a huge problem with over 400 million people off the grid getting no electricity and 60% of businesses maintaining back-up generators on site to cope with power outages...Coal burning power plants supply a majority of the energy with over a third of these plants not meeting emission standards designed for pollution control and India intends to triple its production of coal by 2030...

In 2012 India experienced the largest power outage in history that left half a billion people in the dark...

The social caste system in India is ancient, ignorant, illegal and still observed and practiced by this 'modern' democracy...In this system that has existed for more than 2,000 years, you are born where you should be born to do what you should do from cradle to grave with no way out as a reflection of your previous life...The 'outcasts', as many as one out of every five Indians, are below the bottom and called 'Untouchables' who cannot speak to others, cannot touch the belongings of anyone else, can be humiliated and raped without consequence and killed if they make any real effort to change their status or fate...They still collect the garbage, human waste and the dead as they always have...

In December 2013, India's supreme court overturned a law that had decriminalized gay sex reinstating an ignorant and discriminatory law that dates back to Colonial times...What many opponents to human equality do not understand is that 1: Gender preference is not a choice...It's a hereditary link believed to be associated with epi-marks (information controlling certain genes) that can be "erased" between generations or passed on to the offspring, father to daughter or mother to son...2: If every homosexual on Earth were allowed the freedom to live their life and be with whomever they choose, it would represent less than 3% of the planet's population...This inborn variation of gender preference will never threaten a species that is overwhelmingly heterosexual, a necessity for survival...

To combat the problem of gender preference and infanticide India made it illegal for a doctor to reveal gender of an unborn baby but the introduction of ultrasound technology to monitor the health of the fetus is instead being used by many to determine the sex and the fate of the unborn child...The government has offered to pay poor families to give birth to daughters and raise them to adulthood with additional money at age 18 if unmarried...In 2013, there were about 45 million more men in India than women with statistics reporting a rape every 21 minutes in this nation where ignorant men turn a blind eye to the subjugation of females at every level of society without restraint or consequences...

Every five years in Nepal is held the Festival of Mass Slaughter, a Hindu festival in honor of the Hindu goddess of power, Gadhimal, where more than 200,000 animals (goats, chickens, buffaloes) are killed in the belief that the killing will bring prosperity and eating the meat will ward off evil...

In 2008, India successfully launched a lunar probe that made over 3,000 orbits of the moon and released an impact module that crashed in a controlled way in the South Polar region...India is the 4th nation to reach the moon...

INDONESIA
(253 million humans)
Military Forces = 470,000 active duty / $6.9 billion per year
Indonesia is unique in it's diversity and history as a nation archipelago in the Indian Ocean with a large population geographically located on tectonic plate boundaries and a history of earthquakes and tsunamis that have killed thousands from the eruption of Krakatoa in 1883 to a recent tsunami in 2004 that killed more than 220,000 humans...A Dutch colony until after World War II, today's Indonesia is a

thriving economic power in southeast Asia, the world's largest Muslim majority country and a global player in energy with almost three trillion cubic meters of proved reserves in natural gas and 3.6 billion barrels of crude oil reserves...Government is a presidential republic with 31 provinces and a president elected by majority vote for a five year term...Ethnic tensions are high as expected in a country with hundreds of languages and a government continually dealing with differing cultures in this sprawling archipelago (over 13,000 islands) that seek independence or a recognition of sovereignty...Median age is 30 with a high literacy rate and a life expectancy over 70 years of age...It's estimated 20% of children under age five are malnourished, over four million children in the labor force and around 12% of the population living below the poverty line...Indonesia has strong military ties with the U.S. and has sustained an economic growth through the world recession despite corruption, high unemployment, a challenged infrastructure and labor unrest...Challenges include growing violence from separatists rebels and militant groups, deforestation and multiple borders disputes with surrounding states...

JAPAN
(127 million humans)
Military Forces = 240,000 active duty / $43 billion per year

Japan is a highly technological country with a space program that is pushing boundaries, a very old culture that exists in the modern world and a physical infrastructure that keeps improving through innovation, an evolution in technique and material and well planned implementation of services that support the society...The *Hayabusa* mission was the first time that an attempt was made to return an asteroid sample to Earth for analysis and *Hayabusa* was the first spacecraft designed to deliberately land on an asteroid and then take off again...(*NEAR Shoemaker* made a controlled descent to the surface of 433 Eros in 2000 but was not designed as a lander and was deactivated after arrival)...Technically, *Hayabusa* was not designed to "land": it simply touches the surface with it's sample capturing device and then moves away but still it was the first craft designed from the outset to make contact with the surface of any asteroid and despite it's design, Hayabusa did land and sit on the surface for about 30 minutes...The success of the mission, despite alterations and necessary adjustments both routine and difficult, prove the capability of our species to target any of these thousands of asteroids for analysis, landing, recovery and even mining in a future when the resources of the solar system will be valuable and necessary to support the population of Earth...

In April 2014, Japan's prime minister (Shinzo Abe) removed a self-imposed nationwide ban adopted in 1967 that prevented the sale and export of all weapon systems to foreign nations...The move is seen as both strategic to counter the growing influence of China and economy based to allow manufacturers to sell military equipment to allies in the region and assist less developed countries resist growing disputes involving territorial claims...

Median age in Japan is 46 years with the 2nd highest life expectancy in the world at more than 84 years...Devastated by the Second World War, Japan has made a dramatic recovery to become the world's third largest economy with investment in the electronic industry and automobile manufacturing during the last decades of the 20th century that brought great wealth to a nation that now finds itself with an aging population, enormous debt and uncertainty in it's nuclear industry following the tsunami of 2011 that destroyed a nuclear reactor...The radiation released at Fukushima found its way across the Pacific and still poses a hazard to many people in the area around the reactor...This island nation sits on one of the most volatile parts of the Pacific 'ring of fire', an area of intense tectonic plate activity with frequent earthquakes (over 20% of the world's quakes) and the constant threat of tsunami in a country where 75% of the population live in coastal regions...After the Fukushima disaster Japan stopped the planned building of more than a dozen new reactors and shut down almost all operating plants leaving just two online...In 2014, Japan announced they will restart 48 commercial nuclear plants after they pass safety tests with plans to build more reactors in the future...Since the shutdown, carbon emissions from electricity generation have doubled along with higher prices for power from oil and gas imports...

Current plans are to let the damaged reactors at Fukushima "cool down" until 2020 before any attempt is made to clean them up...Tepco, the giant energy company that owns the plant, admitted in 2014 that 100 tons of highly radioactive water (7 millions times the radiation level safe for disposal into the ocean) leaked out of a storage tank...Misinformation and delay has been the pattern of Tepco's response to this accident, an unaccountability clearly related to the size and power of one of Japan's largest corporations whose political ties and political donations affect regulatory agencies and government oversight...Five years later and radiation levels are still unknown...

KAZAKHSTAN
(17.3 million humans)
Military Forces = 110,000 active duty / $2.4 billion per year

The world's largest producer of uranium with substantial mineral, oil and gas deposits in a large country (2.7 million square km) located south of Russia with China on the eastern border, Uzbekistan, Kyrgyzstan and Turkmenistan on the southern border and the Caspian Sea to the west...Kazakhstan has been ruled by one leader since the nation emerged from the collapse of the Soviet Union in 1991, Nursultan Nazarbayev, a former member of the Communist Party who centralized government power in his hands winning reelection through manipulation of the ballot and a backing parliament who has voted to allow the president to serve an "unlimited" number of terms...The inland Aral sea is heavily polluted and growing smaller due to water diverted for irrigation and there are multiple hazardous regions in the country as a result of the Soviet Union's defense industry and testing of weapon systems in the past...Median age is almost 30 years with a life expectancy of more than 70 years...An estimated 60,000 children in the labor force with around 4% of children under age 5 malnourished...Over 85% of electricity generated is from fossil fuels with the remainder coming from hydroelectric generation...The country has 30 billion barrels of crude oil reserves and 2.4 trillion cubic meters of natural gas reserves with large exports of both commodities through multiple pipelines carrying oil and gas to China and Russia from western oil fields and the Caspian Sea...Almost all media is controlled by the government with restrictions on any criticism of the state, newspapers are owned by the state or by entities in the government and restrictions and filtering is common on the Internet...Nazarbayev considers a move to democracy "too rapidly would create instability" while crushing opposition groups, jailing dissenters, changing the constitution every few years to increase his power (parliament passed a bill in 2010 giving the president immunity from prosecution) and manipulating the electoral process by jailing opponents and controlling the ballot box...Not a single election has been recognized as free and fair by monitors...

KYRGYZSTAN
(5.7 million humans)
Military Forces = 16,000 active duty / $300 million per year

South of Kazakhstan and west of China, this nation also gained independence with the collapse of the Soviet Union in 1991 and has experienced a history of instability and uncertainty since that time with ethnic tensions in the south erupting into violence multiple times...President Almazbek Atambayev is closely allied with Russia supporting a Russian military presence after shutting down a U.S. airbase in Manas critical in America's invasion of Afghanistan and subsequent occupation of Iraq...Another geographically stunning region of Earth that could be a lucrative source of tourism revenue in a stable environment...Media is restricted in some ways but remains a viable voice with private networks broadcasting from Russia and in country...Clashes between the Kyrgyz and Uzbek ethnic communities are commonplace (over 150,000 displaced humans within the country) and anti government protests are frequent...Exports in gold, tobacco and cotton support a poor country with a restless population seeking self-determination and employment...Over 500,000 children in the labor force with more than 30% of the population existing below the poverty line...Challenges include border disputes with Kazakhstan and Uzbekistan, continuing ethnic violence, incursion of terrorist elements, pollution, corruption, drug use and drug trafficking...A member of the Eurasian Customs Union...

LAOS
(6.7 million humans)
Military Forces = 29,000 active duty / $18.7 million per year

In 2012 the government of Laos announced they approved the construction of a dam costing $3.5 billion on the lower Mekong river to generate power to meet it's energy needs and capacity to sell surplus electricity to neighboring countries (Thailand)...Laos insists it has not broken agreements set forth in the 1995 Mekong Agreement between Cambodia, Laos, Thailand and Vietnam that set forth objectives and principles specifically to address sustainable development of the Mekong River Basin, but environmentalists who have studied the project are concerned about fish migration upstream and sediment flow that could affect the economic security of thousands of humans who depend on the river for their livelihood...Laos is bordered by Thailand, Burma, China, Cambodia and Vietnam and remains among the few communist nations in existence with the Lao People's Revolutionary Party controlling the government with a single candidate for president (Choummaly Sayasone)...The economy is rice and foreign aid and investment from China, Japan, Thailand and Vietnam...The government owns all media in the country with severe consequences for anyone who defames or criticizes the government...A poor country with over 25% of children malnourished and some 175,000 children in the labor force...A large tropical rainforest and little infrastructure outside the capital city of Vientiane, Laos is helped by the IMF and various UN food programs to meet the needs of the population...

MALAYSIA
(30 million humans)
Military Forces = 111,000 active duty / $4.3 billion per year

A federation of two regions separated by the South China Sea north of Indonesia, Malaysia is a multi-ethnic society of rich rain forests, beautiful beaches and a Muslim majority...An odd and unique government with a paramount ruler appointed every five years from one of the nine Malay kingdoms in a constitutional monarchy headed by the prime minister with an elected Parliament and an appointed Cabinet...The economy is supported by high tech industry with substantial foreign investment, oil and gas exports and rubber and palm oil exports that increasingly threaten the rain forests...Life expectancy is around 75 years with less than 5% of the population below the poverty line...Almost 4 billion barrels of crude oil reserves and over 2 trillion cubic meters of natural gas reserves...Malaysia is a player in the South China Sea disputes over boundaries, borders and services and is involved in the trafficking of humans, mostly women and children, for forced labor and the sex trade...The government is mostly ineffective in reducing the frequency of this activity...Authorities reserve the right to censor and block all media and internet access citing morality and national security...

MONGOLIA
(2.9 million humans)
Military Forces = 10,300 active duty / $70 million per year

Surprisingly one of the world's growing economies due to foreign investors anxious to mine the enormous quantities of coal, copper, gold and uranium (exports are over 50% of GDP)...It's estimated this country has enough coal to supply China's needs for over fifty years, another clear indicator that world powers and multinational corporations still intend to burn Earth's most damaging fossil fuel for many more years without regard to the environmental consequences...Despite foreign money, most of the population of almost three million humans live in poverty (less than $2,000 a year) as the powerful and corrupt take for themselves...Located on the Central Asian plateau with large pastures, deserts and extreme weather, many Mongolians rely on livestock to exist, a lifestyle that is changing dramatically as climate change brings drought and cold that devastated the livestock in 2010 requiring the United Nations to supply aid for affected families...A nation with a history beyond the 13th century, Mongolia is ruled by a coalition of competing parties with the prime minister and parliament controlling political power and a president who is commander of the armed forces and holds veto power...Religion is mostly Buddhists with a high literacy, a free, unrestricted media and a move to urbanization as more citizens seek opportunities in a growing infrastructure infusing the country with more wealth...Over 30% of the population exist below the poverty line, life expectancy is around 70 years and over 100,000 children are in the labor force...Environmental pollution and deforestation are increasing with little regulation...

MYANMAR
(51 million humans)
Military Forces = 430,000 active duty / $2.2 billion per year

A country ruled by the military for almost 50 years, Myanmar began a transition from military to civilian rule (2011) in a country that has been accused of extreme human rights abuses, the servitude of children in labor camps and forced exile of many citizens...A parliamentary republic with more than 75% of seats held by military representatives or proxies as required by a 2008 constitution, Myanmar suffers from corruption, stagnation, enormous inequality in the sharing of resources and a long, violent conflict with the rebel ethic group Karen (Karen National Union) that continues today...Almost a third of the population are members of minority groups fueling tensions along the Thailand border between government soldiers and rebels that has lasted more than 60 years...A ceasefire was signed by multiple rebel groups and the government in 2015 but the future is unpredictable in this melting pot of many ethnicities...The nation's wealth is generated by mineral deposits (jade, pearls, rubies), offshore oil, gas deposits and some tourism as an ancient center of Buddhist tradition bordering China and the Bay of Bengal...The West has embraced the elections and change in Myanmar with a cautious eye and a lifting of sanctions but the future of this country is still very influenced by the military under the guise of democracy and cooperation...The military is deeply entrenched in society with industrial wealth and long standing control over resources and a requirement that the Minister of Defense, the Minister of the Interior and the Minister of Border Affairs be filled by serving generals with little civilian input...This nation remains a major exporter of opium, heroin and methamphetamine's controlled by the military or ignored with compensation to the military...The media and internet are still censored with 'free news' mostly on radio from sources outside the country...Over 30% live below the poverty line with 22% of children under age five malnourished...Myanmar is ranked third in the world in landmine injuries with estimates of over 100 million unexploded landmines still on Earth...

NEPAL
(31 million humans)
Military Forces = 100,000 active duty / $210 million per year

Bordered on the north by China and on the south by India, Nepal is home to Mount Everest, the highest point on Earth at 29,029 feet above sea level...Almost 4,000 humans have attempted to climb this 60 million year old mountain with over 650 succeeding, the first in 1953...Ruled by a monarchy for centuries, Nepal held it's first elections in 2008 after a decade of turmoil and violence from Maoist rebels supporting the Unified Communist Party of Nepal who won a majority of seats ending the civil war and giving legitimacy to the party...Nepal relies heavily on foreign aid and tourism to continue an economy where over 25% of the population exist below the poverty line with an unemployment rate over 40%, one of the highest in the world...Mostly Hindu with a median age of just 23, Nepal has a literacy rate of just 64% and a child labor force estimated at 2 million...The WFP reports that 41% of children under age five are "stunted", 29% are "underweight" and 11% are "wasted"...Almost 25% of children do not complete primary education with understaffed or non functioning schools, especially in the remote and rural areas of the country...

In 2015, Nepal experienced a 7.8 magnitude earthquake which killed more than 8,000 humans with severe aftershocks continuing for days...In 2016, mostly due to governmental bureaucracy, much of the damaged area remains in ruins...

NORTH KOREA
(25 million humans)
Military Forces = 1 million active duty / $7.5 billion per year

Kim Jong Un is the leader of this communist state controlled system that has brought economic and societal stagnation and widespread poverty to a nation that is essentially isolated from participation in a growing global cooperation...The military controls and runs the country while a population of over 25 million humans rely on foreign aid for food with thousand's dying every year at the hands of a corrupt ineffective government structure that uses fear and brutality to control all of it's citizens...North Korea is the sad result of a post World War II agreement between Communist Russia controlling the northern half of Korea and the U.S. controlling the south...Soviet troops withdrew in 1948 after installing the North Korean Communist Party and North Korea invaded South Korea in 1950 to begin a war that lasted three years and killed more than 2 million humans...The borders at the end of that war are the borders today...China is possibly North Korea's greatest defender and ally because of long standing policies and because the leadership in China is worried of the instability and migration that any change would bring to the region...An estimated 50% of children under five are malnourished...All communications among the people of this country are monitored and internet access is unavailable to all but the elite...Nearly a quarter of a million people labor in camps, suffering torture and starvation while the ruling party dines on imported foods and wines not available to the public...

In 2013, the United Nations investigated human rights abuses in North Korea with North Korea invited to the hearings...A request for a commission of inquiry to go into the country was denied...The investigation included prison camps, abductions of South Koreans, Japanese and other countries' nationals, starvation, torture, and the way North Korea treats handicapped people with testimony from former guards and prisoners...North Korea denies there are prison camps and deny there are human rights problems...Nothing will change...

In 2014, the U.N. Human Rights council released their report stating that officials at the highest levels of the North Korean government are involved in a continuation of "unspeakable atrocities" against their own people, over 100,000 citizens who after being imprisoned for arbitrary crimes against the state are "raped, tortured and forced to dig their own graves before being murdered by the guards"...Impunity reigns with virtually no accountability...The North Korean government "categorically and totally rejects the report"...Nothing will change...

A rogue nation with ties to organized crime and a continuing effort to create instability in this region with internet intrusion programs, military propaganda, counterfeiting, threats and posturing against any and all military or political policies that are adopted by the South Korean government...The military has one of the largest standing armies in the world, limited nuclear capability and a defensive network of tunnels and fortress's throughout the country that would create extreme difficulty for any invading army...Daily control of the people is maintained by restricting opportunities for growth and development through the manipulation of information and resources that has created a population ignorant of their own reality in an extreme way...No citizen can oppose hereditary succession without being opposed to the nation, so like father like son the propaganda and oppression of the population continues...All indicators suggest that North Korea will continue to be an unpredictable international threat steeped in institutional ignorance until it is not...

PAKISTAN
(197 million humans)
Military Forces = 616,000 active duty / $6.9 billion per year

In 2011, over 4,400 people were killed in more than 450 terrorist attacks...Terror groups include Jaish-e Mohammed (Army of Mohammed) who seeks to unite Pakistan and Kashmir, remove foreign troops in Afghanistan and operate openly without fear in many areas of Pakistan...Hizbul Mujahideen is another group seeking unification of Kashmir and Pakistan...Pakistan has long claimed Kashmir and have fought three wars with India over this small majority Muslim border country which remains a controversial flash-point between the two countries today...Pakistan is considered by many security experts as the world's most dangerous nation with a growing nuclear arsenal, poverty, abusive police, corruption, drugs, terrorist groups known throughout the country (some aligned with national security forces) and a government structure that is unpredictable and unstable in multiple ways...A young population with a median age of just 23 and a life expectancy of 67 years, over 31% of children under age five are malnourished with over 30% of the population desperately poor and an educational system so poorly structured and funded that many families send their children to madrasas (over 25,000 in the country) where they are taught and influenced by extremists clerics...The U.S. has given more than $20 billion to Pakistan in the past 10 years with very little going to help the population and much of the money disappearing in unknown (known) ways...Militants and terrorist leaders are routinely set free by a judiciary sympathetic to many of the extreme Islamist ideologies combined with a government that is unpopular and ineffective at dealing with a restless population and an uncertain future...Founded in 1947 as a homeland for Muslims, Pakistan has never really coalesced into a nation with purpose and direction consisting of feudal landlords, wealthy politicians and a ruling class that ignores the needs of many creating opportunities for terrorist organizations to control and influence large populations with protectionist promises, religious schooling and societal control through the brutal application of Sharia Law...Taliban are very active in country launching attack after attack and often acquitted in courts with little or no publicity...Pakistan has one of the world's largest refugee populations with over two million living in refugee camps and no way to be integrated into a society that does not recognize them...

Authorities investigated the killing of five women in the Baluchistan Province who tried to choose their own husbands...The women were shot, thrown into a ditch and buried alive...A prominent lawmaker in the province defended the murders as part of a "centuries-old tradition" 2008

A Pakistani couple killed their 15-year-old daughter by pouring acid on her because she sullied the family's honor by looking at a boy..."She said 'I didn't do it on purpose. I won't look again'...by then I had already thrown the acid...It was her destiny to die this way," said the girl's mother...2012

A striking example of the ignorance and lack of empathy and awareness that permeates governments worldwide is the ongoing animosity between Pakistan and India that has led to the imprisonment of fishermen from both countries who accidentally stray across an invisible maritime border and are apprehended by authorities...With both governments acknowledging that these men engaged in no criminal activities, the fishermen are held for years in prison with no processing or hope of relief...In 2013, an estimated 438 Indian fishermen are in prison in Pakistan and 165 Pakistani fishermen are in India's prisons...Families of the men have little recourse when government agencies and judicial authorities on both sides of the border are dominated by ignorant men and nonsensical policies...An appeal filed in India to address the plight of the Pakistani fishermen has been pending in India's Supreme Court since 2007...A decade or more in a foreign prison for getting lost on the open sea...

RUSSIA
(142 million humans)
Military Forces = 800,000 active duty/ $49 billion per year

Vladimir Putin has now won a third term of president (parliament changed the length of term to six years to keep him in power until 2018) and it's still unknown if he will grow in a progressive way to move Russia toward a viable, transparent, globally cooperative power that has resources and status to influence regional political and environmental progress in invaluable and sustainable ways...Russia is one of the world's largest oil producers (11 million barrels per day) exporting over half to Europe and Asia through a vast network of pipelines that move the oil directly to some of the countries it supplies and to coastal cities for distribution by sea, the 2nd largest producer of natural gas in the world with 33 trillion cubic meters of natural gas reserves and vast amounts of coal reserves (2nd in the world)...A country with nine time zones, Russia's economy is very simply fossil fuels...Currently building the East

Siberia-Pacific Ocean pipeline (3,000 miles long) to supply oil to rapidly developing eastern economies with growing demands, Russia's enormous territory (2% of global population) with an Arctic Sea that is slowly opening up to year round exploration and a growing access to developing countries desperate for energy in every sector, guarantee that Russia will be a world force in energy for many years...

There was hope and desperation in the 1980's and 1990's that *Perestroika* (a restructuring of the Soviet political and economic system) would usher in freedom and democracy, freedom of religion and speech, political and economic reforms not rigid and unyielding and a rejection of the totalitarian system that was stagnating the Soviet empire...But today what has emerged is a more open society with a government that still seeks to control everything and has made steps in that direction with a direct control of media and a rubber stamp parliament...It has not yet had free and fair elections and there is a growing unrest among the population who want a strong country with a strong leader (a sentiment that is very Russian) but want power balanced by a fair economy and a voice in an open government...Some two million Russian's work in businesses related to military spending as Russia plans to spend as much as $700 billion over the next ten years to build or purchase 400 intercontinental ballistic missiles(?), warplanes, ships, submarines and missile defense systems to modernize its military...Russia is the 2nd largest exporter of weapons sending arms to 56 countries (U.S. is #1 sending weapons to 94 nations around the world)...Corruption is a way of life in Russia from the top officials who determine policy to citizens on the street who deal with it in the 'Russian way'...Domestic violence is common with more than 30% of females reporting assaults and violence (not always to authorities in a corrupt and uncaring justice system) and over 13,000 woman killed every year in domestic violence incidents...Median age is thirty-nine with a life expectancy of 70 years...In this century Russia is moving away from global cooperation engaging in Syria's civil war to support Basher al-Assad insisting their focus is ISIS and threatening NATO with retaliation if it moves ahead to strengthen European defenses...Russia was the home of the Chernobyl nuclear meltdown that remains Earth's worst nuclear accident and this country still contains some of the most polluted sites on the planet from military industrial factories that indiscriminately dumped waste in rivers and pits for many years with no regulatory oversight...

In 2014, amid the unrest in Ukraine, Russian troops occupied the Crimean peninsular on the Black Sea and quickly annexed the property to become a part of Russia...The peninsular has been home to an important Russian military seaport for decades and the population in this part of Ukraine are mostly ethnic Russians, a legacy of the collapse of the Soviet Union in 1991...The U.S. and allies have initiated a variety of economic sanctions in protest...

SINGAPORE
(5.4 million humans)
Military Forces = 72,000 active duty / $9.8 billion per year

Independent since 1965, Singapore is a small grouping of islands south of Malaysia populated by a Chinese majority and vital to shipping and commerce in this highly trafficked part of southeast Asia and throughout the world...A single dominant political party (People's Action Party) has governed the nation since before independence from Malaysia (1965)...Government is a parliamentary republic with a president elected by majority vote for a six year term...A stable, free market economy with tourism, major investments in medical technology, pharmaceuticals and a financial services sector that's one the largest in the world...Median age is 34 with a life expectancy of 84 years...A large busy enclave with a stunning skyline and a large migrant population, low unemployment, little poverty and an economy supported by the service sector (84%)...Challenges include ongoing disputes with Malaysia over water rights, territory, maritime boundaries with both Indonesia and Malaysia, urban pollution and a restricted media with government oversight of communications and internet traffic...A hub of illegal activities due to it's geographical location and the availability of transport and services...

Singapore is a city-state (a city whose territory forms a independent state) along with Monaco and Vatican City...

SOUTH KOREA
(48.7 million humans)
Military Forces = 680,000 active duty / $35 billion per year

A remarkably stable society and economy considering that nothing is done without an eye on their neighbor to the north who continue to be a source of tension and unpredictable instability with a new leader (Kim Jong-un, son of former leader Kim Jong-il) who continues policies of the past that starve children and create an illiterate, ignorant and subservient society...South Korea has pursued an open engagement policy with North Korea for years with little to show for it and continues to grow as a major economic force in Asia and across the globe...Current president is Park Geun-hye elected in 2012 for a five year term ruling with a National Assembly elected every four years...A large military and continuing disputes with North Korea's over boundaries in the Yellow Sea...Challenges include the economic realities of a global recession with more demands for an expanded social net and continued posturing from North Korea who are desperate to remain visible on the world stage...Economy is the service sector and industry with 14% of the population below the poverty line...Energy wealth is 5.7 billion cubic meters of natural gas...Education system is excellent, life expectancy is 80 years with a healthcare system that is one of the best in the world...

TAJIKISTAN
(8 million humans)
Military Forces = 11,000 active duty / $100 million per year

Currently referenced as the poorest nation in Central Asia dependent on aid from Russia and China, Tajikistan's proximity to Afghanistan and a large population of young humans without income, direction or purpose allow an increasing Islamist influence, the proliferation of drug smuggling and a growing border insecurity that is escalating as Russia increases it's presence in the country to counter these destabilizing elements, reinforcing military bases and committing a billion dollars to train and equip the Tajik Army...Government is ruled by long time President Emomali Sharipovich Rakhmon (1994) with his People's Democratic Party holding a majority in parliament, but multiple international observers have criticized the legislative and presidential elections in this century for failing to "meet international standards"...Ongoing negotiations with Uzbekistan over borders and minefields...A major crossroad in Asia for the 'black market", Tajikistan is third in the world in opiate seizures...Media is varied with Russian sources being the primary provider of information, internet is controlled by the government as are printing facilities and broadcast stations...More than 14% of children under age 5 are malnourished, life expectancy is 68 years and over 160,000 children work in the labor force...This is a nation rebuilding after civil war (1990's) destroyed infrastructure and killed 50,000 humans...NATO involvement in Afghanistan has helped this nation's economy and infrastructure both on the borders and internally, but in 2016 over 35% of the population still live below the poverty line...

THAILAND
(68.5 million humans)
Military Forces = 305,000 active duty / $5.4 billion per year

Thailand has been transformed in the past 30 years from a primarily agriculture based society to urban development transitioning to industries in computers, automobiles and newly emerging service sectors...A constitutional monarchy with a king and a prime minister appointed by a parliament picked by the military after yet another military coup in 2014...Protests are common in this large population tired of government corruption and nepotism...An ongoing insurgency with Muslim separatists in south Thailand for ten years has killed thousand's and continues today...Conflicts with Cambodia over boundaries and territory...Over 120,000 refugees from Burma and half a million "stateless" humans with no access to voting, property, education, driving, healthcare or employment...Human trafficking, forced labor and drug abuse continue to be problems in this nation with an estimated child labor force over 800,000 and almost half a million people living with HIV/AIDS...The media's somewhat free with an exception for criticism of the military, monarchy or the judicial system...A leading rice exporter with almost ten million tonnes exported in 2015 from over thirty million acres of rice paddies,,,Farming supported and subsidized by the government with price controls, pensions and welfare guarantees to create a secure network for the future...King Bhumibol Adulyadej died in late 2016 bringing another crisis with the king's son (Crown Prince Vajiralongkorn) seen by many as unfit for office resulting in a 'regent' being appointed pending the coronation of the new king...Impossible to know when the political crisis will be resolved with trials, protests and a military still in charge...

TURKMENISTAN
(5.1 million humans)
Military Forces = 23,000 active duty / $203 million per year
A former Soviet republic on the Caspian Sea that gained independence in 1991, Turkmenistan has large reserves of natural gas and new pipelines to export gas to China and Iran...Identifies itself as a secular democracy but the Supreme Court is appointed by the president, the Cabinet of Ministers is appointed by the president and the president is both chief of state and head of the government (won 97% of the vote in 2013 in elections widely dismissed by observers as fraudulent)...Religion is Islam with the majority of the population Muslim...There is corruption at all levels of society with resources needed for education and reform disappearing into a secretive government that strictly controls the media with widespread censorship and threats of reprisal...Crude oil reserves are 600 million barrels and natural gas reserves are over 7 trillion cubic meters (4th largest) amid discussions about drilling rights in the Caspian Sea with Iran, Russia, Azerbaijan and Kazakhstan...

UZBEKISTAN
(29 million humans)
Military Forces = 66,000 active duty / $70 million per year
Almost 29 million humans with a median age of just 26 years make Uzbekistan the most populous country in this region with a large military force and authoritarian rule...Resources in oil (600 million barrels of crude reserves), natural gas (1.8 trillion cubic meters) and the world's 6th largest cotton industry should support a vibrant economy but government corruption has kept growth slow and 18% of the population exist below the poverty line...Strict censorship of internet and repression of all media leads to many citizens seeking news from outside of the borders, mostly from Russia...Majority Sunni Muslim dominated by Islam teachings and Islamic law, this region has a storied history being located on the major overland trade routes of ancient times represented by the architecture, traditions and their culture...Government is highly criticized for human rights abuses with Human Rights Watch reporting more than a million citizens, including children taken out of school and women who were beaten if they resisted, forced to harvest the state controlled cotton crop in 2012...Another government who abuse citizens and take the nation's wealth while receiving military aid ($) from the west and working with the United States involving levels of security, commerce and cooperation...

VIETNAM
(94 million humans)
Military Forces = 440,000 active duty / $3.4 billion per year
Inflation rate over 20% was the highest in Asia in 2011...A single party Communist government long accused of suppressing dissent, curtailing religious freedom and violating human rights of those who criticize government policies...Vietnam is scheduled to bring their first nuclear power plant online in 2028 to offset energy imports to a grid that is mostly powered by hydroelectric plants...Located south of China with over 3,000 km of coastline on the South China Sea, this is a country that was defined by war and conflict in the mid 20th century fighting France, a divided Vietnam and the United States...In

2016, Vietnam is a member of the World Trade Organization and the world's 2nd largest coffee exporter with an increasing population of young averaging 29 years of age...Over 15% of children under age 5 are malnourished and an estimated 2.5 million children are in the labor force...12% of the population exist below the poverty line in a economic reality where government owned industry is responsible for almost 50% of GDP...A communist state with a president elected by the National Assembly and the secretary-general appointed by the Communist Party...Energy wealth is 4.4 billion barrels of crude oil reserves and 700 billion cubic meters of natural gas...Challenges are long running disputes with Laos and Cambodia over maritime boundaries and the territorial sovereignty of offshore islands, internal drug abuse, unexploded mines and ordnance left over from war (government estimates nearly 40,000 people have been killed by leftover munitions), rising inflation and a weak currency...Media is heavily controlled by the government with imprisonment for criticism of political leaders or political policies...

In 2016 the U.S. government lifted it's arms embargo to Vietnam with no restrictions to benefit America's defense industry by opening a new sales market and continuing to make the world a more dangerous and violent place...

Over a 10 year period from 1961 to 1971 the U.S. sprayed over 20 million gallons of Agent Orange, a highly toxic defoliant, across Vietnam which is still polluting environments and causing humans health problems that number in the millions...An estimated 150,000 birth defects (no arms, no hands, no legs), widespread cancers, heart disease and Parkinson's within the Vietnamese population and thousands of military and support personnel affected in the U.S. armed forces still disputed by authorities in America today...Over 4,000 villages in Vietnam were sprayed with this agent and more than five million acres of farmland, almost half of it still unusable...In 2007, the U.S. agreed for the first time to help fund a study into removing Agent Orange and in 2012 the U.S. government initiated it's first direct involvement in decontaminating problem areas...

ENVIRONMENT

Kyoto Protocol was the 1ˢᵗ global pact enforcing carbon cuts ratified by 192 nations with binding provisions for carbon reduction...In 2012, the treaty was extended to 2020 with a revamp in 2015...This weak agreement among the governments of the world came only after two weeks of intense disagreements, posturing, and concessions by the nations who want a stronger binding agreement that could make a real difference...It only addressed 15% of global warming and did not include third world developing countries like China? The United States did not ratify the Kyoto Treaty (dropped out in 2002) and Canada withdrew from the treaty in 2011...

"The State of the Climate in 2012" report released by NOAA in August 2013 confirm the continued rise of greenhouse gas emissions and dramatic changes in the northern latitudes with almost 97% of Greenland's ice shelf melting, Arctic sea ice at it's smallest summer minimum and sea levels reaching a record high in 2012, "Climbing 1.3 inches per decade since satellite tracking of sea levels began in 1993"..."Earth continues to heat up without cessation"

At the world climate conference in Lima, Peru (December 2014) there were no major breakthroughs and little agreement among differing nations who finally agreed on the framework and ground rules for the next climate conference in Paris at the end of 2015 to draft a new global climate treaty...Over 190 nations were represented...

COP21 in Paris (December 2015) is the latest attempt by world powers to define a plan that would address long-term greenhouse gas emissions (net zero emissions after 2050) and global temperature "well below a 2C increase with an effort to limit it to 1.5C" while putting pressure on developed nations to help finance developing nations (somewhere above $100 billion a year after 2025) and laying out a framework of transparency, mitigation and adaptation as well as 'dialogue' in 2018 to check progress...Also meetings every five years starting in 2023 to submit new target goals adjusted to measurable progress and current scientific projections...The agreement will be effective once at least 55 nations representing 55% of global emissions have signed the deal but it is not a binding agreement and there are no penalties for missing deadlines, targets or goals...195 nations agreed to the final draft...It is almost certain the ambiguous "net zero goal" after 2050 cannot be met and given the data that exists today it will not be possible to keep global temperatures below a 2C rise from pre-industrial levels...

In a history of life on Earth lasting 3.5 billion years the human species has evolved over the last few million years in global climate conditions that have remained generally stable and supportive for the colonization of an entire planet and the survival of various plant and animal species essential to human survival...In a century with advanced science, worldwide communication and a long history of analysis and measurement it's expected a majority of people on Earth would understand the importance of maintaining climate conditions and the balance of atmospheric gases essential for life to survive but there is no majority...Even in this age of technology where the sustainability and limitations of Earth's ecosystem are being researched and studied in real time to understand boundaries, restrictions and climate trends everywhere on the planet, there is still no majority...Today what is proven and obvious is not recognized by billions of people who are seemingly unaware of a reality threatening the existence of species, cities, islands and food chains as human global activity rapidly changes Earth's climate with increasing populations, the unchecked burning of fossil fuels and a global ignorance of causality that is stunning...Developing nations and growing industries burn coal and gas without real alternatives for clean, sustainable energy production while new fracking technology, growing access to remote regions of Earth, increasing energy demands by emerging nations and the corporate power structures focused on profit without concern for impact guarantee the policy changes and planning needed now to stabilize Earth's atmosphere will not happen...With military posturing and territorial concerns being a priority in almost every nation on Earth, the global political dynamic necessary to mandate real change is clearly ineffective after two decades of climate "talks" with far more questions than solutions and entrenched interests continuing to control energy markets with influence and money...

Why would a sentient species with a critical knowledge of climate patterns and climate history ignore the dangers of an increasing activity that threatens the ecosystem that supports all life on their planet?

When science talks about the rapid injection of CO_2 into the atmosphere and ocean they are referring to events which can take hundreds or thousands of years to manifest having consequences that can last for millions of years and despite growing evidence of the impact of human activity with studies and conclusions repeatedly verified by analysis, observation and modeling, real science continues to be challenged in obvious and increasingly transparent ways by influential interests, powerful corporations and the politicians supporting these interests...It's unknown if there is a 'tipping point' when 21st century energy policies will cross a threshold and increase exponentially in a way to threaten the existence of vital species and it's clear a global understanding of the long-term effects of human activity is essential for survival on the only planet known to support carbon based life...In a long history of unpredictable threats Earth's ecosystem has recovered from all extinction events without evolving into a Venus type environment (it may be the mean distance of Earth's orbit from the Sun or the rotation and tilt of our planet that prevents this possibility) but the rapid pace and increasing volume of change in this century is an unknown equation and a growing ignorance of consequence is of great concern...

Satellite observations have shown that the amount of autumn sea ice in the Arctic has declined by more than 25% over the past 20 years...Summer ice in the Arctic could disappear in the next 25 years...

Based on what is known about the capacity of Earth's oceans to absorb carbon dioxide and the cumulative atmospheric effects on climate warming, scientists estimate that every 1 trillion metric tons of carbon from carbon dioxide emissions will increase global temperatures between 1.3 and 3.9 degrees Celsius...The optimal cap of 1.5 degrees Celsius is already unrealistic according to David Stocker at the University of Bern in Switzerland in a report published in the journal Science (Nov 2012)..."carbon emission reductions would need to begin by 2027 to meet a cap of 2 degrees Celsius and a 2.5 degree Celsius cap becomes unrealistic after 2040"...At climate talks in 2009 cooperating nations agreed to keep warming below two degrees Celsius citing possibilities that included island nations disappearing, coastal cities flooded, widespread drought and unpredictable weather in a future of continued warming, but seven years later despite talk of major change by developing nations, by industrialized nations, by China and by India, coal-fired power plants are still being built and a growing industry is focused on harvesting shale gas and burning it for the next 100 years with an exploding population that will reach 11.5 billion by 2100 (impossible to stop or even slow down) while every climate conference scheduled to meet this problem is efficient at scheduling future conferences but accomplish little else...

Science can measure the past levels of CO_2 in the atmosphere directly for approximately 800,000 years from air bubbles trapped in the ice in Greenland and Antarctica...What they found is a very close relationship between the amount of carbon dioxide in the atmosphere and the temperature of the planet and it's clear that putting large amounts of carbon dioxide into the air at the present rate is going to lead to an increase in global temperature...This happened in the past with "rapid changes" meaning just thousands of years and many species could not survive...In the 21st century human activity is emitting large amounts of CO_2 into the atmosphere on a geological time scale that's essentially instantaneous...An experiment with no precedent in the known geologic records...

All five mass extinctions were caused by climate change...An increase of CO_2 in seawater increasing ocean temperatures, increases in ocean acidity and increased global temperatures...The events that initiated these changes include worldwide volcanic eruptions, impacts from space, seafloor venting, orbital perturbations, but the result was the same, climate change over a long time period that in each case eliminated over 50% of known species from the planet...In 2016, human activity is creating very similar conditions over a much shorter time period...

PERMIAN-TRIASSIC EXTINCTION EVENT = 252 MILLION YEARS AGO

The most widespread and devastating of the five major extinction events believed to be caused by massive volcanic eruptions near the Arctic (Siberia?) lasting thousands of years in combination with unknown geologic events...Even possible is a gamma ray burst from space or an impact event that resulted in 95% of marine species and an unknown percentage of land species disappearing...Recent scientific research determined the disappearance of almost all life in the ocean was due to an excess of carbon dioxide, increases in ocean acidity, a lack of dissolved oxygen in the seawater and rising water temperatures...*These are the same conditions modern humans are creating in today's oceans with pollution, a lack of global unity and a dependence on fossil fuels*...In this century coral reefs and marine species are disappearing, sea levels continue to rise, ocean acidity is increasing and the temperature in the atmosphere and in the ocean is rising but the future is unknowable with the ability of humans to interact and intervene to affect change in multiple ways...Recent estimates are this extinction event may have took 50,000 to 60,000 years to unfold, a very brief period on geological scales..Estimated time before life on land and in the oceans fully recovered is 20 to 30 million years...

In 2012, global carbon dioxide emissions totaled 31.6 billion tonnes, the highest of any year to date...In a report from the International Energy Agency(IEA), significant reductions in emissions from developed nations using new technology and an increased supply of natural gas were offset by an increase in emissions from China and developing nations who rely on coal as a primary energy source to fuel industry and generate electricity...

Nature Climate Change (December 2015) predicted global CO2 emissions for 2015 will total 35.7 billion tonnes...

TRIASSIC-JURASSIC EXTINCTION EVENT = 200 MILLION YEARS AGO

Over 50% of all known species on Earth went extinct over a 10,000 year period just before the single land mass Pangaea started to separate into the known continents...Again the cause could have been massive volcanic eruptions over extended periods that changed the climate in a way that was not survivable by the species who perished (it's estimated that 30 to 40 super volcanoes exist on Earth and erupt on average every 100,000 years...Last known eruption of a super volcano was approximately 74,000 years ago)...Some explanations of these eruptions speculate that an enormous lava bubble from the mantle of the earth broke through the crust in a high pressure situation that initiated catastrophic climate effects unlike anything modern life forms can imagine...Due to the rapidity of this extinction event many scientists believe it was an impact event from space...It's very difficult to find definitive evidence for either idea...It took millions of years for ecosystems to recover...

CRETACEOUS-PALEOGENE EXTINCTION EVENT = 65 MILLION YEARS AGO

The well-known mass extinction event that killed over 50% of the world's species and ended the existence of the dinosaurs was initiated by a large impact event over 65 million years ago estimated by scientists to have been an asteroid six miles in diameter creating a crater 110 miles wide and over a mile deep...Very few large land animals survived this event that increased the temperature in the ocean more than 15 degrees F above current readings and raised sea levels almost 1000 ft above present day sea levels...Many scientists believe the initial event was so catastrophic that many species died within days or weeks and it created a climate that was very hostile to the life that survived...*Evidence in all of these mass extinction events, regardless of the cause, point to a rapid increase of CO2 emissions in the ocean and atmosphere as a prime factor in the rising temperatures and the accelerated extinction of multiple species*...There is a delay of decades to hundreds of years before the oceans start to respond to the effects of a rapid rise in CO2 emissions but ongoing measurements and a growing body of science confirms this change is already starting to happen this century...As is the case in all of these extinction events, Earth's ecosystems took millions of years to recover...

Scientists believe in about 100 million years the continents will fuse back together and form a single continent in a clear pattern that has determined the orientation of the past subcontinents Pangea, Rodinia and Nuna...

Sea levels around the world have risen more than seven inches since 1900 and are predicted to rise as much as three feet by 2100...This will affect coastal environments and infrastructure worldwide...

Global levels of CO2 have increased from 280 ppmv to 400 ppmv since the beginning of Earth's industrial age...

Greenhouse Gases

Since the beginning of the industrial revolution in the 18th century concentrations of greenhouse gases in Earth's atmosphere have risen almost 40% to levels above 400 parts per million, levels that continue to rise in this planet's "industrial age", a period of activity that began 200 years ago, activity that created the world we live in today and activity that increases globally every year to support a growing human population without constraint...Humans are industrious, creative and busy when given the opportunity to do what humans do best and what we do best is *build better tools*...In every culture on Earth archaeologists define and categorize human evolutionary history based on historical evidence of an evolving ability to conceptualize, design and build increasingly complex tools determined by new materials, new techniques, new energy sources and a need...The last two centuries have seen this skill rise to a level unimaginable 200 hundred years ago but there is an increasingly negative impact in the use of these energy sources (fossil fuels) and today the impact of 200 years of unparalleled growth is in full focus...In this century there exist complex flying machines to carry people through the air and there are enormous machines with the ability to rip a planet open to recover resources or terraform dirt and rock into urban habitats that support millions...It's easy to imagine a future when humans will build solar collectors the size of small moons to feed energy to Earth and a distant future when humans will build machines in space not restricted by aerodynamics or gravity, spacecraft miles in length able to transport our species throughout the solar system for exploration and resource recovery to support life on Earth, Mars and maybe even Venus, but first we must transition to new energy sources to power this civilization for thousands of years and burning a planet for energy is a limited, finite equation...

The short history of global energy production is fascinating as you realize the widespread use of fossil fuels has coincided with a control of electromagnetism and the widespread use of electricity, the energy that will power this planet for thousands of years...Fossil fuels are simply the easiest way to generate electricity but there are different ways and they become more convenient and more available daily...If every vehicle on the Earth were powered by electricity, even if the electricity were supplied by fossil fuel generating facilities, it would make an enormous impact on greenhouse gas emissions...And if every power generation facility were equipped with 21st century technology able to filter and capture dangerous gases, humans could turn this cascading, unstoppable climate event around but this will not happen in the political reality that defines this century...Every indication is the unchecked burning of fossil fuels will continue and even increase during this century as developing nations seek wealth and convenience equal to developed nations and new technology is rapidly making it possible to recover dirty energy sources anywhere and everywhere on Earth...Even with clear evidence of change and causality, human activity continues to produce a variety of greenhouse gases that are rapidly changing the balance of an atmosphere essential for biological life...*Carbon dioxide,* which made up over 80% of the greenhouse gases measured in 2010, has an atmospheric life of 100-200 years and will have the most significant long-term impact on Earth's climate as humans generate tons of it every single day to drive their cars, fly their planes and power their lives *(estimates in 2015 are more than 80 million tons of CO2 released in the atmosphere every day)...Methane,* released from coal mines, numerous petroleum activities, landfills, animals and agriculture is better at trapping heat than CO2 but is short-lived with an atmospheric life of only 10-12 years...*Nitrous-oxide* is released when burning fossil fuels and during specific industrial activities including soil management in agriculture and in the production and use of fertilizers with a atmospheric life of around 120 years...*Fluorinated gases*, synthetic greenhouse gases created to replace the ozone depleting gases being phased out, are generated by industrial processes, are very potent and extremely long-lived gases with an atmospheric life of *thousands of years*...F-gases are highly concentrated and are recognized as High Global Warming Potential gases, the most potent gases created by human activity...These greenhouse gases all mix together well and are found in relatively equal ratios everywhere on our planet without boundaries or borders...

May 2016...NOAA says greenhouse gas emissions are accelerating...CO2 emitted in 2015 = 35-40 billion tons...

The World Meteorological Organization (WMO) annual greenhouse gas bulletin (a publication that measures gas concentrations in the atmosphere, not emissions on the ground) reported in November 2013 that the levels of methane (1,819 parts per billion) and nitrous oxide (325.1 parts per billion) reached record highs in 2012...

It's estimated in 2016 that over nine million tons of methane are being released in the U.S. every year...

Human Adaptation, Greed and Ignorance

The UN assigns carbon credit values to these multiple gases based on longevity and global warming impact with carbon = 1, methane = 21, nitrous oxide = 310 and the dangerous F-gases, primarily HFC23 = 11,700...These carbon credits are sold and traded to allow the increased release of emissions but an industry driven by profit quickly realized they could create an abundance of the highly dangerous HFC23 waste gas keeping market prices low (reduces incentive for manufacturers to find a less harmful alternative) and then destroy this waste gas to receive maximum carbon credits which they can sell on the global market...Since the UN program began, 46% of all carbon credits have gone to 19 coolant companies, mostly in China & India, for destroying HFC23 waste gas created by the companies to take advantage of this program for maximum profit...In 2012, 18% of credits will still go to these 19 coolant companies while only 12% will go to more than 2,300 wind power plants and only 0.2% going to 312 solar power plants...Efforts to change this program is hampered by corporate threats to release the waste gas into the atmosphere (still legal in China and India) if awarding carbon credits for this type of activity is reduced or eliminated...Extortion...There are currently companies in China that freely vent HFC23 into the air and the atmospheric levels are rising...

In May 2013, measurements of CO2 at the Keeling Lab Station in Hawaii exceeded 400 parts per million for the first time since the Keeling station began taking continuous CO2 measurements in 1958 (the first measurements taken in 1958 were around 315ppm) and a report from the WMO (World Meteorological Organization) in November 2013 predicted the global annual average of CO2 concentrations would exceed 400ppm by 2015/2016...In 2015, NOAA (National Oceanic and Atmospheric Administration) announced global carbon dioxide (CO2) concentrations reached a new monthly record of 400 parts per million (401.25), the first month the entire globe broke 400ppm...Ice core samples as far back as 800,000 years record no CO2 levels higher than 300ppm and scientists believe the last time CO2 emissions were regularly above 400ppm was over 2 million years ago...

Mainstream scientists around the world believe a dangerous tipping point for the world's climate may be an average CO2 level of 450 parts per million and to avoid reaching that number industries would have to stabilize emissions by 2020 and reduce emissions more than 50% by 2050...*This will not happen*...Every major country in the world is on a path to consume more fossil fuels and developing countries are demanding equal opportunity to add to the pollution with more cars, more factories, more power plants and little regulation or even an intellectual awareness of the need for regulation...There is little doubt the world's climate will continue to change in rapid ways due to human activity...Evidence from multiple sources including arctic ice bores and tree rings prove the Earth's temperature has risen more in the last 100 years than the previous 2,000 years but many politicians, corporate leaders and the financial markets who set policy and influence governments continue to stare into the headlights too entrenched in familiar patterns of corruption, nepotism, greed and denial to recognize overwhelming scientific evidence and obvious causality...There will be rising oceans, increasing drought, famine, war and eventually a reorganization of world governments as the 1st World Civilization fragments into self-serving territories desperate for their own survival...The most uplifting, positive sounding, value driven advertisements on U.S. television in 2012 were produced by Big Coal urging citizens to remove EPA regulations so they can use their 'clean' power source to provide energy to millions...There is no 'clean' coal...The American Petroleum Institute (API), a national trade association representing America's oil and gas industry (over 550 corporate members), is a continuing presence on U.S. television to convince the public that fracking is safe and everything that oil and gas companies do is in the pursuit of clean, affordable energy, good for the economy and good for the planet...

With new technology coal plants could produce 90% less nitrogen oxide, 90% less sulfur and significantly reduce the amount of other pollutants but most corporations operating coal-fired power plants fight mandated pollution controls and regulation with political influence and money instead of recognizing this reality and creating the most efficient plants possible...Across the world in industries and governments that could make a huge difference with the technology available today, there is little awareness of responsibility and no understanding of the balance necessary for the human species to survive and flourish...Almost all corporate polices focus on a singular, short-sighted motive of profit and when there is a political conflict of economic growth policies vs policies reducing harmful gas emissions the policies promoting economic growth are almost always chosen over environmental concerns and this is true across the globe, even in the most responsible countries...If the world stopped emitting greenhouse gases today global warming will continue to increase for hundreds of years and this world will change in many ways...And the world will adjust...It is unknown how this unstoppable process will affect geopolitical and military strategies on a future Earth with 13+ billion humans...

Federal records obtained by CPI and NPR under the Freedom of Information Act show thousands of coal miners were exposed to excessive levels of silica in each of the last 25 years...Since 1987, coal mining companies and government inspectors turned in more than 113,000 valid mine dust samples and approximately 52 percent of those samples exceeded federal standards...In 1998 alone around 65 percent of the valid silica samples violated the standard...The Federal Mine Safety and Health Administration (MSHA) proposes toughening the coal dust standard by slashing exposure limits by half (to 1 milligram per cubic meter of air) but the industry still opposes the new standard because it would broadly apply to the coal dust in every mine in operation...

In April 2014, the U.S. Department of Labor issued their new rules for coal dust exposure that will improve sampling in mines, lower exposure and close loopholes in current law...The coal industry and political allies managed to keep the exposure limit at 1.5 milligrams of dust per cubic meter of air (as opposed to the expected reduction to 1 milligram) and Murray Energy, a large coal mining corporation, issued a statement vowing to challenge MSHA in court saying the rule "seeks to destroy the coal industry and the thousands of jobs it provides with absolutely no benefit to the health or safety of miners, whatsoever."

Studies in 2013 of methane gas leakage at fracking sites in the U.S. concluded the EPA had underestimated the amount of leakage by as much as 50%...New research in 2015 estimates over 9 million tons of methane is released in the U.S. every year with little oversight...Methane has a greenhouse gas potency many times that of CO2...

Ignored by many is the incredible number of small power engines used for maintenance and lawn care that emit pollution without controls and diesel engines, kilns and cook stoves used in developing countries that create a 'black carbon' which contributes to the early deaths of millions of humans from particulate pollution and contributes to the melting of Arctic sea ice by coating the ice and absorbing more sunlight...

*Even the most optimistic forecasts predict the global market for all electric plug-in vehicles will be less than 5% in 2025 and the market for hybrids which have a small gas engine to recharge the batteries and require no charging source may reach 25% by 2025...**Motor vehicles (over a billion worldwide) are collectively the largest contributors to the increase of greenhouse gases in the atmosphere** and the continued use of internal combustion engines to power the machines humans feel are essential in life will continue to have an extremely negative impact on the environment...In the future there will be advances in battery technologies capable of powering all machines but in this century humans continue to dig large holes in their planet and burn their world for energy...*

Petroleum Coke
A byproduct of the Canadian tar sands refining process, it's hard to overestimate the negative impact of *petcoke* on local environments and global emissions which are changing this planet...It's already known the extraction process of tar sand bitumen produces emissions up to 3x's the emissions produced extracting conventional oil and when burned for fuel, tar sand oil emits 15% to 30% more greenhouse gases than the same amount of fuel burned from conventional oil...When you factor in the impact of *petcoke*, a byproduct of refining that can be burned as coal (producing 50% more CO2 than

the same amount of conventional coal), it's clear to any intelligence in a closed ecosystem that the tar sand oil reserves should be left in the ground...This will not happen...Energy corporations with deep enough pockets to buy the political systems of the world will burn everything they can without regard to the impact on future generations or the immediate local impact to communities of humans located near these refining facilities...A BP plant in Indiana produces over 2,000 tons of *petcoke* daily with plans for expansion (to over 2 million tons a year) to take advantage of Canadian tar sand oil extraction while storing much of the petcoke in Illinois to subvert regulation in their own state...Frequently stored outdoors without restraints, the petcoke (essentially a black carbon powder) is spread by wind currents to neighborhoods and houses many miles away coating houses and humans with a highly dangerous substance that can cause sickness and chronic health conditions...In America the burning of *petcoke* is regulated and the majority of this product is sold to foreign markets in China, India and Mexico to be burned without regulation...

In 2011, the 59 refineries in the U.S. refitted and structured to produce petcoke made over 61 million tons of this product with more than 60% of the total being exported abroad to be burned for energy...

Oxbow Corporation, the largest trader of petcoke on Earth, sells more than 11 million tons annually while spending over $1.2 million on lobbyists in Washington and donating over $4 million to GOP super PACs to avoid "complications and unnecessary regulations"...Oxbow Corporation and KCBX Terminal (the operational arm of the open air petcoke storage sites in Illinois) are owned by one of the Koch brothers, highly influential billionaires deeply embedded in the American political system...

When burned for energy in coal-fired power plants around the world *petcoke* can emit between 30% to 80% more CO_2 per ton depending on the coal's quality...It is priced lower than conventional coal and blended with conventional coal to make burning cheaper and dirtier while producing a coal ash byproduct that's packed with heavy metals...The Potsdam Institute has estimated that industry must limit CO_2 emissions in the first half of this century (2000-2050) to a maximum of 953 billion tons to prevent the average global temperature on Earth from increasing more than 2 degrees centigrade...In the first decade (2000-2010) total emissions of CO_2 were approximately 331 billion tons worldwide, a global accumulation of CO_2 that will exceed the maximum by 2030...

A report published in early 2014 by Proceedings of the National Academy of Sciences concluded the emission levels of polycyclic aromatic hydrocarbons (PAH's) are one hundred to a thousand times greater than reported by the mining companies environmental impact assessments...The unreported amounts are contained in the tailing ponds that collect wastewater from the shale extraction process...PAH's are priority pollutants and have been shown to affect the immune system and the reproductive system causing tumors in lab animals...

Coal Ash
A part of the coal for energy process that coal companies promoting 'clean coal' do not discuss is *coal ash*...Coal ash is the toxic leftover from coal-fired power plants and is filled with arsenic, lead, mercury and other hazardous chemicals...It poisons lakes, rivers, streams, underground aquifers and drinking water supplies...It causes cancer, organ damage and even death...At least 185 coal ash dump sites in the U.S. have contaminated local water supplies and the U.S. Environmental Protection Agency admits that at nearly 504 sites, a major spill will likely cause loss of human life...The largest toxic waste spill in U.S. history occurred in 2008 in Kingston, TN when over one billion gallons of coal ash burst from an earthen dam sweeping away homes and covering 300 acres in toxic sludge...In America there are over 900 coal ash storage sites operating in 35 states while power plants continue to generate more than 110 million tons of coal ash every year...In four years since the 2008 spill no regulation was passed addressing this issue with multiple lawmakers in the U.S. Congress actively trying to promote legislation to prevent the EPA from ever regulating the production and waste storage of coal ash...

In 2014, the EPA agreed to finalize rules on the disposal of coal ash by the end of the year as part of a settlement with a consortium of public interest groups who filed a lawsuit following the Kingston coal ash spill...The consent decree requiring an enforceable deadline does not dictate the content of the final regulation wording by the EPA...The coal industry estimates over three billion tons of coal ash has been generated since the beginning of the 20th century...Coal ash disposal has never been regulated to ensure long-term safety...

In February 2014, a broken pipe at a Duke Energy coal plant in North Carolina spilled an estimated 35,000 tons of coal ash and 27 million gallons of contaminated water into the Dan River., contamination that was found more than 70 miles downstream...The company owns 13 more coal plants in the state and all of them have coal ash storage ponds built many years ago without safety liners...North Carolina's Environmental Agency has close ties with energy corporations and has been criticized for not acting on polluted groundwater until civil actions were heard in court, criticized for issuing small fines and giving energy companies in violation years to conduct further studies before doing any cleanup and criticized for refusing to require Duke energy to stop coal ash seepage from 32 ponds at their 14 coal-fired plants...The governor of North Carolina, Pat McCrory, worked for Duke Energy for 28 years and the company has been a major political donor to the governor's campaigns...

If this is happening in America what is happening in China, a nation with less regulation and a history of corruption and collusion with energy industries? In China's active coal belt entire villages sometimes collapse because of the large number of tunnels dug in the Earth to recover the coal and the mining process uses an enormous amount of water diverted from drinking supplies and irrigation for agriculture...Huge 'coal dust storms' consisting of coal ash and dust are blown hundreds of miles across open country from winds up to 20 mph depositing the ash far away from the source...Approximately one ton of coal ash is leftover from every four tons of coal burned and China burns over 3 billion tons of coal every year with predictions of an increase to over 5 billion tons by 2030...The population of China has no voice on the environmental effects of this activity...

Even if charged by the dirtiest power plants on the planet, electric cars would be better for the environment in every possible way than the internal combustion machines used now...

In America thousands of "drive-thru's" dot the landscape with lines of cars emitting unseen pollution so overweight humans don't have to walk to feed themselves and in every empty parking lot is someone with an internal combustion 'leaf blower' with no exhaust requirements blowing clean the pavement whether necessary or not because this is how Americans live their lives of expectations, thoughtless actions and over-consumption...

Coal ash contains concentrations of arsenic, boron, cadmium, chromium, lead, mercury and other metals...

Water
4% of all deaths on Earth are attributed to water-related diseases...

The most essential resource on Earth is water...It is everywhere, part of the atmosphere, locked up as ice in both polar regions, running through vast underground river systems called aquifers, lying on the surface in collection areas called lakes and covering over 70% of the planet in one ocean that circles the world...There is water in all plant life and all animal life and it is without question the most vital substance necessary to keep what is alive, alive...In the 21st century, distribution of water resources around the world are being challenged by human activities in dramatic and critical ways that will have multiple effects on geographical and political regions everywhere on the globe...Earth's ocean has been a dumping ground for anything unwanted by the human race for thousands of years and the waste of 7.4 billion humans is challenging the survival of many species who live in the oceans, freshwater rivers and lakes everywhere on Earth...There are few enforceable international laws that govern the oceans and no real way to prevent the impact of short-sighted human activity that fails to recognize the balance necessary for humans to exist...Waste runoff continues from cities and industrial sites around the world with little or no regulation and ships routinely dump water polluted with oil and gas in the sea, refilling tanks for ballast and dumping it everywhere in an ocean that is big, but not endless...

Today almost 70% of the freshwater used in the world is for agricultural purposes with the rest used for industrial purposes and human consumption...There exists more than 250 trans-boundary river basins that cover nearly one-half of the Earth's land surface and account for an estimated 60% of global freshwater flow...A stunning growth in population (4 billion people in 60 years) has brought some new cooperation between nations and negotiated international agreements concerning trans-boundary basins and sharing of freshwater sources that suggest some awareness of the growing problem of availability but there are still numerous conflicts, violent and non-violent, in localized regions where the lives and the work of indigenous people and their very existence is determined by how water is controlled and distributed...Innovations in agriculture help use less water by controlling seepage, leveling agricultural land to prevent runoff and utilizing drip irrigation and in a future where humans produce large amounts of solar/wind energy, desalination plants and reverse osmosis could provide agriculture and industry with large amounts of fresh water from the ocean...But this is not the future and in this century the water wars are just beginning...In a changing climate with vast areas of the planet experiencing drought and growing populations, many nations are focusing more and more on control of the water within their boundaries with some seeking cooperation with countries who share the same water source and some seeking to divert water flow for their use creating tensions and hostilities that will only grow as water supplies become increasingly limited...Earth is a water world, human beings are mostly water and how this species manages this vital resource for thousands of years will determine the sustainability and survival of every creature on this world...

Human Blood is 80% water...The Human Brain is 70% water...

EPA has stated they will tighten limits on four specific contaminants proven to cause cancer...Tetrachloroethylene and trichloroethylene used in industrial and textile processes and acrylamide and epichlorohydrin, routinely discharged into the water during industrial treatment processes...

More than a billion humans do not have access to safe drinking water and over 2 billion humans have little access to sanitation facilities...Millions of people globally live on less than 3 gallons of water a day while every American (on average) uses over 70 gallons a day and the unpredictability of climate change is already presenting challenges with changing patterns of water distribution that will increase rationing, crop destruction, violence and force the migration of entire populations...More than 30% of Europe's populations live in nations with acute water problems and in the developing countries almost three million humans die every year from disease related to diarrhea and water contamination, more than half of them children under age five, over 4,000 every day...

Current technology for desalinization, the process of turning salt water and waste water into drinking water, is called the reverse-osmosis spiral modal currently used by more than 12,000 plants around the world...Primary challenges are the destruction of marine life and efficiency of transport...

Humans need a minimum of two liters of drinking water per day to survive...

Chemical Pollution

*In 1976, the U.S. passed The Toxic Substances Control Act (TSCA) banning just five chemicals and grandfathering in over 62,000 chemicals already in use without any requirements for testing the safety of these products on biological life or the environment...Since that time the introduction of new chemicals into industry by corporations were exempted from testing unless there was evidence of harm...More than 21,000 new chemicals have been registered since 1976 with less than 15% submitted with health-and-safety data (voluntary) and the EPA is limited to testing only around 200 new chemicals in a corrupt corporate environment that uses claims of confidential information and trade secrets to discourage and prevent scientists outside the EPA from challenging corporate conclusions concerning the safety, use and disposal of these unknown substances...(In 2016, the U.S. Congress passed and the president signed the **Chemical Safety for the 21st Century Act**, the first update of TSCA in 40 years)*

In 2002, the EPA drafted a strong, comprehensive chemical safety law that was watered down by lawmakers to exempt thousands of chemical facilities in America from any regulation at all and in 2011 a Congressional Research Service identified 483 facilities in the U.S. where a chemical accident could endanger more than 100,000 humans...Efforts continue to produce legislation to regulate these industries in a fair and safe way but fierce opposition to any change by Congress will hamper and probably defeat any attempt to make improvements as millions of humans around the world continue to be exposed to industrial chemicals, pesticides, heavy metals, radionuclides and other toxic chemicals produced by industry in mining operations, agriculture practices, industrial manufacturing and shipping and transport facilities...In Russia and former member states of the Soviet Union there are old weapon storage sites, abandoned aging infrastructure and manufacturing wastelands polluted with multiple hazardous chemicals that continue to contaminate groundwater while in China, India and many other developing nations there are thousands of hazardous waste sites from rapidly growing economies with little government oversight or regulation of industry...

The air in many large cities in Asia is becoming increasingly polluted as populations grow with more energy demands and more internal combustion vehicles sitting in traffic jams for hours while the humans inside these vehicles enjoy a climate controlled environment with little thought to the consequence of this activity...

The most polluted cities on Earth today are in China and India with atmospheres of ash, carbon monoxide, sulphur dioxide and arsenic with an alarming percentage of children suffering bronchitis, pneumonia, cancer and lead poisoning from the unchecked burning of coal...*Lead* smelting around the world that uses recycled metal or ore is estimated to endanger over 20 million humans with careless, unregulated contamination of water, air and soil...*Endosulfan* is a pesticide used widely in India, West Africa, Asia, Australia and is slowly being phased out after killing and damaging millions of humans with neurological disease and cancer...A chemical weapons site at Dzerzhinsk in Russia is said to have toxic levels of the chemical *phenol* at more than 17 millions times the safe limit...Millions of people around the world suffer toxic *mercury* poisoning from activities around poorly regulated gold mines where the mercury is burned to separate the gold, polluting the air, and then discarded into water sources and the ground causing neurological disorders in children...And a documented incident in 1984 in Bhopal, India killed over 3,000 humans when a chemical plant released 40 tonnes of *isocyanate gas* into a population of almost 2 million people creating medical problems that still kill and disable people more than 26 years later...Union Carbide denied responsibility and the site is still being cleaned up...

The EPA in America is required to regulate the storage of oil in above-ground tanks but is not required to inspect storage tanks containing thousands of chemicals that are not oil...In the U.S. an ineffective hodgepodge of "agencies" (Homeland Security, Chemical Safety Board, The U.S. Labor Dept) are expected to provide oversight of the thousands of chemical operations that produce pesticides, fertilizers and chemicals used daily by industry...

The U.S. and other industrialized countries around the world continue to send electronic waste and garbage to third world countries to be melted, recycled and sold releasing chemical toxins into the environment, sickening local populations and eventually finding a way to other nations through the continuous cyclical nature of the Earth's atmosphere and water while the mercury released in toxic smoke from every coal-fired power plant on the planet eventually finds it's way to the ocean polluting many species of fish affecting a food chain essential to the survival of life on land...

There are some 500,000 abandoned mines (copper, silver, gold, lead, etc) in America on federal land with dangerous levels of arsenic, lead, mercury and other contaminants that would cost more than $70 billion to clean up...The Bureau of Land Management, part of the Interior Department, administers over 250 million acres and denies any lack of oversight in dealing with the problem...Politicians in Washington have failed for years to rewrite an 1872 law that lets prospectors stake claims with no responsibility for cleanup...What's in the rest of the world?

The Basel Convention

A study released in late 2013 by the Blacksmith Institute and Green Cross from data collected at more than 3,000 sites in 49 countries worldwide concluded that more than 200 million humans are at risk from toxic waste exposure...#1 in the study was the Agbobloshie dump site in Ghana on the West African coast where millions of tonnes of waste from disposable electronics are processed from countries worldwide...The Basel Convention is an international treaty intended to control the movement of hazardous waste between nations with specific wording to prevent the transfer of hazardous waste to less developed nations coming into force in 1992 ratified by more than 180 countries (The U.S., Afghanistan and Haiti have never ratified the agreement)...The U.S. is the leading exporter of this e-waste using legal maneuvering and loopholes to avoid laws and regulations designed to protect human health with illegal containers of e-waste arriving at Agbobloshie daily, mainly from the U.S., Western Europe and Japan, to be processed by children and unprotected workers polluting water, air and the land...2nd on the list is Chernobyl in Ukraine and 3rd is the Citarum River Basin in Indonesia with textile factories and other industry polluting the river with lead, arsenic, mercury and various other toxins...Heavy metals are particularly hard to remove from the environment and in many places the only option is to move populations and seal off the contamination area...Another example of greed and profit prioritized over all else with the humans who control this process and who understand the "intent" of treaties, laws and regulations actively finding multiple ways to maximize profit and appease a primitive instinct without awareness or concern for the health of the Earth or the species who depend on the resources being destroyed or the children who will be affected for a lifetime by ignorant, uncaring actions...

It's estimated more than 5 billion pounds of pesticides are applied to crops annually in the United States with thousands of farm workers experiencing the symptoms of pesticide poisoning...

A study published in Pediatrics Journal in 2010 that measured the levels of six pesticide compounds used on fruits and vegetables in a sample of 1,139 children found evidence of the compounds in 94% of the participants...

Over 1.2 billion people exist without access to electricity and almost 3 billion people still use solid fuels (wood, charcoal, coal, dung) for cooking and heating resulting in the deaths of more than 1.5 million people every year from indoor smoke responsible for emphysema and other respiratory diseases...

Volatile Organic Compounds

VOC's are chemicals that evaporate and turn to gas when exposed to the air...Many VOC's, including benzene, butadiene and toluene are on EPA's list of toxins or hazardous air pollutants because they are identified or indicated as a carcinogen which are known to cause cancer and other serious health problems...The sunshine chemical reactions between VOC's and nitrogen form ground-level ozone (smog) that's been indicated in multiple health problems including chest pain, coughing, throat irritation, emphysema, asthma and bronchitis...

The 2nd largest refinery and petrochemical plant in America, a 104 year old facility in Louisiana owned by Exxon Mobil, released almost 4 million pounds of volatile organic compounds (VOC's) between 2008 and 2011 beyond the limit approved by the EPA through uncontrollable releases, leaks and spills (EPA permits the release of millions of tons of chemical pollution in the U.S. every year) leaks disclosed by the company when confronted with information from a public records request...In June 2012, Exxon Mobil reported an accident that had leaked over 1,000lbs of toluene, the tolerance threshold for this chemical...In August the company revised the amount leaked in June to 31,000lbs of benzene, 13,000lbs of toluene and 18,000lbs of other VOC's...Exxon Mobil reported there was no evidence that these harmful levels of pollution threatened the safety of the humans who live around the plant...

A subsequent investigation of the Exxon Mobil plant found pipe corrosion, duct taped valves, overdue inspections on 1,500 underground pipes and repeated failures to report leaks for three years...Many experts say this is a reflection of a universal problem in an industry that operates in dangerous cost cutting ways to maximize profit when it's clear the human species has the technology and the intelligence to coordinate and operate these facilities with effective safety protocols and updated equipment without risk...Greed, collusion, corruption and ignorance...

Type	All Emissions	Unauthorized Emissions
Statewide (pounds)	*471.6 million*	*9.5 million*
Refining & Chem. Industries	*208.5 million*	*6.7 million*
Exxon Mobil Baton Rouge Facilities	*28.4 million*	*3.6 million*

In 2013, the Environmental Protection Agency asked 17 Gulf Coast refineries and chemical plants to participate in a voluntary program to reduce pollution...All 17 plants refused to participate...Why is this voluntary?

Because of the way Louisiana regulators categorize pollutant emissions, the "unauthorized" designation is an underestimate...The category doesn't include other unauthorized emissions plants release when they're maintaining, starting up or shutting down equipment...The Statewide category includes all regulated industrial facilities...
NPR analysis of state emissions data

Texas petrochemical plants released nearly 30 million pounds of excess VOC's between 2008 and 2012...

A Global Inventory of Volatile Organic Compound Emissions from Anthropogenic Sources

The results of this study shows total global anthropogenic VOC emissions of about 110,000 Gg/yr...This estimate is about 10% lower than global VOC inventories developed by other researchers...The study identifies the United States as the largest emitter (21% of the total global VOC), followed by the (former) USSR, China, India, and Japan.... Globally, fuel wood combustion and savanna burning were among the largest VOC emission sources accounting for over 35% of the total global VOC emissions...The production and use of gasoline, refuse disposal activities, organic chemical and rubber manufacturing were also found to be significant sources of VOC emissions...
Stephen D. Piccot, Joel J. Watson, Julian W. Jones...Published online: 21 SEP 2012

Ecuador and Oil

Ecuador's a small country on the Pacific coast of South America and home to one of the densest concentrations of biological diversity among species found anywhere in the world...The country is also desperately poor and sits atop a reservoir with an estimated volume of almost a billion barrels of crude oil that companies are eager to harvest...Yasuni National Park is a reserve of nearly 4,000 square miles in the Amazon basin that would be impacted in an extremely negative way if these oil companies were allowed access after a clear history of previous activities and operations by energy corporations over 40 years polluted vast stretches of the Amazon through negligence, indifference and corruption...

In 2011, a judge in Ecuador ruled that Chevron was responsible for polluting the rainforest and ordered the company to pay $9.5 billion to farmers and others damaged by Chevron's activity...In 2014, a judge in the U.S. ruled the verdict against Chevron was unenforceable because the plaintiff's lawyer had used fraud and corruption to influence the original judgment...This allegation seems to have validity but what is stunning is the oil pollution clearly caused by Chevron's activities has never been cleaned up...There has been oil on the ground for over thirty years and the corporation responsible for the careless destruction of indigenous peoples livelihoods and the careless pollution of an extremely valuable and fragile rainforest still refuses to clean up their mess while continuing to use legal maneuvering to avoid responsibility...Chevron's profit in 2011 was $26.9 billion...

In a concerted effort to prevent more destruction to regions of the Earth that are not replaceable, environmentalists and scientists reached out to the president of Ecuador to consider other options and a plan was put forward to leave the oil in the ground, conserve the biodiversity of the region and prevent the release of almost a million tons of carbon dioxide in the atmosphere...The plan, the *Yasuni-ITT Initiative*, asked the international community to replace 50% of the estimated market value of the oil with donations that would be administered by the UN Development Program for social development and investment in alternative energy projects to strengthen Ecuador's economy and allow the world to move forward on a global environmental policy that could be a blueprint for a changing awareness of what is necessary to maintain a global environmental balance...This one area on Earth has the highest concentration of insect species in the world, more woody-tree species than in all of North America, over 500 species of birds, nearly 30 species of endangered vertebrate and about 30% of all the reptile and amphibian species found in the Amazon...There were serious risks with this plan in a nation with political instability and uncertain governance but it could have been a first step in educating other nations to recognize the importance of limited resources and sensitive areas where careless, destructive activity could have long-term irreversible effects impossible to offset with future policies...

In August 2013, President Rafael Correa approved oil drilling in the Ishpingo-Tambococha-Tiputini area of Yasuni National Park and canceled the trust fund (The Yasuni-ITT Initiative) that would have compensated the country for leaving the oil underground...Oil extraction operations are already scattered across more than half of Yasuni National Park...1% of the Yasuni Biosphere Reserve amounts to 20,000 hectares and each of those hectares houses more than 100,000 species of animals, many unique to this area and found nowhere else on Earth...

In the south of Ecuador there's more than $200 billion in copper and gold ore slated for extraction through "environmentally sound" mountain-top removal that will devastate local populations and native species...How is it possible to remove the top of a mountain in an "environmentally sound" way?

Deforestation

The rate of deforestation in the Amazon Basin increased by more than one-third in 2013 compared with the rate of forest destruction in 2012...Primary motivation is the exploitation of resources for agriculture, mining and hydroelectric power...In almost all cases the local populations do not benefit as profits from these activities are taken out of the region leaving local cultures to cope with less water, industrial waste and agricultural pollution...

It's estimated the Amazon rainforest stores over 100 billion metric tons of carbon...Reducing Emissions from Deforestation and Forest Degradation (REDD) is a set of rules designed through the United Nations to develop projects and invest money into rain forest stabilization to reduce slash and burn deforestation that accounts for over 15% of human carbon dioxide emissions and these efforts are slowly working in South American nations containing a majority of the world's rainforests...More than 30 millions acres of Earth's rainforests have disappeared every year for the past 20 years from forest destruction and have made Indonesia and Brazil the third and fourth largest emitters of greenhouse gas globally (U.S. and China are number 1 and 2)...Brazil has lost over half a million square kilometers of forest in the last 30 years mostly due to cattle ranching to supply a growing world population and making Brazil the world's largest producer and exporter of beef in the world, over 80 million head of cattle annually...Almost 2 billion humans rely on forest products for their livelihood and saving the trees essential for human survival is a growing problem with no easy solutions...Subsistence farmers, commercial agriculture, logging and road construction are ongoing activities contributing to rainforest depletion as an exploding human population engulfs the planet...

The Amazon basin contains almost 50% of the world's remaining rainforest, a diversity of trees and plants vital to the circulatory ecosystem of the entire planet...Rainforests produce 20% of Earth's oxygen...

Industrialized countries consume over twelve times more wood products per person than non-industrialized countries...The U.S. has less than 5% of the world's population yet consumes more than 30% of the world's paper...In 2011, participating nations at the conference of the 17th United Nations Framework Convention on Climate Change made progress toward the creation of the Green Climate Fund to help mobilize $100 billion a year from private and public sources by 2020 to help developing nations adapt to climate change and convert to clean energy...Little else was accomplished and even the Green Climate Fund is just a proposal without details...

Deforestation is responsible for approximately 20% of the world's CO2 emissions...

The United Kingdom has delayed setting carbon emission targets for 2030 until they hold elections in 2016...Critics say this will make it extremely difficult to meet the UK goals on climate change...

A Warning System for Earth...*With no centralized system to monitor and report change in Earth's life-support systems, scientists in 77 nations have established GEO BON (Group on Earth Observation Biodiversity Observation Network) which integrates existing data streams into a single platform to provide a global warning system for Earth's biological and social systems...The data is structured around eight working groups focused on genetics, terrestrial species monitoring, terrestrial ecosystem change, freshwater ecosystem change, marine ecosystem change, ecosystem services, remote-sensing integration and data integration and interoperability...*

Global Warming Effects

In 2012, the United Nations Environment Program released a report highlighting 90 important environmental issues and found that meaningful progress is being made in only four areas...Lead free petrol, ozone layer depletion, access to clean water, and research on marine pollution... *"lack of action on climate change, overfishing, deforestation and a growing demand for fresh water makes humanity's path unsustainable"* the report warns *"If current trends continue, if current patterns of production and consumption of natural resources prevail and cannot be reversed, then governments will preside over unprecedented levels of damage and degradation."*... This is despite over 700 international agreements designed to tackle specific aspects of environmental decline and agreements on alleviating poverty and malnutrition...*"Population growth, unsustainable consumption in western and rapid industrializing nations and environmentally destructive subsidies all need urgent action"*...*"Air pollution indoors and outdoors is believed to cause over six million premature deaths each year...Greenhouse gas emissions are on track to warm the world by an average of three degrees Celsius by 2100...The majority of river basins contain places where drinking water standards are below WHO standards and only 1.6% of the world's oceans are protected...The last time in Earth's history when it was hotter than 2012 was 135,000 years ago based on ice core samples, fossil records and computer modeling...In 2100, if humans continue on a trajectory of global warming measured over the past decade, Earth's average temperature will be hotter than it has been for *25 million years*...The 1st World Civilization is accelerating past the two degree Celsius limit set in Brussels in 2007 and continuing to ignore the actions needed to cut worldwide emissions by up to half of 1990 levels by 2050...Earth's weather is determined by a chaos equation that make it impossible to predict individual weather events occurring year to year, but Earth's climate fluctuates in normal and long-term patterns *determined more by the pace of change than the amount of change* and that is the uncertainty recent human activity has created...This planet's climate is changing very quickly in an exponential way that will be catastrophic to many species if the unchecked burning of fossil fuels continues and there is no indication it will not continue through the 21st century...

New oil and gas from shale in America has industries talking not of the impact of burning these resources but how quickly they can produce the new gas and oil in a fracking process that requires tens of thousands of holes to be drilled in the ground with unknown impacts...India and China continue to build coal-fired power plants at a record pace with large reserves of coal available in America and Mongolia that has corporations willing and anxious to harvest and sell all of it to meet energy demands in growing third world countries who insist on internal combustion transportation and unlimited power for industry...Just one or two degrees of average global temperature could turn huge areas of the Earth used for agriculture into desert and current estimates are for an increase in average temperature of two to four degrees...Sea levels will rise, islands will disappear and many coastal cities around the world will face difficult, expensive decisions to determine how to protect or move millions of humans...Many energy experts believe even if humans stopped burning everything on the planet immediately, it would still be too little, too late to stop the unpredictable changes already set in motion and the global policies existing in this century indicate humans will continue to harvest and burn this planet for years...It will be interesting...*A sentient self-aware species burning their planet with no place else to go...*It will take all of our inventiveness, adaptability and intelligence to deal with the changing dynamics of Earth that are now inevitable in a future that is unknowable...

November 2012: The UN reports global attempts to curb emissions of CO2 are well short of what's needed to stem dangerous climate change...The Emissions Gap Report 2012 compiled by 55 scientists from 20 countries says that without action greenhouse gases in the atmosphere will be the equivalent of 58 gigatonnes of carbon dioxide per year by 2020, 14% above where they need to be in 2020 for temperature increases in this century to stay below 2C...

March 2014: The International Panel on Climate Change (IPCC) released the second of three reports on the impact of human activity on Earth's climate and it's predicted impact on civilization in the near and distance future...The document titled "Climate Change 2014: Impacts, Adaptation, and Vulnerability" emphasizes the increase in the amount of new scientific evidence on warming impacts available in 2014 (twice that of 2007) with some 12,000 scientific studies making up the basis of this report...The reports says climate change is evident in every area of the world with sea ice melting, coral reefs disappearing, negative impacts on fresh water and acidity rising in the ocean leading to the elimination of some fish species and the migration of many more...The continuing unchecked pollution will lead to a growing food insecurity with crop yields that could drop more than 25% by 2050 leading to new "pockets" of poverty worldwide as frequent flooding and drought will be catalysts for the migration of entire populations with conflicts growing between nations competing for resources in a world growing more crowded (9 billion humans by 2050) and desperate...Over 600 million humans will be directly affected by rising sea levels and the report cited a World Bank estimate of over $100 billion yearly (per nation) to help developing nations in dealing with the devastating effects of climate change...The authors admit almost no real progress has been made globally after more than 20 years of international efforts to limit greenhouse gas emissions...These reports by IPCC were commissioned by 195 nation states representing almost every government on the planet to inform policy and decisions impacting climate change...

Something rarely spoken of is the proliferation of millions of internal combustion convenience devices (mowers, leaf blowers, trimmers, etc.) used by humans with no filters or regulatory restrictions on exhaust fumes...All these devices should be powered by batteries charged by power plants with 21st century technology of the highest standard...

James E. Lovelock "Gaia Theory" states that the Earth itself, with its atmosphere, geology and biological organisms, is a self-regulating system maintaining the conditions that allow it's perpetuation...He suggests that the planet as a whole could be viewed as a single living organism...

THE 9 BILLION NAMES OF GOD

Jeebo, Jengu, Waaq, Eostre, Frige, Frigg, Ingui Fréa, Tupa, Seaxneat, Thor, Tiw, Wéland, Wóden, Brekyirihunuade, Kwaku Ananse, Anansi. Asase. Ya, Bia, Nyame, Altjira, Bagadjimbiri, Bajame, Bamapana, Bobbi-Bobbi, Bunjil, Daramulum, Dilga, Djanggawul, Eingana, Galeru, Gnowee, Julana, Juunggul, Karora, Kidili, Kunapipi, Mangar-kunier-kunja, Numakulla, Pundiel, Ulanji, Walo, Wawalag, Wuriuppranili, Yurlungur, Ayya, Vaikundar, Sivan, Nathan, Thirumal, Arumukan, Chalchiuhtlicue, Cinteotl, Coyolxauhqui, Ehecatl, Huehueteotl, Huitzilopochtli, Ilamatecuhtli, Itztlacoliuhqui-Ixquimilli, Mayahuel, Mictlantecuhtli, Ometeotl, Quetzalcoatl, Tepeyollotl, Tezcatlipoca, Tlahuizcalpantecuhtli, Tlaloc, Toci, Toncatecuhtli, Xipe, Totec, Xochipilli, Xochiquetzal, Bahaullah, Buddha, Jesus, Krishna, Bangputtis, Diwas Dievas, Kurche, Laima, Mahte, Mara Potrimpos, Zeme, Abellio, Agrona, Alaunus, Fin, Ambisagrus, Ancamna, andarta, Anextiomarus, Artio, Aveta, Belatu-Cadros, Belenus, Belisama, Borvo, Brigit, Brigantia, Camma, Camulus, Cernunos, Cissonius, Cocidus, Condatis, Coventina, Dagda, Damara, Eduardo, Epona, Esus, Fagus, Glanis, Grannus, Gwydion, Loucetios, Lugh, Lyr, Manannan, Maponos, Morrigan, Nantosuelta, Nemain, Nemetona, Nodens, Nuadha, Ogma, Rhiannon, Robur, Rosmerta, Rudianos, Segomo, Sirona, Smertios, Sucellus, Sulis, Tammesis, Taranis, Toutatis, Cai-Shen, Chang'e, Guan Yin, Guan Yu, Jade Emperor, Matsu, Shangdi, Tian, Tu Di Gong, Zao Jun, Na Tuk Kong, Na Tok Kong, Tnee Kong, Teh Choo Kong, Tu Di Gong, Matsu, Choy Sun Yeh, Kwan Kong, Kwan Yin Ma, Di Zhu, Gao Yao, God The Father, El, Yahweh/Jehova, God the Son (Jesus, Yeshua), God the Holy Spirit (Spirit of Christ/God or Holy Ghost), Bendis, Gebeleizis, Zamolxis, Age, Ayaba, Da, Gbadu, Gleti, Gu, Lisa, Loko, Mawu, Nana Buluku, Sakpata, Sogbo Xevioso, Zinsi, Zinsu, Eris, Abassi, Atai, Amun, Anubis, Apep, Atum, Bast, Geb, Hapy, Hathor, Heget, Horus, Imhotep, Isis, Khepry, Khnum, Maahes, Ma'at, Menhit, Mont, Naunet, Neith, Nephthys, Nut, Osiris, Ptah, Ra, Sekhmet, Sobek, Set, Shu, Taweret, Tefnut, Thoth, Wepwawet, Peko, Pikne, Tharapita, Vanemuine, Uku, Ilmarine, Alpan, Aplu, Menrya, Nethuns, Tinia, Turan, Uni, Voltumna, Ahti, Jumala, Loviatar, Mielikki, Otso, Pekko, Perkele, Rauni, Tapio, Tuonetar, Adonis, Aphrodite, Apollo, Ares, Artemis, Athena, Chaos, Cronus, Demeter, Dionysus, Enyo, Eos, Eris, Eros, Gaia, Hades, Hebe, Hekate, Helios, Hephaestus, Hera, Heracles, Hermaphroditus, Hermes, Hestia, Hypnos, Pan, Persephone, Poseidon, Selene, Thanatos, Ouranos, Zeus, Aeons, Archons, Christ, Sophia, Yaldabaoth, Yao, Saklas, Samael, the Demiurge, Abaangui, Jurupari, Vishnu, Shiva, Brahman, AdiShakti, MahaSaraswati, MahaLakshmi, MahaKali, Brahmâ, Lakshima, Parvati, Vishnu, Garuda, Ananta, Shesha, Narada, Dattatreya, Havagriva Nataraja, Dakshinamurti, Mahadeva, Hanuman, Parvat, Ardhanarishwar, Ganesh Kartik, Veer Bhadra, Ayyappa, Nandi, Sati, Annapurna, Lalita, Shashti...

- The Warrior Manifestations of Parvati
  - Kali - The Goddess of Time and Death
    - The Ten Great Wisdom Manifestations of Kali
      - Kali - The Goddess as Time
      - Tara - The Goddess as Space
      - Chinnamasta - The Goddess as The Cycle of Life and Death
      - Bhuvaneshvari - The Goddess as Perfection
      - Tripura Sundari - The Goddess as the Most beautiful
      - Bhairavi - The Goddess as the Most frightful
      - Bagalamukhi - The Crane headed Goddess as upholder of Universal Order
      - Dhumavati - The Widowed Goddess as Chaos and Misery
      - Matangi - The Goddess as Leftovers and Salvage
      - Kamala - The Goddess as Perfection

- Durga - The Goddess of Power and War
  - The Nine Manifestations of Durga
    - Shailaputri
    - Brahmacharini
    - Kushmanda
    - Skanda Mata
    - Katyani
    - Chandraghanta
    - Siddhi Dhatri
    - Maha Gauri
    - Kaal Ratri
  - Maya - The Goddess of Illusion and Mystery
- The Adityas
  - Indra - god of weather and war
  - Mitra - god of honesty, friendship and contracts
  - Ravi, Surya - the Sun gods
  - Varuna - god of the oceans and rivers
  - Yama - god of death

Some of the most important <u>Devas</u>:

- Agni - god of fire
- The Asura - Demons, Anti gods
- The Aswini - gods of sunrise and sunset
- Dyaus-pitar - ('Heaven-father') cognate of the Roman god Jupiter
- Ganesh - personified with the head of an elephant, god of wisdom, intelligence, education and prudence
- Parjanya - god of wind
- Parvati or Parvathi, wife of Shiva
- Prithivi - the Earth goddess
- Purusha - the Cosmic-Man
- The Rudras - the storm deities
- Soma - the lunar deity
- Ushas - The goddess of sunrise
- Vasus
- Vayu - god of wind
- The Visvedevas
  - Ishvara - One who gives prosperity.
  - Hari - One who destroys sins (obstacles on the way to Moksha (liberation from the cycles of birth-death-birth)).
  - Narayana - The final destination towards which all individual souls are traveling.

Igbo

- Aha Njoku
- Ala
- Chukwu

Incan

- Apo - mountain god
- Apocatequil - god of lightning
- Chasca - goddess of dawn, twilight, and Venus
- Coniraya - moon god
- Ekkeko - god of hearth and wealth
- Illapa - weather god
- Inti/Punchau - sun god
- Kon - god of rain and wind from the south
- Mama Allpa - multi-breasted fertility goddess
- Mama Cocha - goddess of sea and fish
- Mama Pacha - dragoness fertility goddess
- Mama Quilla - moon goddess
- Mama Zara - goddess of grain
- Manco Capac
- Pacha Kamaq - creator god
- Pariacaca - water god
- Supay - god of death
- Urcaguary - god of metal and jewels
- Viracocha - creator of civilization
- Zaramama

Islamic

- Allah (Monotheism).

Isoko

- Cghene

Jehovah's Witnesses

- Jehovah

Judaic

- Adonai/Ehyeh-Asher-Ehyeh/El/Elohim/Shaddai/Shalom/Yah/YHWH/YHWH Tzevaot (God) ("YHWH" is often transliterated "Jehovah" or "Yahweh", but only by people outside of Jewish tradition)

Khoikhoi

- Gamab
- Heitsi-eibib
- Tsui'goab

Korean

- Dangun - the grandson of the god of heaven.
- Hwanin - the grandson of Hwang-gung, one of the Four Men of Heaven and considered a direct ancestor of the Korean people.

Xũ

- Mantis
- Prishiboro

Lotuko

- Ajok

Latvian

- Auseklis
- Dēkla
- Dievs - a god (the name was later used for the Christian god)
- Jumis
- Kārta
- Laima
- Māra
- Pērkons
- Saule
- Ūsiņš

Lugbara

- Adroa
- Adroanzi

Lusitani

- Ares Lusitani
- Atégina
- Bandonga
- Bormanico
- Cariocecus
- Duberdicus
- Endovelicus
- Geleshus
- Mars Cariocecus
- Nabia
- Nantosvelta
- Runesocesius
- Sucellus

- Tongoenabiagus
- Trebaruna
- Turiacus

Lydian

- Annat
- Anax
- Artimus
- Asterios
- Atergätus
- Atys
- Baki.
- Bassareus
- Damasēn
- Gugaie/Guge/Gugaia
- Hermos
- Hipta
- Hullos
- Kandaulēs
- Kaustros
- Kubebe
- Lamētrus
- Lukos
- Lydian Lion
- Mēles
- Moxus
- Omfalē
- Pldans

Maya

- Ahaw Kin - Sun God
- Bacabs - Gods of the 4 directions
- Balac - War God
- Balam - Protector God
- Bolon tza cab - Ruling God of All
- Chaac - Rain God
- Hunah Ku - Creator God
- Itzamna - Reptile Creator God
- Ix Chel - Moon Goddess
- Kukulcan - Feathered Serpent God
- Xbalanque - God of the Jaguar
- Xi Balba - God of the Death
- Yum Kaax - Corn God

Mesopotamian

- Anshar - father of heaven
- Anu - the god of the highest heaven
- Apsu - the ruler of gods and underworld oceans
- Ashur - national god of the Assyrians, thought by the Assyrians to be king of the gods
- Damkina - Earth mother goddess
- Ea - god of wisdom
- Enlil - god of weather and storms
- Ereshkigal - Goddess of Darkness, Death, and Gloom
- Hadad - weather god
- Ishtar - goddess of love and one of the highest-ranking deities in Mesopotamian myth
- Kingu - husband of Tiamat
- Kishar - father of the earth
- Marduk - national god of the Babylonians, later thought to be king of the gods
- Mummu - god of mists
- Nabu - god of the scribal arts
- Nintu - mother of all gods
- Ninurta - god of war
- Nergal - god of war, disease, death and destruction; ruler of the underworld
- Shamash - god of the sun and of justice (Shapash in Ugaritic, Shamsa in Sumerian)
- Sin - moon god
- Tiamat - dragon goddess slain by Marduk

Mormonism

- Heavenly Father
- Jesus Christ
- The Holy Spirit

Native American

Abenaki

- Azeban - trickster
- Bmola - bird spirit
- Gluskab - kind protector of humanity
- Malsumis - cruel, evil god
- Tabaldak - the creator

Haida

- Gyhldeptis
- Lagua
- Nankil'slas
- Sin
- Ta'axet
- Tia

Ho-Chunk

- Kokopelli

Hopi

- Aholi
- Angwusnasomtaka
- Kokopelli
- Koyangwuti
- Muyingwa
- Taiowa

See also: kachina.

Huron

- Airesekui
- Heng
- Iosheka

Inuit

- Igaluk - lunar deity
- Nanook - master of bears
- Nerrivik - sea mother and food provider
- Pinga - goddess of the hunt, fertility, and medicine
- Sedna - sea goddess, ruler of the underworld
- Torngasoak - sky god

Iroquois

- Adekagagwaa
- Gaol
- Gendenwitha
- Gohone
- Hahgwehdaetgan
- Hahgwehdiyu
- Onatha

Kwakiutl

- Kewkwaxa'we

Lakota

- Canopus
- Haokah
- Whope
- Wi

Navajo

- Ahsonnutli
- Bikeh Hozho
- Estanatelhi
- Glispa
- Hasteoltoi
- Hastshehogan
- Tonenili
- Tsohanoai
- Yolkai Estasan

Pawnee

- Pah
- Shakuru
- Tirawa

Salish

- Amotken

Seneca

- Eagentci
- Hagones
- Hawenniyo
- Kaakwha

Norse

- Balder - god of beauty and light, slain by the trickery of Loki
- Bragi - god of bardic poetry
- Freyja - goddess of fertility
- Freyr - the brother of Freyja and a fertility god
- Frigg - goddess of marriage, household management, and love, Queen of Heaven, and wife of Odin
- Heimdall - god of the rainbow, a bridge to heaven. His job is to blow his horn if danger approaches.
- Hel - daughter of Loki and the giantess Angrboda, Queen of the Dead
- Hodur - brother of Balder and tricked by Loki to kill him
- Idunn - guardianess of the Apples of Youth that kept the gods young
- Loki - trickster-god, giant, blood-brother of Odin, will eventually lead the forces of evil against the gods in Ragnarok
- Niord - god of sailors and fertile seaside land
- Odin - king of the gods, god of wisdom and runes
- Sif - the wife of Thor
- Thor - god of war and storms, famous for his hammer, Mjolnir
- Tyr - god of war and glory

Philippine

- Amanikable - God of Hunters.
- Amihan - North Wind.
- Anitan - Guardian of lightning.
- Anitun Tabu (Anitong Tao) - Goddess of wind and rain.
- Apolake - God of war, guardian of the sun.
- Bakonawa - Celestial Serpent, devourer of sun and moon.
- Bathala - Supreme god of the ancient Tagalogs.
- Dian Masalanta - Goddess of love.
- Hukluban - Goddess of death.
- Idianale - Goddess of agriculture and husbandry.
- Ikapati/Lakan Pati - Goddess of fields, fertility, and lands.
- Kalinga - God of Thunder.
- Kan-Laon - Ancient Visayan god, king of time.
- Lalahon - Goddess which resides in Mt. Kanlaon. Bringer of famine if unappeased.
- Manggagaway - Goddess of sickness.
- Mangkukulam - God of fire.
- Manisilat - God of broken homes.
- Maria Makiling - Protector of Mt. Makiling.
- Mayari/Bulan - Lunar goddess.
- Magwayen - Ferryman of the dead.
- Tala - God/Goddess of the stars.
- Mandangan- God of War
- Mabuyan/ Bai Bulan - Godess of the underworld.

Polynesian

Hawaiian

- Atea
- Ina
- Kane Milohai
- Lono
- Maui
- Papa
- Pele

Māori

- Haumia-tiketike - deity of uncultivated food, especially fern root
- Hine-nui-te-pō - deity of death
- Hine Tītama - deity of the dawn
- Hine Raumati - deity of the summer
- Papa-tū-ā-nuku - deity of the earth
- Rā - deity of the sun
- Ranginui - deity of the sky

- Rongo - deity of cultivated food
- Ruaumoko - deity of earthquakes
- Tāne Māhuta - deity of the forest and its creatures, man
- Tangaroa - deity of the sea and its creatures
- Tāwhiri-matea - deity of the weather, especially storms
- Tū Matauenga - deity of war
- Uenuku - deity of rainbows

Prussian

- Ukapirmas - the chief god, creator of the world
- Perkūns / Perkuno - the god of thunder
- Pikullos (Pikuls; Pickollo) - god of war and death (in Christian times, he was vilified as the devil)
- Kaūks - a deity
- Swāikstiks / Suaixtix / Swaixtix - a deity
- Dēiwas - a god (in Christian times, the name was applied to the Christian god)
- Zempat - god of the earth and of cattle

Pygmy

- Arebati
- Khonvoum
- Tore

Roman

- Acis - river god near the Etna, son of Faunus and the nymph Symaethis
- Aesculapius - god of health and medicine
- Apollo - god of the sun, poetry, music, and oracles, and an Olympian
- Aurora - goddess of the dawn
- Bacchus - god of wine and sensual pleasures, not considered an Olympian by the Romans
- Bellona - war goddess
- Caelus - god of the sky
- Carmenta - goddess of childbirth and prophecy, and assigned a Flamen Minore
- Ceres - goddess of the harvest and mother of Proserpina, and an Olympian, and assigned a Flamen Minore
- Consus - chthonic god protecting grain storage
- Cupid - god of love and son of Mars and Venus
- Cybele - earth mother
- Diana - goddess of the hunt, the moon, virginity, and childbirth, twin sister of Apollo and an Olympian
- Discordia - goddess of discord
- Fama - goddess of fame and rumor
- Faunus - god of flocks
- Febris - goddess who prevented fever and malaria
- Flora - goddess of flowers, and assigned a Flamen Minore
- Fortuna -goddess of fortune

- Hercules - god of strength
- Hespera - goddess of dusk
- Hora - Quirinus' wife
- Janus - two-headed god of beginnings and endings and of doors
- Juno - Queen of the Gods and goddess of matrimony, and an Olympian
- Jupiter - King of the Gods and the storm, air, and sky god, and an Olympian, and assigned a Flamen Majore
- Juturna- goddess of springs
- Juventas - god of youth
- Libitina - goddess of the underworld
- Lucina - goddess of childbirth
- Luna- moon goddess
- Lupercus - god of shepherds
- Mars - god of war and father of Romulus, the founder of Rome, and an Olympian, and assigned a Flamen Majore
- Mercury - messenger of the gods and bearer of souls to the underworld, and an Olympian
- Minerva - goddess of wisdom and war, and an Olympian
- Mithras - god of Mithraism, a separate religion
- Morpheus - god of dreams
- Nemesis - goddess of revenge
- Neptune - god of the sea, earthquakes, and horses, and an Olympian
- Orcus - a god of the underworld and punisher of broken oaths
- Pluto - King of the Dead
- Poena - goddess of punishment
- Pomona - goddess of fruit trees, and assigned a Flamen Minore.
- Portunes- god of keys, doors, and livestock, he was assigned a Flamen Minore.
- Priapus - god of fertility
- Proserpina - Queen of the Dead and a grain-goddess
- Quirinus - Romulus, the founder of Rome, was deified as Quirinus after his death. Quirinus was a war god and a god of the Roman people and state, and was assigned a Flamen Majore.
- Saturn - a titan, god of harvest and agriculture, the father of Jupiter, Neptune, Juno, and Pluto
- Silvanus - tutelary spirit of woods
- Sol Invictus - sun god
- Somnus - god of sleep
- Suadela- goddess of persuasion
- Terra - goddess of the earth and land
- Terminus - the rustic god of boundaries
- Trivia - goddess of magic
- Venus - goddess of love and beauty, mother of the hero Aeneas, and an Olympian
- Vesta - goddess of the hearth and the Roman state, and an Olympian.
- Victoria - goddess of victory
- Volturnus- a god of water, was assigned a Flamen Minore.
- Voluptas - goddess of pleasure
- Vulcan - god of the forge, fire, and blacksmiths, and an Olympian, and assigned a Flamen Minore

Sardinian

Sardinian deities, mainly referred to in the age of Nuragici people, are partly derived from Phoenician ones.

- Janas - Goddesses of death
- Maymon - God of Hades
- Panas - Goddesses of reproduction (women dead in childbirth)
- Thanit - Goddess of Earth and fertility

Semitic pagan

- Adad
- Adonis
- Amurru
- An/Anu
- Anat
- Anshar | Asshur
- Asherah
- Astarte
- Beelzebub
- Baʿal/Hadad
- Dagon
- El
- Enki/Ea
- Enlil
- Ereshkigal
- Inanna/Ishtar
- Kingu | Kishar
- Lahmu & Lahamu
- Marduk
- Mot
- Mummu
- Nabu
- Nammu
- Nanna/Sin
- Nergal
- Ninhursag/Damkina
- Ninlil
- Tiamat
- Utu/Shamash
- Yaw

Gods of Ur of the Chaldeans (only appearing in the LDS Book of Abraham)

- Elkenah
- Libnah
- Mahmackrah
- Korash

Shinto

- Aji-Suki-Taka-Hiko-Ne - god of thunder
- Amaterasu - sun goddess
- Ama-no-Uzume - fertility goddess
- Amatsu Mikaboshi - god of evil
- Chimata-No-Kami - god of crossroads, highways and footpaths
- Ho-Musubi - god of fire
- Inari (mythology) - god of rice
- Izanagi - creator god
- Izanami - creator goddess
- Kagu-tsuchi - god of fire
- Kura-Okami - god of rain
- Nai-No-Kami - god of earthquakes
- O-Kuni-Nushi - god of sorcery and medicine
- O-Wata-Tsu-Mi - god of the sea
- Sengen-Sama - goddess of the Mt. Fujiyama
- Seven Gods of Fortune
    - Benzai-ten or Benten - goddess of money, eloquent persuasion, and knowledge
    - Bishamon-ten - god of happiness and war
    - Daikoku-ten
    - Ebisu (also spelled Yebisu)
    - Fukurokuju
    - Hotei-osho
    - Jurojin
- Shina-To-Be - goddess of wind
- Shina-Tsu-Hiko - god of wind
- Sojobo - king of the tengu
- Susanoo - god of storms and thunder, snakes and farming.
- Taka-Okami - god of rain
- Take-Mikazuchi - god of thunder
- Tengu - minor trickster deities
- Tsukuyomi - god of the moon
- Uke-Mochi - goddess of food
- Wakahiru-Me - goddess of the dawn sun

Sikhism

- Waheguru.

Slavic

- Belobog - god of light and sun (speculative)
- Berstuk - evil god of the forest
- Cislobog - moon goddess
- Crnobog - god of woe (speculative)
- Dajbog - sun god
- Dziewona - equivalent of Diana

- Flins - god of death
- Hors - god of the winter sun
- Jarilo - god of vegetation, fertility, spring, and associated with war and harvest
- Juthrbog - moon god
- Karewit - protector of the town of Korzenica
- Lada and Lado
- Marowit - god of nightmares
- Perun - god of thunder and lightning
- Podaga - weather god, and god of fishing, hunting and farming
- Porewit - god of law, order and judgment
- Radegast (god) - possibly a god of hospitality, fertility and crops
- Rugiwit
- Sieba - fertility and love goddess
- Siebog - god of love and marriage
- Stribog - god and spirit of the winds, sky and air
- Svarog - god of fire
- Svetovid - god of war, fertility and abundance
- Triglav - three-headed god
- Veles - god of earth, waters, and the underworld
- Zirnitra - dragon god of sorcery

Sumerian

- An
- Enki
- Enlil
- Inanna
- Nammu
- Nanna
- Ninhursag
- Ninlil
- Sin
- Utu

- Zibelthiurdos

Ugarit

- Anat - war goddess
- Athirat - mother goddess
- Athtart - goddess of fertility
- Dagon - god of earth
- El - the father god and head
- Haddu - commonly titled Ba'l, a rain god and storm god.
- Mot - the god of death
- Yam - the god of the sea

A simple internet search for God...God's are ubiquitous...Every society in human history has had many gods for many things or one God for everything...Gods are constructs created by a connective, developing, self-aware intelligence...Gods are not real, but they are very human...

Religion is the product of a developing intelligence in a hostile world...

All the stories, myths, legends, gospel and writings that have tried to explain and understand human reality since the beginning of human history are historical, entertaining, insightful, contradicting and necessary progressions in the developmental evolution of human awareness and human intelligence and throughout recorded history these writings on philosophy, religion and spirituality are equally brilliant explorations of the nature of human existence or just plain silly...It's easy in retrospect to wonder how Greeks might have believed in the gods we now know as the Greek myths, but 2,500 years ago humans did not understand laws of nature or the laws of science known today...One definition of intelligence is a need to understand and explain the nature of the reality each intelligence experiences through sensory input and conceptual thought and the existence of gods, demons and the many variations of speculative supernatural phenomenon were rational constructs to a curious emerging species seeking to explain the forces of nature and understand the mystery of existence...Life was a mystery without scientific order and life is still a mystery today but the questions are bigger, stranger and structurally organized around an evolved intelligence and awareness stretching billions of light years in all directions...The human species has been looking out from Earth and asking these existential questions for thousands of years and the telescope, a fundamental tool of human exploration and analysis, was only invented some 400 years ago...Since that time the advancement of science and unbounded human curiosity has proven to our species the nature of our existence in three-dimensional space and the natural wonder and real mystery of a reality that is now definable and increasingly understood...Humans now *know* the answers to many of the questions asked by every civilization throughout time and it's changed the way intelligence explores this continuing question of existence...

Almost no human being on Earth today believes the Greek gods actually existed and all religions on Earth will someday suffer the same fate...

Humans are tiny biological creatures living on a rock circling a black hole with billions of stars, comets, planets and moons, everything moving incredibly fast in a mysterious emptiness growing larger every second, a universe with a chaotic and violent beginning in unending motion that will continue to 'grow' for trillions of years...The *why* of human existence and the *why* of existence itself is the essential question our species will explore for the next billion years and the reality of what might be possible is mostly unknowable at this time simply because the human race is so young...Is there a vast intelligence beyond the boundaries of human awareness and beyond all human science that understands the reality of this 'eternity'? Are there unknown multiple dimensions of existence in exotic possibilities not yet discovered, ways to travel faster than the speed of light, ways to teleport matter, cross-dimensional travel, molecular re-cohesion or modification that could extend a human lifespan thousands of years?

The answers to these questions are unknown but it's clear the reality of human existence and the why of everything is a much bigger question than any religion ever imagined on Earth...Today it's increasingly clear that all the religions, myths and fantasy which have influenced and dominated human culture and civilizations since the beginning of human history are simply the result of a self-aware, sentient species exploring and recognizing the reality of their own existence...Human stories of man's 'relationship' with deities and gods from every culture, every civilization and every religion are a remarkable mixture of memories, history, metaphor, illusion, dreams, story-telling, observable reality, unexplained phenomenon and the need to understand *why* we exist...Just to understand *where* humans exist in physical space and time has taken thousands of years of discovery and it's unknown if we can ever know why this universe exists or the nature of reality beyond the Big Bang but it's clear in this century, after a chaotic, unscripted, creative, violent and enlightened 200,000 year history of human expansion and occupation, it's clear that all of Earth's religions are the invention of man...

The famous philosopher Descartes who tried to understand human reality without God still believed in God and how could he not, knowing something existed somewhere beyond Earth...Still, he believed physical phenomena must be explained by the laws of nature and God could not change nature after creating these laws in the beginning...

An advanced alien race visiting this planet today would view every church, every temple, every mosque, every single religious structure on Earth as a clear indication of the level of human intellectual development, a planet full of monuments reflecting dogma, ignorance and the absence of a shared intellectual awareness on an overpopulated world consumed with selfish ideologies and violence...At this time in the history of human evolution, with war, widespread ignorance, millions of babies dying, children neglected and discrimination, greed and abuse on a planet-wide scale, it's clear that the many idealizations of a deity or deities and their imagined involvement with life on this world are without foundation...There is no evidence for even a single deity, no miracles, no resurrections, no spells, no witchcraft and no magic...The reality of human religion is clear to all intelligent creatures but the question of eternity and the reality of origins and existence are questions that might never be fully answered...*How can answers be known to questions this big?* How does life exist and why does life exist and what will existence be like in a trillion years of unending expansion? Are such vast, endless time scales possible and how can they not be possible? Where does physical reality end on the scale of the very large or the impossibly small? Humans exist in a continuum of size and spatial awareness that has no solution in either direction...How can this be? Think of the speed of light, the speed limit of the universe...How can there be a speed limit if we don't exist in a reality of laws and restrictions and why are there laws and restrictions? How can there be forever and how can there not? As a species humans can no longer attach superstitious meaning or superficial reasoning to these ancient questions...We must wait for answers...It's likely our species will continue to make scientific progress for a billion years and still not know what exists beyond our measurable, observable reality...Or will we? A sentient species who seek to know the reality of their existence will not accept conjecture without proof...Humanity will continue to ask these ancient questions, explore their reality, observe, measure, survive and wait...

There is no mystery of God because there is no evidence of God...

The North Carolina Constitution states: "The following persons shall be disqualified for office: First, any person who shall deny the being of Almighty God." Arkansas, Maryland, Pennsylvania, South Carolina, Tennessee and Texas have similar provisions in the constitutions of their states...

The Reality of Religion
(in no particular order)

MONESTARY OF THE ANGELS - A group of Nuns who trace their history back 800 years...The response to why they don't actively help the needy? "We sit and pray...God will help the needy"

"For most of it's existence, Christianity has been the most intolerant of the world faiths doing its best to eliminate all competitors with Judaism a qualified exception" Diarmaid MacCulloch "Christianity"

In 1996, Pope John Paul II declared evolution was "more than just a hypothesis"...

CATHOLIC CHURCH – A religious organization that traces it's history back 2000 years with a global following estimated at over a billion humans and a single supreme leader...The recent history of this organization has been a revelation of secrecy, abuse, intolerance and injustice on a global scale...

The only way so many abuse cases can be from so many different places is if the abuse has been ingrained for centuries, a long standing culture and history of sexual abuse on woman and children that has continued unquestioned for over a 1,000 years...New evidence of this systemic abuse in nearly every nation with a Catholic hierarchical structure has been uncovered recently due to the emergence of technology, the beginning of a world civilization and a global spotlight impossible to avoid when an institution has such a sordid history cloaked in deliberate secrecy...In the 21st century the bureaucracy of this church is using every defense possible to turn the spotlight away from the reality of their history but their attempts are weak and transparent with 'leaders' and policies entrenched in the Vatican hoping beyond hope the masses will be too unaware to sort things out and believers will continue to support waste, excess and wealth without reason...With this knowledge of the church and a history that is unforgivable by any standards, no intelligent human would support the ideas of this organization but over a billion members of the Catholic faith still believe that these men, who wear costumes and molest innocent children while hiding a culture of the worst kind and destroying lives without remorse, these men are the chosen representatives of God...An all-powerful, all-knowing deity with the best interests of all of Earth's men, women and children in mind...Imaginary nonsense...

In 2013, the new pope Francis enraptured the world with his humility and concern for the poor while maintaining that women are inferior and have no place on the dais, reaffirming that birth control and a female's right to control her own health are always secondary to God's ordained role for women (subservient and obedient to man) and that all homosexual acts are unacceptable, "a moral disorder" (including the acts forced upon children with no voice by trusted representatives of the church) while stating that being homosexual is ok ("who am I to judge?")...Pope Francis seems to understand in some kind of way the diversity, the commonality of all humans and the negative impact of unequal resource distribution on the extreme scales currently experienced in almost every nation on Earth and he talks of a global waste culture without limit, the perversity of wealth accumulation driven by primitive instincts and a lack of awareness or responsibility on a single world of 7.4 billion humans where over two billion people have no access to sanitation and many millions go hungry every day...Yet his refusal to judge 'homosexuals' does not prevent him from judging 'females' every day convinced that they are inferior to males, unsuitable for the responsibilities his imaginary God has assigned only to males and obligated to be subservient to males in any and all situations continuing the same ignorance of reality with the same self-serving rhetoric that has been the foundation of this organization built upon a myth...

In a recent tour of the United States, Pope Frances continued to treat women as 'pets', patting their head and telling them what a 'good job' they are doing, essential to the church, but you are still just women...He is incapable of recognizing that men and women are equal while discouraging birth control on an overcrowded planet...

In 2010, over 60% of American Catholics polled stated they still viewed the pope in a favorable light which represents over 65 million Catholics in America with many variations of what they choose to believe in...The Catholic Church has criticized the removal of funding in it's effort to help sex trafficking victims but is steadfast in refusing to refer these abused women to health organizations who provide birth control...(Throughout the history of the church a woman's health and well-being has always been secondary to the Catholic 'rules of God')...In April 2012, after four years of investigation by the 'Congregation of the Doctrine of the Faith', the Vatican admonished the largest organization of nuns in America (The Leadership Conference of Women Religious) saying they promoted "radical feminist themes" by challenging the church on homosexuality, contraception, male-only priesthood and accused the nuns of spending too much time on 'poverty' and 'economic issues' and not enough time voicing concerns of the Vatican's social morals...In keeping with the mantra 'the pope's word is final' the Vatican appointed a man (archbishop) to rewrite the rules for the sisters and approve or reject every single item, every project, every speaker, all literature and any statements at his sole discretion...The sisters who head this organization criticized these invasive actions and it's a rare sign of independence by women who give their lives helping others who are less fortunate creating some hope that this ripple could eventually challenge the authority this small group of men believe they have over millions of humans... 'The Catholic Committee on Laity, Marriage, Family Life and Youth' has looked into the Girl Scouts of America for "problematic relationships with other Organizations" criticizing their links with groups like Doctors Without Borders, the Sierra Club and Oxfam because they support family planning and emergency contraception...In 2013, Pope Benedict resigned (the 1st resignation in almost 600 years) with a statement "my strengths, due to an advanced age, are no longer suited to an adequate exercise of the ministry"...The backdrop of his resignation is a Vatican riddled with financial inconsistencies, sexual misconduct at the highest levels and a hierarchical system that's been broken and ineffective for many years...The Catholic church is a religion trapped in a myth of their own making supported by huge sums of money from the very poor and the powerful exhibiting all the symptoms of a religious organization that is primitive, redundant and hopelessly mired in policies of the past, dedicated and beholden to an ideology from the Dark Ages...

In 1854, Pope Pius IX revealed that Mary, the mother of Jesus, was conceived without sin...

In a speech in Aparecida, Brazil in 2007, the 80 year old pope defended the Catholic Church's bloody campaign to Christianize indigenous people...

In one of his first actions since becoming pope in 2013 Pope Francis released a statement supporting the church's crackdown on the Leadership Conference of Women Religious, an umbrella group of U.S. nuns who help the poor and needy...The nuns have been accused of "radical feminist themes incompatible with the Catholic faith." and are being disciplined and restructured by a group of old men including Archbishop Gerhard Mueller who insists that a 'program of reform' is necessary to keep these women in their place...In the latest effort to stop free thinking women from doing what is needed and necessary to improve the lives and circumstances of the world's oppressed, the Vatican appointed Seattle Archbishop Peter Sartain and two other bishops to oversee a rewriting of the conference's statutes, to review its plans and programs, approve speakers and ensure the group properly follows Catholic prayer and ritual in an ongoing effort to control their females and keep them subservient the way God intended...A group of 60,000 women being restricted, supervised, admonished and controlled by a small group of self-appointed men...

In 2014, the UN committee on the Rights of the Child (CRC) denounced the Catholic church for adopting policies the allowed priests to sexually abuse thousands of children with concealment and collusion among the Catholic hierarchy in the Vatican to "manage" a problem strongly believed by the church to be a matter of "Canon Law" (church law) without regard to criminality or the welfare and safety of the children...The Vatican said it would "examine" the report while accusing the world of interference in church matters...

SOUTHERN BABTISTS – Founded in 1845...16 million humans...This group claims to be the largest Protestant body in the US and the largest Baptist denomination in the world...Emphasis on the International Mission Board to enable the members to "respond faster and more efficiently to disasters worldwide *as a way* of introducing their religion to non-Christians" "It helps us take the hope of the gospel to the ends of the earth, literally"...The church uses disasters as membership drives and who wouldn't join when the church approaches in the middle of a disaster with the resources to help...They support a male-only pastorate and passed a resolution to ensure the offices requiring ordination be restricted to men...They have over 40,000 churches in the U.S. and the funds collection and distribution program for 'ministry support' claimed over $10 billion in collections in 2008...

CREATION MUSEUM, *Kentucky, USA...$29.95*
"The state-of-the-art 70,000 square foot museum brings the pages of the Bible to life casting it's characters and animals in dynamic form and placing them in familiar settings. Adam and Eve live in the Garden of Eden, children play and dinosaurs roam near Eden's Rivers. The serpent coils cunningly in the Tree of the Knowledge of Good and Evil. Majestic murals and great masterpieces brimming with pulsating colors provide a backdrop for many settings."

UNITED METHODIST CHURCH – The largest mainline Protestant denomination in the U.S. with close to 8 million members in America and over 4 million in the rest of the world, mostly Africa, voted in 2012 to keep intact their proclamation in the church's "Book of Discipline" that homosexuality is "incompatible with Christian teaching" and there would be no support or tolerance for humans who are born with a different orientation...

JUDAISM – Based on Abraham (leader of the Hebrew people almost 4000 years ago) and the teaching of the Torah (the Jewish holy book) the practice of Judaism as a religion is believed by many to be the obligation of every Jewish individual but there are many variations of belief and structure even within this strict one faith view of a supreme god...There are currently around 16 million humans worldwide who identify themselves as Jewish and maintain a group identity through a core ancestry and a shared belief of history and religion...Any human on Earth can be Jewish simply by 'converting' to Judaism and accepting Jewish religion but this doesn't make one 'Jewish' to all Jews because of the radical interpretations of God's word among differing Jewish populations...Ethnically, Jewish people are identified as belonging to one three major groups, the *Ashkenazim* from central European descent (the majority of Jewish people worldwide), the *Sephardim* of Portugal and Spanish descent and the *Mizrahim* of mostly middle-eastern and African descent...The divisions, similarities, tensions, beliefs, practices, restrictions and diversity among Jewish ethnic groups are generally confusing and certainly at odds with their stated belief in one defining god of wisdom and power...The *Ashkenazi* group represents ultra-Orthodox Jews, radical, strict religious practitioners of Judaism and only the Ultra Orthodox all male Rabbatical court can give permission for a woman to divorce a man which is not allowed unless the man says 'yes' even if the husband is in a coma, in jail or missing...Ultra Orthodox Jews are very extreme in their views on religion and social morals within their community and in this misogynist, closed society women are not allowed to sing and are encouraged to marry very young and bear as many children as possible...These men restrict the education of their children and believe in a Hebrew bible that states all daughters of Eve should suffer divine punishment forever believing that woman are unclean and untouchable during childbirth and during their menstrual cycle, a biological function of more than three billion humans and billions of other mammals on Earth...Within the Ultra Orthodox community, two groups, the Ashkenazi and the Sephardic, are frequently forcibly kept separate to avoid conflict, another example of ignorance, racism and a lack of understanding in a highly discriminatory society...The Jews are not 'the children of god' or 'the chosen ones' anymore than any human on this planet...The Jewish people are a fascinating cultural group of long descending bloodlines and ethnicity who have a sad history of persecution by ignorant, misguided groups of humans with little awareness of human reality...It's impossible to discuss the Jewish state, the Jewish people, their politics or their society outside of their religion...Israel is basically a religious state without exception...

GOD, YAHWEH, ALLAH
Christians, Jews and Muslims all worship the same god, only by different names...This object of worship is vengeful, insecure, jealous, confusing, vain, angry, a war monger and a product of vivid imaginations, ancient storytelling and legends based on the occurrence of natural phenomenon in all three religions...All scientific evidence for a supreme being, if written out in any language, would not color the point of a needle...There is no evidence...

BUDDHISM – a belief system established around 2,500 years ago by Siddhartha Gautama, the son of a royal family in India who developed a philosophy called Dhamma, Enlightenment or the Truth...Questionable as a religion because there are no Gods associated with Buddhism, it's teachings are followed by over 500 million humans worldwide based on texts in three categories, the *Mahayana Sutras*, the *Abhidharma* and the *Vinaya*...A complex philosophy with many teaching and interpretations, followers believe happiness can only be attained by modifying behavior concerning desire and aversion and accepting that life is suffering and the ultimate state of happiness, *Nirvana*, can only be reached by giving up wants and desires and living life one day at a time...Siddhartha Gautama was raised as a baby in a rich environment of constant attention and most likely grew up aware of the reality of his world, the injustice to the poor and the unhappiness of the rich despite their wealth...He spent his last 50 years of life living simple, poor, teaching and practicing enlightenment and simplicity to his followers...

In 2013, archaeologists unearthed a 6th century Buddhist Shrine at Lumbini in Nepal, long believed the birthplace of Siddhartha Gautama...These remains are the earliest known evidence of Buddhism...

"You only lose what you cling to"...Siddhartha Gautama

HINDUISM – Recognized as the oldest religion on Earth based on the four Vedas, *Rig, Sama, Yajur and Atharva,* Hinduism is a deep chasm of beliefs with many gods, subdivision of 'types' and two schools, Vedanta and Yoga...Followers are allowed to believe in one god or many gods according to interpretation...The Rig Veda and the Sama Veda are collections of more than 1,000 hymns and songs subdivided into ritualistic sections that tell histories and stories of ancient societies based on the Rig-Vedic civilization...The mantras in the Veda's are some of earliest forms of Sanskrit dating back past 1300 B.C...The Yajur Veda is a liturgical guidebook for ceremonial rituals performed by Hindu priests and the Atharva Veda is a controversial collection of songs, spells and charms from a later time describing historical and societal practices...Hinduism has many gods and goddesses for many things which are all considered different lower incarnations of a single Supreme Being...The Hindu belief system is rooted in reincarnation with rebirth dependent upon past life behavior and on 'karma', human actions that produce an effect that is cyclic in always coming back to the originator of the action or 'actions'...The religion of over 800 million humans in the second most populous nation on the planet...

In 2007, Pope Benedict XVI affirmed that 'Hell' is real and it is 'Eternal'...

ISLAM - The Qur'an is the central religious text of Islam which Muslims consider the verbatim word of God (Allah) and is often misinterpreted and used by radical Islamists to spread fear and enforce their agenda (By example according to scholars, the Qur'an states that men and women are equal, a truism not accepted in any way by the majority of Muslim and Arab males)...The origin of this religion is the birth of the said prophet Muhammad in Mecca, the holiest site of Islam, in 570 A.D. to bring the word of Allah to the people...Over 1.4 billion humans are followers of the Islamic religion, the majority being Sunni Muslims totaling more than 85% of worshipers with Shiite Muslims being the second largest group...A continuing source of social stratification and tension in the Middle East is the ongoing separation and strife between these two groups who are essentially the same people...Both claim Muhammad as their prophet and both claim Allah as their God so the perceived differences and the continuous violence between Shiite and Sunni are a perfect example of two identical groups of humans with similar beliefs, different ideologies and an inability in the majority of either group to understand the reality of their existence and consequently, the benefits of a non threatening coexistence that would be profitable and peaceful for both populations...For more than 1,300 years thousands have died and continue to die over this imaginary difference which splits governments, ostracizes individuals and threatens the destruction of entire nations and their people...

The Five Pillars of Islam Central to the Islamic Religion...
Bearing witness that there is no God but Allah and that Muhammad is the Messenger of Allah...
Observance of Prayer (5x's daily while facing in the direction of Mecca)...
Paying Zakat (money every adult Muslim should pay to support specific categories of people)...
Fasting during Ramadan (the ninth month of the Islamic calendar)...
Pilgrimage to Mecca, called the 'hajj' at least once during a lifetime...

It is believed the Prophet Muhammad gave his last sermon at Mount Arafat east of Mecca over 1,400 years ago...The hajj is an experience that includes a visit to the Grand Mosque, an animal sacrifice to mark the beginning of Eid al-Adha and a symbolic stoning of the devil...

WAHHABISM – a form of Sunni Islam that is puritanical and strict in it's interpretation and application of Islamic Law, followers of this brand of Islam seek to 'purify' Islam of any practices that do not date back to the 7th century teachings of the prophet Muhammad...The state religion of Saudi Arabia, Wahhabi beliefs have dominated the educational and judicial policies of the Saudi's resulting in one of the most repressed, controlled populations in the world...Woman have absolutely no voice in this society where men rule everything and dictate all policy, another example of an Arab country whose male population is frightened of their females...Human males in this nation, who cannot understand causality or the definable reality of human existence, are scared to talk to women, interact with women, live with a woman, let women drive, let women work, let women play, a completely stagnant society with one source of income (oil) and 50% of the population unable to participate because of an ignorant fear that has frozen the male population in the 7th century...By that standard one could say their religion works...Also called Salafiyya (preferred by a majority) this is the adopted conservative Islam promoted by Al Qaeda who embrace the 7th century and wish to govern every nation under the strict application of Shari'a Law, a very violent and repressive variation of Islamic Law which advocates beheading, dismemberment, murder and the stoning of women to control behavior...Saudi Arabia has denied supporting religious extremism through education or culture but they are challenged by the facts...The majority of Muslims around the world who practice Islamic religion in a rational, progressive and beneficial way do not agree with or adopt the teachings of this extremist religion...

TAOISM – one of the great ancient religions of China, Daoism (the modern form of Taoism) dates from before the 4th century B.C. (although references go back much further in Chinese history) and embraces a philosophical view with various interpretations as a guide to mysticism, transcendent monism, ethical intuitionism and supernaturalism with an emphasis on 'living in harmony' with forces both mystic and naturalistic...This religion has a long complex interaction with all of Chinese culture, shaping industry, medicine, invention, martial arts, astronomy and astrology and has never been recognized as a singular unified religion but rather as a methodology based on teachings arising from original revelations...The cosmological beliefs of Taoism are often cited by scientists who study the fate of the universe with it's description of the two forces in play, entities of *yin* and *yang,* which promote a cyclic evolving universe that is continually reinventing itself...There are multiple deities and estimates of the followers of Taoism range from 20 million to more than 300 million...

Most religions are complex stories of philosophy, history, tradition, sacrifice, ambition, calculation and visions of an unseen afterlife as a continuance after death...

SIKHISM – the fifth largest organized religion in the world based on the teachings of ten progressive Guru's with a following of more than 30 million humans, mostly in India...Many historians believe this religion to be an amalgamation of many independent beliefs and practices, possibly from

149

Hindu and Islamic teachings, but this is disputed by Sikhs who believe Sikhism is a direct revelation of God...Founded in the 15th century by the first Guru, Sikhs believe in a single formless God with many names, 'karma' (accumulated good and bad deeds), a Code of Conduct that prohibits intoxicants and 'reincarnation' a belief in rebirth after death until God judges one worthy to sit at his side...They also reject the *caste system* believing all humans are equal, a challenge in the nation of India whose very foundation is rooted in the *caste system* (a belief that one is born into their fate)...As in many religions newer generations reject and modify the fundamental practices of older generations causing violence and dissent within the religion...The Hindus, the majority religion of India, clash with the Sikhs on a continuing basis over perceived inequality, religious intolerance, governmental control and resource allocation although there are many, many similarities in the cultural and ethical views of these two religions...Thousands have died in the past 150 years from confrontations begun in confusion and misunderstanding when humans from both sides, in a frenzy of group ideology unchecked by rational thought and considerations, kill others without purpose or reason in a primitive mindset intended to avenge and satiate a god or gods that do not exist...Despite the differences in deities, idol worship, life after death beliefs, dietary practices, origins and destinations, the Hindus and Sikhs are the same people on the same planet with the same needs and concerns as every human in this world...Future generations born and raised with love, security, nutrition and stimulation will realize this simple reality one day and the differences between the communities will seem trivial and ancient...

CONFUCIANISM – another fundamental philosophy/religion that's permeated Chinese culture since the 5th century B.C., Confucianism is a non-threatening way of celebrating the power of human nature and the multiple possibilities of human achievement through self examination, teachings and practice...Believed to be actively embraced by an estimated 6 million humans with a perspective of *Heaven* as a force rather than a place, Confucianism is rooted in the principle of *ren* (benevolence) comprising of *li* (ritual), *zhong* (loyalty to the self), *shu* (reciprocity) and *xiao* (filial piety) which together comprise *de* (virtue)...Embodied in the sacred text of the Analects and the Confusion Canon, this religion is an example of the human brain seeking clarification of the natural state of things with questions of a higher existence and the possibilities of human achievement...

BAHAISM – an interesting religion founded in 19th century Persia (Iran) by Mirza Hoseyn 'Ali Nuri' which teaches that God is omniscient, inaccessible, unknowable, and that all the prophets and religious leaders throughout the history of man are manifestations of God's ultimate plan to unite all religions and all of humanity into a single belief system and establish a universal faith...The Baha'i religion is a child of the Babi faith whose founder and Ali Nuri's mentor, Mirzi 'Ali Mohammad, was executed in 1850 by a Shiite Muslim clergy and government who were afraid of his teachings...In an effort to purge the followers of the Babi teachings from the population, more than 20,000 humans were killed or imprisoned for their beliefs...Currently Bahaism is practiced on multiple continents including North America and Europe by an estimated 7 million humans with no ceremonies or clergy, abstinence from all substances that affect the mind, an annual Nineteen Day Fast and a recurring Nineteen Day Feast where all religious scripture from all religions is examined and discussed...The nation of Iran does not tolerate the Baha'i religion with all adherents subject to persecution and imprisonment by an intolerant Islamic government...

JAINISM – a nonviolent religion devoted to the well being of the Universe and all beings that exist within that Universe...A belief in reincarnation in a continuous cycle until a creature obtains perfection and escapes the cycle to exist forever in a state of bliss...Jains believe the path to perfection is a life of harmlessness and renunciation with minimal impact on world resources and three guiding principles (the three jewels) *right belief, right knowledge and right conduct*...Coalesced and interpreted by Mahavira around 540 B.C. and embodied in the text Agamas, Jains believe in no Gods or spiritual beings that interact with humans and they seek truth through vows of celibacy, truth and honesty...There are many stages of spiritual development in Jainism but the great guiding principle of the religion is nonviolence (ahimsa) with the teachings and interpretation of Jainism left to an extended community of monks and nuns...It's estimated 5 million humans follow these teachings, mostly in India...In parallels surprisingly close to knowledge being discovered by modern science the Jains believe this universe was not created by a God but instead is cyclic and self regulating...They believe nothing in the universe is created or destroyed but simply changes form in a universe that is eternal and unending...

Research by archaeologists have determined the Hebrew Bible was written in the 6th century BC and had many authors and contributors of unknown origins...

SPOTLIGHT ON RELIGION
Ipsos/Reuters Poll 2011
(18,829 people participants across 23 countries)

51% believe there is an afterlife and a divine entity
18% believe there is no god
17% aren't sure if there is a god
28% believe god created humans
41% believe in evolution
31% don't know what to believe
7% believe you are reincarnated
2% believe in heaven but not hell

Creationism belief was strongest in South Africa followed by the United States, Indonesia, South Korea and Brazil...

This is Life with Lisa Lang on CNN on 11/28/14...She's talking to a priest and the priest says the "molestation of children is not just a physical failure or an emotional failure, but a spiritual failure and this spiritual failure is what concerns the church the most" completely dismissing what the victim is dealing with which is extreme physical and emotional trauma that will last a lifetime...But the spiritual failure is most 'concerning'...A religion completely out of touch with all aspects of humanity and reality...

The National Council on Bible Curriculum in Public Schools in America
"To date our Bible curriculum has been voted into 687 school districts (2,235 high schools) in 38 states. Over 550,000 students have already taken this course nationwide on the high school campus, during school hours for credit..." (Studies and reviews by independent groups have found this curriculum to be devotional as opposed to educational with mostly Protestant Christian perspectives)

CONCLUSION

On Earth today in a 21st century global environment that exposes dogma and the ignorance of stagnant institutions without change, there is a growing awareness that all the religions and deities of Earth were created as a normal part of the evolutionary process from a very limited intelligence and awareness to a highly evolved intelligence and awareness...Although no scientific evidence exists on Earth to suggest any of the many gods, myths and demons invented by humans have ever existed, *is there evidence anywhere in this universe that this reality was created or initiated by an unknown intelligence beyond comprehension?* It's a question without an answer but it's as possible and as serious a question as any explanation seeking to explain human existence and it's conceivable our species may be one of a select few in this reality with the intelligence and science to find evidence if any evidence exists...This question of origin and the question "why?" is a necessary question for every sentient species and the continuing search for an answer is what defines self-aware intelligence...

Pew Research estimates that around 16.5% of Earth's population (1.1 billion humans) do not identify with any religion making "non-religious" the third largest religion behind Christianity and Islam...Psychology Today 2015

The sad reality of a species with technology to manage a planet is that religion still determines behavior in billions of individuals on Earth who continue to exist within rigid separatist structures that divide populations to promote rigorous, unyielding belief systems, organizations who have dominated human culture for thousands of years helping millions in need, creating great wealth for themselves and controlling many governments and rulers responsible for the deaths of millions of people...

There are far too many people in this world who wrap their perception of reality around a non-existent God and are oblivious to the millions of infants and innocents dying every year, the billions of humans without food or shelter and the uncounted thousands who die violently in places convenient to ignore while insisting "It's God's Will" a common response from the religious to release themselves from responsibility...Why after thousands of years has religion not brought peace, enlightenment and stability to Earth? Why do humans who live in the same reality on the same world not understand the same physical processes and proven realities that define human existence? And why do humans who control the resources on this planet not understand the causality of unequal resource distribution on the daily lives of billions of creatures who are identical to them? The answer to all three of these questions is the same...This is a very small book in a very big world but the hope is there are millions of people who are disillusioned and disappointed by the failure of religion to end violence anywhere on Earth, the failure of religion to protect innocent children who are still abused by the very people they are asked to 'trust' and the failure of religion to help the poor and feed the hungry as the leaders of these large worldwide organizations live comfortable, opulent lives while focusing on money and expansion with little or no tolerance for new ideas after a thousand years of proclaiming themselves to be the "voice of truth"...Hope that the humans who are beginning to understand human reality will leave these religious organizations or change them with a new realization that *the singular cause of human suffering and death is human activity and human behavior*...It's not "god's will" and it's not 'ok' to allow human policies, human activities and human ignorance to permit the premature deaths of millions of innocents while the majority of people on Earth practice their faith and "look the other way"

To survive as a species humans must move forward and embrace life and the 'nature of man' rather than the tenuous existence of a belief without substance while practicing all the good aspects of faith by aiding any individual who is hungry, sick or oppressed and uniting in organized ways to create institutions essential and critical in helping any and every baby born anywhere on Earth experience love, adequate nutrition, abundant sleep, security and positive stimulation for the first 3 years of life...A global focus on initial brain growth, the next evolutionary step for this species...

If large religious organizations focused their efforts on expectant mothers and newborn babies anywhere and everywhere on Earth to provide adequate nutrition and a secure, safe environment for growth and learning, this world would begin to change in just a few generations...Promoting outreach and sensible government intervention everywhere on Earth to lift any restrictions as it applies to babies, these organizations could use their influence and power to persuade political structures everywhere to admit interactive caregivers into all societies, the Palestinian territories, war-ravaged African nations, Middle Eastern countries and the barrios of South America, anywhere on the planet where babies need human love and protection...With a growing intellectual awareness of human possibilities that exceed any religious teachings of the past, these religious structures (who all preach a desire and obligation to help humanity) could use their influence and money to access women in societies across the globe with education, resources and human caretakers to provide whatever is needed to promote maximum brain growth in every baby without concern for any group's philosophy or religion...If these organizations wish to "find God", promote intellectual development in every baby born and the future generations of Earth will find these deities if they exist...

"No one starts a war warning that those involved will lose their innocence...That children will definitely die and be forever lost as a result of the conflict...That the war will not end for generations and generations, even after cease-fires have been declared and peace treaties have been signed...No one starts a war that way, but they should"
Alexandra Fuller

AFRICA

The second largest continental land mass on Earth, 55 independent nations, more than a billion inhabitants and a history of origins...All evidence in anthropological and evolutionary studies indicate that Africa is where primates developed upright mobility some four to six million years ago and started making rudimentary tools some two million years ago...Starting in the 1920's with some of the most compelling findings coming in just the last decade, fossil records confirm an emerging evolutionary history of our species that's exclusive to this region of Earth...In the future Africa's unique diversity will be recognized for its role in the development of life, recognized for the enormous variety of plant and animal species, recognized for variations of climate and topography that allowed different species to evolve and for more than 11.6 million square miles of stunning landscapes with resources that could improve the lives of millions of people...The birthplace of humanity and the home of the largest animal species on this planet, Africa's future, even in the short-term, is very uncertain...

When you view the lights of Earth from space most of Africa is dark with an absence of critical infrastructure, economic stagnation, widespread poverty and endless ongoing conflicts between ruling strongmen and rebel groups which has continued for years and prevented economic growth and development in devastating ways...Still there is hope...In this undercurrent of violence many nations in Africa are discovering new resources in an accepting global environment which is creating some of the fastest growing economies in the world with corporations and governments beginning to recognize the potential of a rapidly expanding African market that could benefit local populations in critical areas of infrastructure, sanitation, communication, healthcare and education...Hope...In this century transparent democracies have a opportunity to improve the lives of all the people in Africa if a unified world can remove existing dictators and strongmen who rule with corruption, brutality and ignorance, a very big "if" in a world where status quo benefits the wealthy and money controls all politics...

The United Nations estimates more than 120,000 children work in the gold mines (bush mines) in Western Africa, some young as four years old, separating gold, using mercury without protections and working 12 to 16 hours a day...

The geographical center of the ancient landmass Pangaea, Africa is very old and incredibly rich in historical diversity with over 90% of the continent mostly unknown to the 'modern' world during the flourishing of ancient Greek culture, the rise and fall of Roman civilization and the great Asian cultures that grew in isolation for thousands of years...When examining Africa in a modern 21st century, few realize that most of Africa was unaware of a two thousand year history in Europe and Asia (400BC to 1600CE) which was revolutionary in the development of new weapons, writing, science, medicine, systems of governance and resource recovery that's become the foundation of the world today...Arab slave traders explored the interior of the continent south of the Sahara over a thousand years ago but little was known until explorers in the 18th and 19th centuries started mapping the interior leading to a "gold rush" by European powers in the late 19th century who embarked not on a human exploration of enlightenment to advance world knowledge but rather an exploration of exploitation and conquest that destroyed the lives of millions...Representative of everything that defines our species, primitive humans with dominating territorial instincts destroyed multiple African societies for power, resources and extreme personal wealth while dividing land among themselves without concern for indigenous people who were treated with cruelty and a complete absence of human rights, labeled as 'savages' who must be controlled or killed...An ignorance responsible for the death and enslavement of millions of innocent people and an ignorance that still dominates the heart of Africa in this century...

The home of humanity's closest relatives (chimps and gorillas), the survival of Africa's native species is uncertain with the encroachment of civilization and a disregard for native habitat in a global society still seeking to exploit this continent for it's mineral resources and recently discovered gas and oil reserves without concern for the negative impacts on humans or the environment...An arena of war, slavery, murder and rape since its invasion and exploitation by "civilized" humans less than 200 years ago, millions of humans in Africa continue to exist in unimaginable poverty and violence, victims of a desperate infancy that leaves many of them surviving on instincts in an impoverished violent reality without the stability needed for change...Millions of babies are born uncared for, dying and not dying, growing up unable to grasp their reality in any way beyond the need for survival with children given modern weapons and encouragement from warlords to kill, maim and abuse at will with no sympathy or empathy for others, young men growing up into a violent future and present day leaders who take Africa's wealth for themselves while forcing millions of people to survive in desperate ways...Ignored by the world for most of history, Africa is home to 16 million refugees displaced internally or otherwise by ethnic slaughter, climate change, conflict and African strongmen who lead in name only...

After years of transition from colonialism to self-governing states that promised the growth of democracy across sub-Saharan Africa, the trend is reversing and the democratic policies and promises of a new cooperation across the continent are fading...Almost every nation in Africa hold elections but many of the 'free democratic' elections in Africa today are rigged with power gained through bribery, violence, intimidation and manipulation at the voting booths...Today less than 20 independent nations in Africa are recognized as full electoral democracies but there is some hope with a new transparency from the proliferation of mobile communication devices and new technology that can provide people with multiple sources of information and make it difficult for corrupt governments to keep citizens uninformed through the control of local media...Younger populations, a growing awareness of global practices, an awareness of the advantages gained from a fair political process and a growing realization of the failed, violent histories of Africa's strongmen who promised prosperity but enriched themselves at the expense of everyone could energize many citizens in Africa to participate in change and use the resources rich and abundant on this continent to improve the lives of every citizen...

This is possible in Africa if the multilateral powers of the world would support real change and refuse to turn a blind eye to the ignorance, corruption and brutality of the many African leaders who are allowed to continue a systematic abuse of populations and self-enrichment in exchange for profit and a false sense of security, but there is little evidence this is happening or will happen this century...There is an enormous shadow economy in this continent with estimates of $4 billion in resources exported every year and almost none of it gets to the populations who continue to exist year after year in desperation and poverty while 'leaders' supported by governments and financial institutions in the developed world ignore the misappropriation and outright theft of billions of dollars that could help stabilize the lives of millions...The global paradigm for corporations is maximum profit without regard for environmental or human costs and the power structure's of 21st century Earth are all too willing to turn a blind eye to any and all consequence if the bottom line is increased...This indifference to suffering across all of Africa, the enslavement of females throughout the Arab world, the brutality of Islam extremists, the brutality of drug cartels, the insatiable greed of Wall Street and a growing ignorance of consequence in almost every nation on Earth continue to reflect a reality six billion humans are unable to understand...

A stark example of the entrenched hopelessness which is embraced by western corporate structures and their governments is Nigeria where almost 70% of the population live beneath the poverty line and over 25% of children under age five are malnourished existing amid broken infrastructure in a violent environment of despair while foreign energy corporations make enormous profits in collusion with a corrupt government and deny any responsibility for the environmental damage that has made the drinking water unsafe and affected the health of millions...Status quo...

ALGERIA
(38 million humans)
Military Forces = 500,000 active duty / $10.5 billion per year

In northern Africa on the Mediterranean Sea, Algeria gained independence from France in 1962 after more than a million Algerians were killed...The economic engine for this country is mostly natural gas and oil exports with estimated reserves of over 12 billion barrels of crude oil and 4.5 trillion cubic meters of natural gas...In 2012, the Algerian gas and electric company set a goal of 22,000 MW from solar and wind farms by 2030 and with over 75% of the country part of the Sahara Desert, this is a perfect environment to maximize this green technology...Religion is Sunni Muslim and the country is still recovering from a civil war which erupted after the military canceled elections in 1992, violence between the Algerian army and Islamic insurgents that lasted most of the decade and killed more than 140,000 people before stability was restored...Terrorist group is Al-Qaeda in the Islamic Maghreb, a Sunni Muslim group who is dedicated to violently removing the Algerian government and establishing an Islamic caliphate...The government is a constitutional republic with a strong military presence, a parliamentary body, a prime minister and the president as head of state since 1999 having amended the constitution in 2008 to remove any term limits and remain in power (very common in Africa)...Poverty, unemployment and corruption are ongoing problems...An estimated 300,000 children are in the labor force...Almost 5% of children under age five are malnourished and almost 25% of the population exist below the poverty line...Women are better tolerated and accepted by men in this society with females representing a high percentage of lawyers and judges...

ANGOLA
(22 million humans)
Military Forces = 87,000 active duty / $3.9 billion per year

Oil revenues generate over $60 billion per year...The constitution altered in 2010 by the current regime to maintain the power of the president (ruler for over 35 years) states that the president will be selected by the ruling party and not by popular vote...One of the poorest countries in the world due to ineffective government, entrenched corruption and a president who enriches himself and his cronies while ruling through intimidation and violence...Twenty million people with a life expectancy of just fifty-three years, crumbling infrastructure, unexploded land mines, refugees, armed civilian population (the legacy of civil war that destroyed much of the country) and continuing instability...The civil war lasted 25 years and killed more than a million humans...Median age is 18 with a child labor force of almost one million...Over 15% of children under age five are malnourished and 40% of the population exist below the poverty line in a nation with 9 billion barrels of crude oil reserves and over 275 billion cubic meters of natural gas reserves, third largest economy in Sub-Saharan Africa...A single newspaper and a state controlled media that limits all information to citizens with restrictions enforced by an inept government who want no dissent...Electricity is available to only 40% of the population and hunger is widespread with many of the citizens dependent on foreign aid for their survival...Major oil companies Chevron, Total, BP and ExxonMobil are in country developing deep-water oil fields and despite obvious human rights abuses and corruption, the U.S. buys Angola oil to supply energy needs...

BENIN
(10.3 million humans)
Military Forces = 4,900 active duty / $65 million per year

A colony of France until 1960, Benin is a relatively stable democracy with the president as head of the government (and the military) dealing with a weak economy that could grow substantially from the recent discovery of oil offshore near the Nigeria-Benin border...Less than half the population are literate, median age is just 18 with a child labor force estimated at more than a million...Over 35% of Benin's people live below the poverty line and over 20% of children under five are malnourished...One of Africa's large cotton producers, Benin is still a poor country with entrenched corruption, challenging sanitation, border clashes with neighboring states, subversive alleged attempts challenging leadership and a life expectancy of sixty years for both male and female...

BOTSWANA
(2 million humans)
Military Forces = 9,000 active duty / $350 million per year

A moderate African nation gaining independence in 1966 bordered by Zimbabwe, Namibia and South Africa, a multiparty democracy that has been very successful for the Botswana Democratic Party winning all elections since independence with the current president in power since 2008...Botswana is the world's leading producer of diamonds, home to the Kalahari desert, home to one of the world's largest protected wildlife reserves and home to one of the most aggressive and effective HIV treatment programs in Africa...Once known to have the world's highest rate of HIV infection, Botswana has made enormous progress in the treatment and prevention of HIV with free drug distribution, treatment and awareness campaigns but it's still estimated as many as 1 in 3 adults are infected with HIV...Over 11% of children under age five are malnourished, an estimated 45,000 children employed in the labor force with over 30% of the population living below the poverty line...Challenges include the availability of arable land, the availability of freshwater and an ongoing dispute with native bushmen of the Kalahari whose homeland became the focus of profitable diamond mining operations threatening the lifestyle of native bushmen, inhabitants of this region for over 20,000 years...Diamond mining accounts for over 30% of GDP and 33% of government revenues...Tourism is a growth industry with multiple wildlife reserves which protect some of the most endangered animals on Earth...The media is mostly free of censorship in a stable African nation that is party to the Kyoto Protocol, the Endangered Species Act, the Law of the Sea and Ozone Layer Protection Agreement making Botswana reflective of what's possible with a stable civilian government and a rational road-map for dealing with the challenges of global diversification and limited resources...

BURKINA FASO
(18 million humans)
Military Forces = 11,000 active duty / $140 million per year

Earth's third poorest nation (UN), Burkina Faso is a landlocked country bordered by Mali, Ivory Coast, Ghana, Togo, Benin and Niger...Widespread corruption under President Compaore who stepped down in 2014 (in power since a 1987 coup) after protests resulted in a military takeover that eventually installed an interim president in a troubled part of Africa where smuggling, frequent rebel uprisings, poverty and human rights abuses are a way of life...Main export is cotton (leading cotton producer in sub-Saharan Africa) and gold (substantial reserves)...Life expectancy for both men and women is under 55 and it's estimated over 70% of girls and women suffer from genital mutilation...Sanitation facilities are poor with a high risk of infectious disease...26% of children under five are malnourished, literacy rate is less than 40%, child labor force is estimated at more than 1.5 million and 45% of the population exist below the poverty line...After two decades of demands by the people for transparency, economic stabilization, higher wages and lower food prices, Burkina remains one of the poorest countries in the world with a corrupt ineffective government, high fertility rate (5th highest in the world) and over 70% of the population unemployed...

Despite the abject poverty of Burkina, Ex-President Compaore is enormously wealthy with increasing wealth evident within his family, a common pattern in Africa that's growing in multiple nations worldwide...

BURUNDI
(10.5 million humans)
Military Forces = 23,000 active duty / $47 million per year

Burundi's western border is Lake Tanganyika (the 2nd deepest freshwater lake in the world) and the Democratic Republic of Congo with Rwanda and Tanzania to the east and south...Ethnic violence between the Hutu and Tutsi lasting more than 50 years has resulted in hundreds of thousands of dead humans who couldn't understand they're the same species...It's exhausting history includes assassinated leaders, government takeovers, Tutsi government forces killing over 115,000 Hutus in the 1970's, Tutsi soldiers sparking a conflict that killed more than a quarter of a million people in the 1990's, Hutu and Tutsi slaughtering each other in 1994, women and children, innocent and guilty, friends and enemies, cease-fires signed and cease-fires broken, senseless loss of so many because of ignorance, superstition and the complete absence of growing babies in trusted environments to stop this cycle of generation after generation stumbling through life without any awareness or purpose...President Nkurunziza was first elected in 2005, again in 2010 (amid accusations and boycotts), again in 2015 after a constitutional court ruled he could run for the third time, a poor decision that plunged the country into a new cycle of violence...One of the poorest countries in the world (70% of the population live in poverty), Burundi is still a violent country bordered by violent nations and cross-border rebel groups...Only 50% of children ever go to school, less than 3% of the population have access to electricity, one out of fifteen adults has HIV/AIDS, the median age in country is just 17 and life expectancy for both men and women is about 57 years...420,000 children in the labor force and a fertility rate ranked 2nd in the world...Over 30% of children under age five suffer from malnutrition and generations of children who have known little else but war, conflict and continuing distress are the adults who will take this nation into the future...

A nation with over 500 denominations, the official religion of Burundi is Roman Catholicism...About 67 percent of Burundians are Roman Catholics, 23 percent practice indigenous beliefs, and 10 percent are Muslims...

CAMEROON
(22 million humans)
Military Forces = 14,000 active duty / $370 million per year

In 2013, Cameroon took full sovereignty of the Bakassi peninsula after the International Court of Justice (ICJ) awarded them oil rights in a dispute with Nigeria, a decision handed down in 2002 with four years passing before Nigeria acceded and began the process of transfer...Over 300,000 Nigerian citizens still live on the peninsula with uncertainty about their future...Cameroon's President Paul Biya is another long-term African leader (1982) changing the constitution to remove restrictions that would prevent him from continuing as president...Cameroon is recognized globally as one of the most corrupt countries on Earth enriching Biya and his cronies with elections in 2004 and 2011 cited for widespread fraud, intimidation and irregularities...As is the case in too many African nations, it makes little or no difference...European powers occupied this country and did as they wished for years profiting from agriculture and slaves until a French government granted independence in 1960...16% of children under five are malnourished with 1.3 million children in child labor...About 50% of the population live below the poverty line with high unemployment rates (30%)...200 million barrels of crude oil reserves and 135 billion cubic meters of natural gas reserves...Life expectancy is mid 50's for male and female...

CENTRAL AFRICAN REPUBLIC
(5.1 million humans)
Military Forces = 4,300 active duty / $51 million per year

A country occupied by mercenaries backed by a fractured Sudanese government landlocked in central Africa with mineral wealth (diamonds) and tropical rainforests...The Central African Republic (CAR) is one of the least-developed nations on Earth with a desperately poor population and some of the largest populations of lowland gorillas, forest elephants and lush rainforests framed in unchecked violence and ignorance...Spilling over the borders from Uganda, the LRA (Lords Resistance Army) is a brutal faction of violent, lawless renegades lacking morality, intelligence or any awareness of human rights, killing and raping women and children with little resistance from an ineffective military and a distracted government...Led by an extremely "damaged" human, Joseph Kony, the LRA has regrouped and rearmed many times in this region of the world without any cohesive force available to stop such madness and in 2016 Kony remains free, a bitter, ignorant man with a legacy of rape, mutilation and the deaths of thousands who represent the failure of global organizations to protect innocent people from violent humans...Many thousands of people have fled this country, another example of European colonialism (France) gone wrong as every attempt at peace, a unity government and stability fail...CAR is full of illegal weapons, unchecked crime and armed rebels who ousted former president Francois Bozize (fled the country in 2013) after Bozize rigged elections and enriched himself while ignoring the widespread poverty that is devastating to five million humans who exist with a life expectancy of less than fifty years...Michel Djotodia then dissolved parliament, suspended the constitution and declared himself ruler until chased from office...Unimaginable is the plight of children and babies existing in the violence and uncertainty of these African nations dominated by humans who care little for these small children unable to care for themselves...These babies will not develop the neural growth necessary for awareness and intelligence in such environments and the cycle of ignorance, brutality and hopelessness continues...Infant mortality ranks the 4th worst in the world and over 20% of children under age five are malnourished....Millions have died from hunger and disease in the past decade and today it is still Christian against Muslim...Potable water and sanitation facilities are available to less than half of the population and over half a million children are in the labor force...

This country descended into religious and ethnic violence in late 2013 with most of the country controlled by armed gangs and former rebels with violence escalating between Muslim and Christian populations who lived side by side until Djotodia took power, the first Muslim leader of the nation...A weak UN Security Council was urged by the Secretary General Ban Ki-moon to impose "sanctions"? against those who commit mass rape and murder...A sad lack of unity, clarity and vision among world leaders...The African Union took charge of 2,500 UN peacekeepers in 2014 with Ban Ki-moon supporting an increase to 12,000 troops..."Sanctions" will not stop rape and murder...Tens of thousands of humans surrounding the airport desperate for safety with no sanitation or potable water...World leaders "urging" religious leaders and government officials to promote reconciliation between the Muslims and the Christians while the powers of the world sit on their hands...France sent 1,600 troops and so far few allies of France have offered forces to stop the murder and rape that goes unchecked and unpunished...

In 2014, Catherine Samba-Panza was installed as the new interim president of CAR in a world effort to bring some kind of stability and peace to a violence which has displaced more than one million people in less than a year...Instead, the consequence of the Muslim leader (Djotodia) leaving office has emboldened Christian "anti-balaka" militia's into revenge killings with reports of French and African peacekeepers standing by while bodies are mutilated and violence continues...After suffering under a Muslim tyranny lasting less than a year, the "Christian" majority is seeking revenge and are every bit as brutal as Muslim militia's in machete killings and abuse of women and children...A graphic expression of the primitive and ignorant with a global audience watching...

A few minutes after the new president (Samba-Panza) insisted the violence "must end" in a speech to the Christian FACA (The National Army), the Christian FACA lynched a Muslim man in front of the international press...Absolutely no fear of consequence from a world that does not know how to react...February 2014

A new president was elected in February 2016 (Faustin-Archange Touadera) to take the nation out of this spiraling sectarian violence that has displaced over 25% of the population and killed 6,000 humans in just three years...It will take years to return to any kind of 'normal' existence...

CHAD
(11.5 million humans)
Military Forces = 30,000 active duty / $240 million per year

Africa's 5th largest nation, landlocked and bordered by Niger, Nigeria, Sudan, Libya, Central African Republic and Cameroon...Home to Lake Chad, a large shallow wetland that is a vital source of water to over 30 million humans, Chad has a violent history of instability that continues...A former French colony that gained independence in 1960, Southern Chad is predominately Christian while humans in the north are mainly Arab-Muslim, a source of ethnic and religious conflict for half a century...After 30 years of revolts, coups, civil war, invasions and chaos, Chad held it's first multiparty election in 1996...President Idriss Deby has been in power since 1990, a military African strongman who has survived multiple attempts to overthrow his government...Life expectancy is around 50 years for both men and women in a country with close to half a million refugees from the Darfur region, Central African Republic and an instability within the country that has forced almost 200,000 people to leave their homes and seek safety elsewhere...Women and girls are routinely treated as less than men due to tribal customs and traditional practices...Domestic violence is widespread and female genital mutilation is practiced on almost 50% of the female population...An estimated 70% of the humans in this nation live in poverty with less than 50% of residents in urban areas able to access potable water and less than 5% with access to basic sanitation...There are large reserves of gold, oil and uranium in the country with billions being spent on oil recovery by western corporations, but the tiny percentage promised to deal with infrastructure and poverty have yet to be realized...One of world's most corrupt nations with incompetent governance, a median age of only 17 years and a ranking of 6th worst in the world for infant mortality...1.5 million children in the labor force, a literacy rate of less than 40% and 30% of children under age 5 are malnourished...With more than 1.5 billion barrels of crude oil reserves and a pipeline to the coast, this country swims in poverty with almost no money or resources going to the population...In the new global paradigm of profit and greed, the people are ignored as the oil wealth fills the pockets of the president and his cronies...A region whose history stretches back thousands of years, Chad is another African nation shaped by colonialism, ethnic violence, poor governance, warring neighbors, religious ideology and cultural traditions, unaware of the needs of their babies and children with inadequate help from the rest of the world...Deby wins a fifth term as president in 2016...

REPUBLIC OF CONGO
(4.5 million humans)
Military Forces = 10,000 active duty / $157 million per year

Half the country is run by militias and warlords with civil wars and conflict common in Congo, a former French colony which experienced the first of two bloody episodes of fighting when disputed parliamentary elections in 1993 led to ethnically-based fighting between pro-government forces and the opposition...President Nguesso is now serving his seventh term as another corrupt leader who steals the nation's wealth for himself while half of the population lives in poverty...One of Africa's primary oil producers...Widespread violence against women results in as many as 1,000 rapes a month, one in five children die before reaching age five, more than 25% before living just one month and the ones who survive are malnourished with almost half receiving no education at all...Oil is the economy with over 1.6 billion barrels of crude oil reserves and an estimated 90 billion cubic meters of natural gas reserves...In 2004, the nation was expelled from the Kimberley Process established to prevent conflict diamonds from entering the world market when Congo could not explain the origin of rough diamonds it's exporting in quantity...IMF suspended debt relief in 2006 for corruption...Median age is 19 with an estimated 12% of children under five malnourished and over 250,000 children in the labor force...The Republic of Congo is rich with tropical equatorial forests that are home to the highest known gorilla densities on Earth and with an increase of population and industry, forest elephants, western gorillas and chimpanzees are increasingly endangered as their natural habitat grows less and less secure...

DEMOCRATIC REPUBLIC OF CONGO
(72 million humans)
Military Forces = 152,000 active duty/ $161 million per year

Another African nation where upheaval, corruption and poverty are an accepted part of daily life, the Democratic Republic of Congo (DR Congo) is a large country with enormous resources and a continuing humanitarian crisis that has seen thousands killed, injured and displaced by the fighting and posturing of ignorant dictators and corrupt governments...The world sits by and watches...A peace deal amid a new government in 2003 did little to stop marauding militias and the army in a war over mineral wealth and natural resources that has killed some six million humans...Rwanda invaded the renamed Zaire in 1996 to seek out Hutu militias which enabled rebels to capture the capital Kinshasa installing Laurent Kabila as president and renaming the country the Democratic Republic of Congo...Estimates of over 2.5 million humans killed in conflicts fueled by fear, greed and widespread ignorance between 1998 and 2001 when Kabila was killed and his son (Joseph Kabila) took control...An estimated mineral wealth of over *$20 trillion* while 63% of the country's population of 72 million live in extreme poverty on less than a dollar a day under an ineffective government riddled with corruption and nepotism...Over 70% of the population does not have access to drinking water and over 25% of children under age five are malnourished...Conflicts over fishing rights turned into ethnic struggles for power in the north with thousands of refugees seeking safety elsewhere...In the northeast, Uganda's Lord's Resistance Army (LRA) are active, raping and killing local populations in Congo and neighboring nations...A country with one of the largest UN peacekeeping forces on the planet...In 2009, there were estimated to be over 8 million orphans and vulnerable children in the country, more than 90% existing with no means of support...Current estimates are over 8.1 million children forced into child labor...Over 40,000 children are homeless in the streets (the majority in Kinshasa) abandoned by their families, refugee children or orphans of war,...Domestic violence against women is widespread with little police protection against sexual violence and laws that permit female genital mutilation...Authorities rarely enforce laws against corruption which is widespread at all levels of government, a government whose laws are words on paper that mean little unless you're poor and unnecessary...The re-election of Kabila in 2011 was highly controversial and the government is still weak and ineffective ruling this enormous country providing little for citizens while enriching the rulers...Kabila has a personal net worth estimated in 2013 of over $200 million...Millions of humans remain threatened by war, disease and famine...

Wars in Congo have killed over six million humans...The world watches...It's about mineral wealth, desperation, the uninterrupted flow of weapons and the ignorance of nations...

In 2014, more than 2 million humans were still displaced by the conflicts in the 1990's surviving in the forests of eastern Congo eating bush-meat that is decimating populations of bonobos (dwarf chimps/apes) in "protected areas" and poaching for food, activity that is now estimated to have reduced the gorilla population in Kahuzi-Biega National Park by half...A recent study of the effects of this displaced population on the forests has shown significant deforestation in the Luo Scientific Reserve and the Iyondji Community Bonobo Reserve, areas of swamp forest and lowlands where many of the remaining animals survive...

DJIBOUTI
(900,000 humans)
Military Forces = 17,000 active duty / $31 million per year

A strategically important country for France and the U.S. with access to the Gulf of Aden and the Red Sea, Djibouti is essentially a clearing house for imports and exports connecting many African countries with the rest of the world...Gaining independence from France in 1977, the first president (Aptidon) installed a one party rule in 1981 in an effort to consolidate his power resulting in resentment and civil war in the 1990's ending with power sharing deals in 1994 and 2000...Aptidon was replaced by his nephew (Ismail Guelleh) in 1999 who won a 3rd term in 2011 (made possible by a constitutional amendment voted in by a parliament without any opposition representatives) and a 4th term in 2016...A restricted media with no private TV or radio, Djibouti suffered extreme drought in 2011 and is heavily dependent on international food aid...Large Somali population, mostly Muslim, live in the capital city...30% of children under age five are malnourished and approximately 14,000 are in forced child labor...25% of the population live below the poverty line with an unemployment rate over 60%...The French and the U.S. have a permanent military presence in the country...

EGYPT
(88 million humans)
Military Forces = 667,000 active duty / $6.3 billion per year

A nation which has existed for more than 4,000 years, Egypt's sovereignty in the modern world only dates back to 1952 when revolution removed occupying British forces and the Egyptian military came to power...A military that has never relinquished control of the Egyptian Republic and a military that's in power again in 2016...The political figureheads in the presidential seat since the revolution in 1952 number just 5 with Hosni Mubarak ruling for almost 30 years until a popular uprising forced him from office in 2011 and Mohammed Morsi, a member of the Muslim Brotherhood and chairman of the Freedom and Justice Party, winning presidential elections in 2012 to become Egypt's first head of state freely elected by the citizens...After a short honeymoon promising new democracy, economic growth and new transparency, Morsi challenged the status quo and the will of the people by pushing an Islamic agenda without concessions until large demonstrations on the anniversary of his election mobilized the military to depose the new president and retake control of the political process and a fragile economy where over 25% of the population live below the poverty line...

Much of the military aid given to Egypt by the U.S. is returned to more than 40 American companies who, under contract, sell mostly military hardware to Egypt...If the U.S. government cuts military aid to Egypt (over $1.3 billion a year) they are obligated to reimburse these companies for profits lost in this odd arrangement which gives U.S. taxpayer money to Egypt that returns as profit for private corporations...

Since Morsi's removal in 2013, the population of Egypt and numerous international partners have waited to understand the military's agenda in re-establishing "civilian" rule...The military's deeply entrenched in the economy of Egypt with unshakable control of resources and vast wealth while real change is slow and opaque in a nation considered an essential partner in the Middle East...Estimates of more than 20% of males and 35% of females illiterate...Over 9% of children under the age five of are malnourished, three million humans live in extreme poverty and over a million children are in Egypt's labor force...With an energy wealth of 4 billion barrels of crude oil reserves and 2 trillion cubic meters of natural gas reserves, Egypt is dependent on tourism, agriculture and foreign investment to provide stability to an economy that's seen billions of dollars in investments withdrawn in the past few years as uncertainty becomes the normal state of affairs...In March 2014, Abdel Fattah al-Sisi (the general who removed Morsi from office) resigned from the military to run for president and won...Egypt in 2016 is a military state without exception with media restricted, citizens powerless and an estimated 40,000 political prisoners locked up and held for years with allegations of widespread torture...

Over a period of just five weeks, a single judge in Egypt sentenced 1,212 people to death in two trials targeting the Muslim Brotherhood...Judge Saed Youssef (The Butcher) looked at no evidence and heard no witnesses before he announced his verdict...Egypt's Justice Ministry said the proceedings were legal...

Since taking office, President Abdel Fattah al-Sisi has proven himself to be a misogynist intolerant leader who runs an opaque government tolerating no dissent...Allies continue to support this government with money and weapons...

Between 1948 and 2011, the U.S. has given Egypt $71.6 billion in aid with the majority going to the military...
(Washington Post)

EQUATORIAL GUINEA
(740,000 humans)
Military Forces = 1,500 active duty / $170 million per year

President Teodoro Obiang Nguema has proven to be another small minded African leader who has ruled for more than 30 years with violent oppression of political opponents, flawed elections and an abysmal record of taking care of his nation's population and infrastructure...After the discovery of large energy deposits offshore in the 1990's, Nguema enriched himself and his cronies while presiding over a nation where 11% of children die before age 5 and over 10% of the remaining children under five are malnourished...Senate investigations found the president has received enormous sums of money from U.S. oil companies and Nguema has been a guest at the White House as western powers turn a blind eye to corruption and human rights violations in a growing global paradigm of profit and greed...With a population under a million humans and Africa's highest wealth ranking, this nation could be the gold standard for healthcare, modern infrastructure, human rights, high-tech expansion and a wealth which could help poor surrounding countries join the global community and raise the standard of living for everyone...But as is too often the case on this continent, power is in the hands of a leader incapable of understanding morality or reality in any definitive way, driven by primitive instincts to acquire as much as he can without regard to the suffering of others...Media is controlled by a government using violence and intimidation to obstruct the flow of information...14th in the world in infant mortality...Half of the population has no access to clean water or sanitation facilities...

ERITREA
(6.3 million humans)
Military Forces = 300,000 active duty / $467 million per year

A one party state ruled by Isaias Afewerki, again, an African leader who rules with violence and oppression and no recognition of basic human rights...On the Red Sea bordered by Ethiopia and Sudan, Eritrea was created as a federation of Ethiopia by the UN in 1952 but became an independent nation after Ethiopia tried to take the territory in 1962 resulting in a "war" which was finally resolved in 1993 with Eritrea joining the UN as sovereign state...Predictably, these two countries have been adversaries ever since that time with more than 60,000 humans killed in the late 1990's in border clashes while the populations of both nations suffer from an extreme ignorance of governance with drought and famine killing hundreds of thousands and military forces being a priority over all else...Median age is just 19 years with more than 35% of children under age 5 malnourished...Clean water and adequate sanitation are chronic problems while 50% of the population live below the poverty line...The only African nation with no private media outlets as the government controls information with intimidation, imprisonment and violence...Another leader frightened and confused by an inability to understand existence in any real way and primitive unfocused violence is a typical reaction in such men...Recent attempts by the UN to investigate human rights violations have been discouraged by government intervention...

ETHIOPIA
(93 million humans)
Military Forces = 182,500 active duty / $340 million per year

One of the few African nations that was never colonized, Ethiopia has almost a million humans in need of food assistance, water management, sanitation and hygiene as droughts, famine and a lack of cohesive governance continue in a violent, unstable country with a very uncertain future...A key ally of the United States...The former prime minister Meles Zenawi (died in 2012) employed a vast security organization that raped and killed humans with no consequence and anyone who protested or reported this violence was imprisoned...Current prime minister is Hailemariam Desalegn, a close friend of Zenawi who promises to carry on in the same manner...A scorched earth policy by the government in the 1980's destroyed hundreds of thousands of acres of food resulting in widespread famine...Median age is only 17 with over 25% of children under age five malnourished...Literacy is less than 50% and a reported 10 million children are in Ethiopia's labor force...Over 24 billion cubic meters of natural gas reserves but less than 500,000 barrels of crude oil reserves...Over 300,00 refugees from Somalia, Sudan and Eritrea and ongoing disputes with Somalia over rebel groups, subversion and boundaries...

GABON
(1.6 million humans)
Military Forces = 5,000 active duty / $70 million per year

Located south of Equatorial Guinea bordered by Cameroon and Congo, Gabon is more stable than most African nations with only two rulers for the first 49 years of independence...The president in 2016 is Ali Bongo, son of former leader Omar Bongo who ruled Gabon for 40 years accumulating enormous wealth for his family and friends...The new president's election in 2009 is still disputed today...Gabon's stability is largely due to a French military presence in country and reserves of crude oil estimated at 2 billion barrels with 28 billion cubic meters of natural gas reserves...Gabon has over 1.5 million hectares of protected land set aside for hunting or protection of large populations of lowland gorillas, chimpanzees, forest elephants and hippos making this nation a popular tourist destination...The distribution of the nation's wealth is unequal and ongoing corruption has resulted in the same problems experienced across all of Africa with a high unemployment rate (21%), and a large population who exist below the poverty line...Median age is 18, life expectancy is less than 60 years and malnourished children under age 5 around 10%...Gabon has a large number of ethnic tribes who still practice art and traditions centuries old and a government controlled media that uses legal tactics, financial pressure and the threat of imprisonment to censor critical reporting of questionable policies...

GAMBIA
(1.8 million humans)
Military Forces = 2,500 active duty / $4.5 million per year

President Jammeh won his fourth term election in 2011 (all the elections have been disputed by international groups), another African strongman who came to power in military dress and tolerates no media freedom or criticism with the threat of imprisonment...The president has threatened to "cut off the heads" of homosexuals, he has reinstated the death penalty, he has threatened to "kill human rights workers" as saboteurs and he has claimed that he can "cure AIDS" (no proof of this)...The nation's only TV station is owned by the government...The economy is reliant on tourism and agriculture with almost half the population living below the poverty line, median age is 20 and life expectancy is around 65 years...More than 16% of children under age 5 are malnourished, literacy rate is only 55% with over 100,000 children in child labor...Supreme court is appointed by the president, the Cabinet is appointed by the president and there are no term limits for the president...All but one opposition party boycotted the 2012 elections to fill the legislative branch...In 2016, the president declared that his nation is now an 'Islamic Republic' with new restrictions on women and freedoms...

GHANA
(25.7 million humans)
Military Forces = 13,500 active duty / $118 million per year

The first sub-Saharan country to break free of colonial power, a functioning parliamentarian democracy since 1992, Ghana is often looked at as a model of reform for the routine dysfunctional state of Africa's many elections and coups...Ghana is Africa's second largest producer of both gold and cocoa which has sustained an economy that is growing with the recent discovery of offshore oil (700 million barrels of crude oil reserves) and a proven reserve of over 22 billion cubic meters of natural gas...The nation has worked closely with IMF (International Monetary Fund) to reduce poverty, maintain good governance and promote private sector competition...Median age is 20 years with a life expectancy of approximately 65 years...Drinking water and sanitation facilities are challenged, about 15% of children under age 5 are malnourished and the child labor force is estimated at more that 1.7 million...Media is relatively free from government interference or manipulation, a rarity on the African mainland...

GUINEA
(11 million humans)
Military Forces = 12,500 active duty / $100 million per year

Guinea is bordered by Liberia and Sierra Leone to the south, Ivory Coast and Mali to the west, Senegal and Guinea Bissau to the north and over 300 km of coastline on the Atlantic Ocean...Located in a violent region of Africa, Guinea has a violent history of almost 3 decades of oppressive socialist rule that witnessed thousands of humans tortured and killed and a violent history from that time with mutinies, rebel clashes on borders, assassinations, internal revolution and military rule whose actions brought international sanctions...Current president is Alpha Conde who took over from the military in 2010 in Guinea's first democratic election since independence in 1958...Maternal mortality rate is 13th in the world and life expectancy is under 60 years...Over 20% of children under 5 are malnourished in a desperately poor population with a literacy rate under 40% and a child labor force of more than half a million children...Almost 50% of the population exist below the poverty line (under $1 per day) in a nation with gold and diamonds and the world's largest reserves of iron ore and bauxite (used to make aluminum)...Guinea is often the destination of refugees from violence in neighboring states with rebel groups and warlords in Mali, Sierra Leone and Liberia...

IVORY COAST
(21.3 million humans)
Military Forces = 18,000 active duty / $350 million per year

President Alassane Ouattara continues to work toward reuniting a divided north and south with the support of UN and French peacekeepers after the 10 year reign of Lautent Gbagbo destroyed vital infrastructure and killed thousands in a civil war that split the nation into a rebel controlled north and a government controlled south...Widespread abuses by military and security forces have been highlighted in multiple reports from human rights groups...Home to what was once the largest forest in Africa now decimated by logging, legal and illegal...Drinking water and sanitation are unreliable with a high risk of infectious disease...15% of children under age 5 are malnourished and over 40% of the population live below the poverty line...Literacy rate is under 50% and 1.8 million children are in the labor force...The world's largest producer of cocoa and an exporter of coffee and oil with 100 million barrels of crude oil reserves and 28 billion cubic meters of natural gas reserves...Ivory Coast (Cote D'Ivoire) is recovering from a decade long period of instability which divided populations and destabilized economic and energy sectors...Border disputes continue with Ghana and life expectancy is below 60 years of age...

KENYA
(44 million humans)
Military Forces = 24,500 active duty / $590 million per year

Kenya is home to the world's largest refugee camp (Dadaab) totaling almost 500,000 displaced humans fleeing war in Somalia or drought and famine with no end in sight...It's estimated over the next 40 years that some thirty million refugees displaced by drought, floods and changing seasonal patterns affecting food growth and harvest will overwhelm African governments already stretched by a lack of resources and cooperation with neighboring nations...Over 1,000 refugees a day pour into these three camps in Kenya which were designed to provide for 90,000 and will soon be over half a million and growing...A new constitution adopted in 2010 added 47 new districts with policies on power sharing and governmental checks and balances...Kenya is one of Africa's most scenic destinations, home to the Great Rift Valley where fossils of humanity's earliest ancestors have been discovered, mountains rising over 15,000 feet and a large variety of native African wildlife...Bordered by Somalia, the Indian Ocean, Tanzania, Uganda and Ethiopia, Kenya is often challenged by drought, famine and refugee populations looking for safety from the continuous violence in this region of the continent...Mostly Christian with a median age of 19 years, Kenya has an estimated 2 million children in the labor force and over 17% of children under age five malnourished...43% live under the poverty line with an unemployment rate of 40%...A majority of the labor force work in an agricultural and service based economy although recent discoveries of energy reserves could accelerate Kenya's economy if infrastructure needed to produce and export the oil is built and maintained in this unstable region of Africa...Corruption and inflation are continuing problems as is Al-Shabab, a violent terrorist group based in Somalia...

LESOTHO
(2.1 million humans)
Military Forces = 2,100 active duty / $48 million per year

One of Earth's poorest nations landlocked in South Africa near the southern tip of the African continent, Lesotho exists in an environment unfriendly to agricultural expansion with less than 10% of land being used in this way surviving on a dam and reservoir system, mining, textiles and the aid and investment of South Africa...Over 50% of the population live below the poverty line with the majority "desperately poor" and almost one million requiring food assistance to survive...Life expectancy is around 51 years for male and female, a mostly rural population with 15% of children under age five malnourished and over 100,000 children in the labor force...Currently ruled by a coalition after years of military coups and political uncertainty...Heavy reliance on South Africa's support...

LIBERIA
(4.1 million humans)
Military Forces = 2,200 active duty / $12.3 million per year

Famous for it's founding by freed American and Caribbean slaves, Liberia today is a legacy of the ruinous rule of Charles Taylor in the 1990's with competing rebel factions, civil war, infighting, war crimes and Taylor's alliance with Sierra Leone rebels which killed thousands of innocents...After U.S. and UN intervention in 2003, Liberia elected the first female African head of state, Ellen Johnson Sirleaf, who won reelection and a Nobel Peace Prize in 2011...Africa's oldest republic is still monitored by a UN peacekeeping force of 15,000 while training security forces to maintain a stable democracy still very dependent on foreign assistance...Rich in resources, including timber, rubber, diamonds and recently discovered offshore oil, Liberia is rebuilding infrastructure and dealing with an unemployment rate of around 80% in an economy where 60% of the population live below the poverty line...Median age is 18 years, literacy rate is less than 50%, life expectancy for both male and female is around 60 years and 16% of children under age 5 are malnourished...Infant mortality rate is 15th in the world and maternal mortality is 8th highest with over 150,000 children in the labor force...With a coastline of 579 kilometers (400 miles), Liberia is very attractive to smugglers, traffickers and surfers in an equatorial climate with mangroves, swamps, tropical rainforests and inland mountains...

Liberia was at the center of the world's largest Ebola outbreak in 2014 which challenged a world humanitarian response and a fragmented medical infrastructure...The number of dead from the virus were almost 5,000...

Africa has the highest fertility rates on Earth (4.7 children per woman) and is predicted to have a population of almost four billion humans by the end of the 21st century with multiple cities inhabited by more than 20 million people...With no serious policies or interventions being considered to implement strategy that might educate a global population of the dangers and challenges on a single world with 13 billion humans, it's almost certain this will be our future...If you consider the reality of Africa's poor in 2016 with security, warring factions, entrenched rulers, wildlife preservation and malnourished children, solutions introduced in the first half of this century targeting the problems in a world of 7.4 billion humans will have no chance of success in a reality of 13 billion humans...

LIBYA
(6.4 million humans)
Military Forces = 36,000 active duty / $3.1 billion per year

Tribal powers, a fight for oil revenues (9th largest proven reserves in the world), a fractured government and rebuilding after Muammar Gaddafi...Less than a year after Gaddafi's fall, the Libyans held their first democratic elections in sixty years to elect a National Congress who appointed a prime minister, Ali Zeidan, to form a interim government and prepare for new elections in 2013...A large turnout and the absence of violence in these initial elections held promise of a successful transition to a democratic society...*This did not happen*...Elections in 2014 resulted in two opposing governments, the CoR (Council of Representatives) and the GNC with an Islamist agenda who selected their own prime minister in a power struggle...The UN pulled out later that year and the world ignored Libya until ISIS started occupying port cities with the extreme violence associated with these ignorant ideologues...The UN has set up an interim government in Tunisia in 2016 but they are not recognized by many and the challenge of creating a unity government is extreme...With Egypt to the east, the Mediterranean Sea to the north, Chad and Niger to the south, Algeria and Tunisia to the west, Libya's in a strategic location with over 1,700 kilometers of coastline on one of the most active bodies of water in the world...Mostly desert with less than 1% of arable land, Libya economy is energy with 48 billion barrels of crude oil reserves and 1.5 trillion cubic meters of natural gas reserves...Over 37% of the population exist below the poverty line in a country that imports 80% of it's food...Unemployment rate is 30%, approximately 6% of children under age five are malnourished in a society where there is no central government able to provide water or sanitation to the majority...Life expectancy is around 75 years...In the global chaos that defines much of the world today, Libya has enormous potential for wealth and prosperity with a new government and a new (as yet unwritten) constitution to provide for it's citizens and create a stable, prosperous society...But there is equal potential and possibility for disruption and fragmentation with the many militia groups who have various demands and powerful weapons, with religious and ethnic tensions found everywhere in this region of the planet and with the uncertainty of a bickering, divided government who cannot share or invest the oil wealth for the people's prosperity, reverting to corruption as so many African nations have done and still do...It continues...

In the continuing fight over oil, the government in the east has set up it's own oil company and shipped it's first cargo of oil that the government in Tripoli claims is illegal to sell...It has appealed to the UN to stop the shipment as tensions continue to rise in this fractured economy...Militia's and rebel groups continue to control large areas of the nation and the self-proclaimed Islamic State (ISIS) continues to attack targets on the coast to take advantage of instability and chaos in a mostly dysfunctional region of Earth where the focus of government is still acquiring personal wealth while populations suffer...Oil production at the start of 2016 is less than 500,000 barrels per day...

MALAWI
(17 million humans)
Military Forces = 5,500 active duty / $47.6 million per year

An elongated nation on the shores of Lake Malawi that is dependent on food aid as drought and floods destroy crops in a society where 80% of the population live in remote rural areas dependent on subsistence farming...Median age is 17 years with a life expectancy of 60 years...Over 13% of children under age five are malnourished with clean water and sanitation unavailable for large segments of the population ...Almost a million children are part of the labor force and over half of the population live below the poverty line...Current president is Peter Mutharika elected in May 2014, brother of former president Bingu Mutharika who is recognized for destroying Malawi's economy and infrastructure through mismanagement and widespread corruption...Primarily an agricultural economy, Malawi is currently in dispute with Tanzania over rights to possible oil and gas deposits in Lake Malawi while this densely populated nation fights homophobia and an HIV epidemic that has Malawi ranked 9th in the world in HIV/AIDS deaths leaving an estimated one million children orphans...

MALI
(16 million humans)
Military Forces = 7,500 active duty / $190 million per year

With a history of government involvement in arms smuggling, drug smuggling, corruption and weakness, with a strong presence of al-Qaeda terrorists and anti-government Tuareg rebels and under extreme pressure from a worried international community, Dioncounda Traore was installed as interim president in April 2012, a very challenging year for Mali...Extremist group Ansar Dine controlled much of the north, including Timbuktu, allied with al-Qaeda in search of a fundamentalist Islamic ideology until the beginning of 2013 when French military intervention took back control of the north in less than four months...While the rebels were in control of north Mali, works of art, ancient statues, tombs and archaeological ruins were destroyed by these Islamic fundamentalists citing some vague claim of idolatry...In August 2013, President Ibrahim Boubacar Keita was elected with promises to stabilize and rebuild the country...One of the poorest nations in the world, Mali is dependent on gold and agricultural exports, primarily cotton, to provide for a population where the median age is only 16 years with a life expectancy of 55 years...Birth rate is one of the world's highest (#2), infant mortality rate is one of the world's highest (#2) and almost 30% of children under age 5 are malnourished...Literacy rate is less than 40% and over 1.5 million children are in the work force...Violence with Islamist extremists is still ongoing in the north, forced labor, sex trafficking, domestic servitude and a prevalence of child labor in the gold mines and agricultural settings is widespread...

MAURITANIA
(3.5 million humans)
Military Forces = 16,000 active duty / $215 million per year

An Islamic Republic on the west coast of Africa with a similar history shared by many former colonies...Independence from France in 1960, Mauritania's been led by strongmen and military leaders with flawed elections, coups, violence and corruption...Current president is General Mohamed Ould Abdelaziz who ousted the sitting president in 2008 by military coup and had himself elected in 2009 in a questionable procedure...Parliamentary and local elections held in November 2013 were boycotted by almost all opposition parties...Median age is under 20 years with almost 20% of children under age five malnourished...Less than 4% of the GDP is spent on education in a nation with a literacy rate of only 50%...Unreliable access to clean drinking water and sanitation with 40% of the population living below the poverty line and an unemployment rate over 35%...Over 120,000 children in the labor force...The country's vast mineral wealth in gold, iron ore, copper and recently discovered offshore oil promises revenue which could help a largely agricultural society cope with the uncertainty of drought and food security, but in a majority of African nations, the wealth of the nation rarely gets to the people...Located on the coast of the Atlantic Ocean with Senegal and Mali to the south, Western Sahara and Algeria to the north, Mauritania is a crossroads for an active slave trade and the illegal trafficking of women and girls to border countries or the Middle East for prostitution and domestic servitude...The government outlawed slavery in both 1981 and 2007 with little or no effect on this illegal activity...An increase in terrorism since a 2009 suicide bombing at the French embassy led to presidential promises to spend more on the military to fight Al-Qaeda and has resulted in an alliance with Mali, Niger and Algeria to tackle the growing threat of terrorism in a cross-border cooperation...

In 2013, the Morocco government was criticized for a penal code allowing a rapist to marry his victim to avoid prosecution after a 16 year old girl who was forced to marry her rapist committed suicide..In 2014, the parliament amended the code to remove this option...25% of females in Morocco are victims of violence in a culture of ignorance where unmarried women who lose their virginity, even forcibly, are a dishonor to the family...

MOROCCO
(33 million humans)
Military Forces = 190,000 active duty / $3.3 billion per year

This mountainous country on the northwest coast of Africa is a constitutional monarchy ruled by King Mohammed VI who's royal lineage dates to the 17th century...The government is a moderate Islamist coalition although King Mohammed is chief of state and can dismiss parliament, is the head of the military and holds veto power over parliamentary decisions...With Algeria to the east, Spain across the Strait of Gibraltar to the north and Western Sahara (still unresolved) to the south, Morocco is allied with western powers in the fight against terrorism and holds a non-permanent seat on the UN Security Council...A challenging economy with large government subsidies supporting most of the population, industry and service sectors providing more than 80% of the nations income and 15% of the population living below the poverty line with a child labor force of more than half a million children...Literacy rate is below 70% with a life expectancy of 75 years...Media is "watched" by government for anti-Islamic reporting or critical views of the monarchy...Energy wealth is less than a million barrels of crude oil reserves and 1.5 billion cubic meters of natural gas reserves...Continual disputes with Spain over territory (islands) and illegal migration across the Mediterranean...Because of it's geographical location, Morocco is a point of departure for trafficking in forced labor, illegal drugs and the sex trade...One of the world's largest producers of hashish and a transit point to Western Europe...

MOZAMBIQUE
(24.5 million humans)
Military Forces = 11,200 active duty / $86 million per year

A country with large deposits of gold, copper and iron yet to be mined, Mozambique is one of the fast growing economies in Africa with new natural gas discoveries and rich coal mines, yet more than half of the population exist below the poverty line...A safe haven for smugglers with inaccessible roads, rugged terrain and corrupt security forces...A republic led by the prime minister with a president appointed cabinet and a parliament of 250 seats...Median age is 17 years and life expectancy is around 52 years...Ranked 13th in the world in child mortality with clean water and basic sanitation unavailable to a majority of citizens...Over 15% of children under age five suffer from malnutrition, literacy is just 60% with more than a million children in the labor force...Bordered on the east by the Indian Ocean, Mozambique has more than 2.8 trillion cubic meters in natural gas reserves and exports seafood, cotton and around 12 billion kWh of electricity from hydroelectric plants on the Zambezi river including the largest plant in southern Africa at Cahora Bassa...High rate of HIV infection (over 1.5 million, 5th in the world) and the unemployment level is near 20%...Floods in this century have displaced thousands and required food assistance from the world community...Challenges include corruption, illegal logging and rhino poaching...Even with a continuing improvement in all areas of economic growth, Mozambique is still dependent on foreign assistance to maintain it's economy...

NAMIBIA
(2.3 million humans)
Military Forces = 9,100 active duty / $420 million per year

Located on the west coast of southern Africa with Angola to the north, Botswana to the east and South Africa to the south, Namibia has over 1,500 km of coastline on the South Atlantic Ocean in a hot, dry climate with frequent drought and unpredictable rainfall...Rich in minerals with diamond deposits, zinc, gold and copper, Namibia is the 5th largest producer of uranium on the planet...Despite this wealth of resources almost 30% of the population live below the poverty line with an unemployment rate of more than 25%...Half of the population (mostly rural) rely on food assistance from foreign sources and 13% of children under age five suffer from malnutrition with a life expectancy of just 52 years...The government is a republic with the president limited to two five year terms in office...This region has a long history of repression and struggle with thousands killed by the Germans in the early part of the last century and many thousands killed in a thirty year struggle for independence from apartheid South Africa...Challenges include a proposed hydroelectric dam on the Angola-Namibia border that is highly controversial with concerns of environmental impact and the wide displacement of people affected by the project...A high incidence of HIV/AIDS among the adult population and a lack of food security for thousands affected by drought and uncertainty in a region that is mostly desert...

A newly discovered water source, a vast aquifer found in Namibia that could last for centuries, may have a major impact on development in the driest country in sub-Saharan Africa...Estimates suggest the aquifer could supply the north of the country for 400 years at current rates of consumption...Scientists say the water is up to 10,000 years old but is cleaner to drink than many modern sources, however, there are concerns that unauthorized drilling could threaten this supply...Natural pressure of the water means it's easy and inexpensive to extract but because a smaller salty aquifer sits on top of the new aquifer, it could be threatened by careless drilling..Non-compliance with strict technical specifications could create a rupture between the two aquifers and lead to contamination...The 800,000 people who live in this area depend on a 40-year-old canal that flows from Angola to supply drinking water and over the past decade the Namibian government has been trying to address the lack of a sustainable water supply in partnership with researchers from Germany and other EU countries...The new aquifer, called Ohangwena II, contains enough stored water to equal the current supply in this area of northern Namibia for 400 years...

NIGER
(17 million humans)
Military Forces = 9,300 active duty / $58 million per year

Landlocked in Africa and bordered by seven nations, Niger is a desperately poor country with a history of military coups and ineffective policies...Frequent drought makes food security a continuing worry, a broken health system with a shortage of doctors and facilities, poor sanitation with almost 40% of children under age 5 suffering from malnutrition, a child labor force of 1.5 million, literacy rate is only 20% and 60% of the population exist below the poverty line...Infant mortality rate is 7th highest in the world and maternal mortality rate 14th highest...Current president is Mahamadou Issoufou elected in 2011 after another military coup in 2010 (re-elected in 2016) with power to appoint the prime minister, the 26 member cabinet and the 7 judges on the Constitutional Court...Mineral wealth in diamonds, gold and uranium (some of the largest deposits in Africa) as well as new oil and gas exploration desperately needed to alleviate extreme poverty and build infrastructure in an unstable region of Africa...Challenges include the influx of refugees from Mali to the west and an unstable Nigeria to the south, the ongoing tradition of slavery (Niger banned slavery in 2003 with little effect), unacceptable level of malnutrition among infants and children who are the future generation of Niger, a birth rate ranked #1 in the world in a country who cannot feed the millions of humans who already live there, corruption in government, Al-Qaeda militants whose presence is increasing in unknown ways and human traffickers who leave people to die in the desert after extracting money for transport...Median age is just 15 years...

NIGERIA
(173.4 million humans)
Military Forces = 130,000 active duty / $2.6 billion per year

Africa's most populous state and another African nation "developed" by the British who merged the Islamic north with non-Muslim communities in the south, a decision that's makes Nigeria one the most violent nations on Earth...Among the world's leading crude oil exporters, there is enough wealth from resources to support the population but over 50% of the people live in abject poverty due to entrenched corruption at the highest levels of government and in the corporations that control the resources, mainly oil...70% of Nigeria's population exist below the poverty line with 37 billion barrels of crude oil reserves and more than 5 trillion cubic meters of natural gas reserves...There is widespread knowledge among the people of a power structure which cares little for the average citizen, funneling the country's wealth to an elite who spend without limits or concern for the societal problems that are evident everywhere...This indifference and corruption allowed rebel organizations like Boko Haram, a militant group who have killed thousands battling Nigeria's government, to operate with broad approval from the many poor and impoverished...Founded just a decade ago, Boko Haram continues to operate in country with a clear goal of creating an "Islamic state" in the Muslim north with a ban on western education and the implementation of Sharia Law...Civil war in the 1960's killed one to three million people from fighting, disease and starvation leaving half a million preschool children malnourished...In 2016 over 30% of children under age 5 were malnourished, infant mortality is 10th highest in the world, maternal mortality rate is 11th and over 11 million children are in the labor force...Life expectancy is just 53 and literacy rate is under 60% while fighting, poverty and corruption continue in a society with enough natural wealth to provide for all citizens in a different reality...

Almost 1,000 humans died in government detention in 2013...The government disputes this number...

In 2010, many children under age 5 began dying unexpectedly of what was later determined to be acute lead poisoning which causes permanent neurological damage in victims it does not kill...The lead contamination was linked to gold mining where primitive methods of extraction using mercury continue with no protection for the environment or workers (many are children)...Rapid intervention by outside aid groups to treat the children and educate miners reduced the number of deaths by 2012 but the mining of gold continues with no real regulation...A sad legacy of a broken government that will continue for decades...Broken infrastructure with broken roads, little electricity, unreliable drinking water and a lack sanitation facilities to meet the needs of an increasingly desperate population...Dense garbage in slums stretching for miles, no schools and no health facilities define a system of corporate collusion with local political powers to increase the wealth of the chosen leaving millions of humans without resources or hope...Nigeria imports much of it's gasoline because of broken refineries, cheap gas for citizens until 2012 when the government removed subsidies and prices more than doubled...A nation defined by violence, ineffective governance and widespread corruption...

In 2014, president Jonathan Goodluck signed a law criminalizing same-sex relationships...

Africa's largest oil producer but there is $16 billion "missing" in revenues from the state owned oil company...

A report by Amnesty International who studied the "oil spill investigation process" in Nigeria for six months concluded there was no legitimate basis for numerous oil company claims that the enormous number of spills devastating the Niger delta were caused by theft and sabotage...Investigations are not transparent in this corrupt country with oil companies and operators often in charge of their own investigations and conclusions...Many spills result from equipment failure, widespread corruption, operator error and the absence of any substantial regulatory structure to insure quality and safety...Major oil producers in Nigeria deny any responsibility calling all of the study's findings "unsubstantiated assertions without merit"...

February 2014...Gunmen from Islamist group Boko Haram stormed a boarding school in Nigeria overnight and killed 29 pupils, many victims dying as the school burned to the ground...The group has killed thousands in their quest to create an Islamic state in northern Nigeria with over 200 people killed in two attacks as militants destroyed an entire village and killed the inhabitants as they tried to escape...The militants have also increased their abduction of girls as civilians blame an ineffective government and military for the violence...

April 2014...Boko Haram kidnaps over 300 girls from a school in northern Nigeria threatening to sell them into slavery unless the government releases jailed militants...The world takes notice and does little...

The Council on Foreign Relations estimates Boko Haram was responsible for over 4,000 deaths in 2014 with more than 10,000 deaths linked to this group in the past 12 years...The attacks in May took place after regional and Western leaders pledged "total war" on the militant group at a summit in Paris...

There is no fundamental difference between Hutu & Tutsi and the biggest obstacle to a reality of cooperation and shared existence is this population's inability to understand that there is no difference between Hutu & Tutsi...

RWANDA
(12 million humans)
Military Forces = 33,000 active duty / $92 million per year

Millions of humans have died in this nation primarily from the ethnic tension between dominant Tutsi minority and majority Hutu...The history is long and reflective of maybe the most violent century in the progression of human development...In 1994, after President Juvenal Habyarimana (Hutu) and Burundian President Cyprien Ntaryamira (Hutu) were killed in a plane crash (shot down), the Rwandan military and Hutu militias began murdering Tutsi with a million humans of both ethnicities brutally killed in just over 100 days in an attempt by one ethnic group to completely eliminate another ethnic group...They are the same species...Some 2 million Hutu fled to the Democratic Republic of Congo joined by rebels who continued the elimination of Tutsi...Rwanda Tutsi invaded the militia camps to kill the Hutu...This caused a conflict involving Zimbabwe, Namibia and Uganda who continued fighting in the Democratic Republic of Congo even after the arrival of UN forces in 2000 until 2002 when the Rwandan troops withdrew to their country...Paul Kagame was elected president in 2003 (still president in 2016) and multiparty parliamentary elections held just months later swept Kagme's party to power (RPF) amid allegations of fraud and intimidation...Since that time some economic progress has been made in a nation where 40% of the population exist below the poverty line in a densely populated society with coffee and tea being primary exports and a heavy reliance on agriculture...Kagame is a ruthless opponent to those who oppose him using total control of the army and controversial legislation to crush dissent...The fallout continues over the 1994 genocide with strained relations between Rwanda and the DRC as well as Britain...Median age is 18 years, life expectancy is 58 years, drinking water and sanitation are problematic in both urban and rural areas, 12% of children under age five suffer from malnutrition and over half a million children are in the labor force...Natural gas reserves of more than 56 billion cubic meters...The state controls a majority of media with a history of jailing journalists or exiling them from the country...Tourism is a growing industry with diverse elevated geography (lowest point is over 3,000 feet in elevation) mountainous, many lakes and savannas and a high interest in the populations of mountain gorillas who live at high elevations in protected areas...

SENEGAL
(13.3 million humans)
Military Forces = 19,500 active duty / $225 million per year

Located on the west coast of Africa with Mali to the east, Mauritania to the north and Guinea (and Guinea Bissau) to the south, Senegal is a mostly stable African economy with a new president, Macky Sall, elected in April 2012...Median age is 18 with a life expectancy of just 61 years...Over 16% of the children under age five are malnourished with more than 40% of the population living below the poverty line and a high unemployment rate of around 45%...This nation is heavily dependent on foreign investment and foreign aid to help support an economy dependent on phosphate mining, commercial fishing, iron ore exports and agriculture...Over 70% of the population work in the agricultural sector in a country where just under 20% of the land is arable and less than 1% of the land is dedicated to permanent crops with seasonal flooding and frequent drought...Senegal's location makes it a popular spot for the departure of illegal migrants traveling to Europe and for the illicit transfer of drugs destined for Europe and North America...Many citizens have left Senegal in the past decade to seek work in the Mediterranean states and many refugees still live in the country after fleeing violence in neighboring countries...Child labor force number around 600,000 and literacy rate is under 60%...

SIERRA LEONE
(6 million humans)
Military Forces = 13,100 active duty / $59 million per year

A legacy of over 50,000 humans killed and tens of thousands deliberately mutilated in civil war led by The Revolutionary United Front (RUF), a rebel group renown for human mutilation, rape and murder on an unimaginable scale (with aid and assistance from Liberian President Charles Taylor) that ended when UN and British forces implemented a ceasefire with a disarmament of all rebel forces in 2002, elected President Kabbah and began over ten years of war crime trials and hearings...Current president is Ernest Bai Koroma who won a second term in 2012...A poor nation with about half of the population working in subsistence farming and over 70% below the poverty line, Sierra Leone recently discovered energy reserves in coastal waters desperately needed to improve an economy dependent on diamond mining, some manufacturing and foreign aid...Median age is 19 with a life expectancy under 60 years...Maternal mortality rate is 5th in the world and infant mortality rate is 11th in the world with a shortage of doctors and hospitals...Over 17% of children under age 5 are malnourished, literacy rate is less than 50% and half a million children in the labor force...The 2014 Ebola outbreak in this region of Africa killed over 400 humans in Sierra Leone...Corruption is widespread in a government hostile to journalists who criticize officials or expose corruption...With miles of unspoiled coastline, Sierra Leone is looking to rebuild necessary infrastructure in an effort to make tourism a profitable growth industry...

SOMALIA
(9.8 million humans)
Military Forces = 12,000 active duty / $59 million per year

2013...Doctors Without Borders (MSF) says it can no longer trust the Somali government nor fundamentalist groups like Al-Shabab to guarantee their safety and they are leaving the country...The government of Somalia says the result could be catastrophic...In 2012, the medical aid group delivered over 700,000 medical treatments to Somalis, everything from vaccines to surgery that will now end completely..."Our criticism goes against all authorities, be it the government, be it Al-Shabab, be it the traditional leaders because as you know the government does not control all of Somalia...There are different authorities in central Somalia, in Puntland and in Somaliland controlled by Al-Shabab, so our criticism is the widespread lack of respect for humanitarian action"...

A violent and impoverished nation without little functioning government outside of the capital Mogadishu...25% of Somalia children die before age 5...Terrorist group is Al-Shabab, a violent military wing of the Somalia Council of Islamic Courts that was 'defeated' in a two week war, they continue to operate in Somalia in uncoordinated and fragmented ways with bombings, assassinations, demands and forcible control of native populations...This country continues to practice genital mutilation on women sewing a female's vagina shut so they will remain virgins and removing the clitoris so there will be no pleasure during sex...Ignorance unimagined with a lack of awareness and a complete absence of any empathy, unthinkable that we still have humans who behave this way and unthinkable that this world still allows this to happen...In the north of the country are thousands of humans suffering drought and famine who need help with water sanitation, livestock diversification and resources for farming...The UN says a million or more humans in Somalia are in need of aid...

A 13 year old girl raped by 3 men was stoned to death in a packed stadium of over 1,000 cheering spectators...2008

The UN extended it's mandate in 2012 to keep forces in Somalia for 12 more months to help in the fight against the Al-Shabab rebels who intimidate with violence, chopping off limbs and starving the population...A new government was installed in late 2012 with President Hassan Sheikh Mohamud winning the first election since 1967...The U.S. decided to recognize the new government in 2013, the first such recognition in more than 20 years with a promise of military assistance to combat rebels and terrorists...Al-Shabab Islamist militants continue to target Kenya forces and Kenya civilians in 2016 for their intervention in 2011, an ongoing attempt to destabilize the new government...8th highest birth rate in the world and 13th highest death rate...Maternal mortality is 3rd in the world and infant mortality is 3rd as well...Median age is 18 and life expectancy is just over 50 years...Sanitation and drinking water are severely challenged with over 30% of children under age 5 malnourished and more than a million children in the labor force...Average school life for children is just three years reflected by the nation's overall literacy rate of less than 40%...Economy is agriculture and livestock with exports of bananas, fish, rice and livestock and there are some natural gas reserves (5.6 billion cubic meters) recoverable in a more stabilized environment...The Indian Ocean to the east has been a source of piracy that has grown less as more countries (NATO) patrol the waters off the coast...Continuing border disputes with Ethiopia and severe drought in 2011 displaced millions and challenged the survival of thousands facing starvation with the world community unable to provide a basic nutritional standard for infants, children and adults who have no options in a country ungoverned and ignored for decades...

NATO has worked with the African Union since 2007 to assist peacekeepers in Somalia with air and sea support...

SOUTH AFRICA
(51.8 million humans)
Military Forces = 88,000 active duty / $4.6 billion per year

Bordered on the north by five countries (Swaziland,, Botswana, Zimbabwe, Mozambique and Namibia) South Africa's coastline (2,798 km) wraps around the bottom of the continent making it the strategic and economic focus of Sub-Saharan Africa...After more than 40 years of apartheid polices that favored a white minority with institutionalized discrimination and imprisonment for the black majority, the first multiracial elections were held in 1994 that ended apartheid and brought the African National Congress (ANC) to power...Two decades after that historic election, South Africa still struggles with the legacy of apartheid, political corruption, labor unrest, increasing violence against migrant workers and the highest percentage of humans with HIV/AIDS in the world...A parliamentary republic with the president elected by a multiparty National Assembly (5 year term)...More than 10% of children under age 5 are malnourished in a nation where over 35% of the population live below the poverty line...High unemployment rate (25%) in an economy dependent on natural resources (world's largest producer of gold and platinum) and a growing service sector...Some energy resources with 15 billion cubic meters of natural gas reserves and 15 million barrels of crude oil reserves...Challenges include border control with neighboring countries to prevent poaching, smuggling and migration, lack of renewable water sources, unpredictable drought, an increase in illegal drug shipments and money laundering controlled by organized crime and mostly ignored by corrupt officials...South Africa is home to almost all of Africa's rhinoceros, the second largest land animal on Earth and a herbivore who is a threat to no one with no natural predators (excepting humans)...1,175 rhino killed in 2015 by poachers for their horns (desirable in Southeast Asia by humans who believe it has mystical medicinal powers to cure cancer, hangovers, possession by the devil and rheumatism)...Some wildlife experts estimate the extinction of rhino within two decades if this trend continues as inadequate protection, high demand and widespread corruption give the rhino very little chance in a future very uncertain...

On May 1st, 2008, South Africa ended a 13 year ban on killing elephants insisting a growing population of the largest land animals on Earth is increasing conflicts between elephants and humans and destroying the environment...

SUDAN
(46 million humans)
Military Forces = 120,000 active duty / $2.3 billion per year

Another British colony separated into the south and north with very negative consequences that continue today...Sudan became an independent democracy in 1956 and less than 2 years later was taken over in a military coup that banned political parties, unions and free press...Democracy was restored in

1964 with the overthrow of the military government but another coup in 1969 brought the military back in power in a fluctuating cycle that is Sudan's history...After decades of civil war between the north and south with declining services, rising fees, high commodity prices and unemployment, Sudan partitioned itself (2011) into Sudan and the Republic of South Sudan in an international effort to bring stability and a more equal distribution of oil wealth to the people...*This has not happened*...The oppressive, sadistic leader of Sudan, Omar al-Bashir, has been in power for more than 22 years and is charged by the International Criminal Court with genocide, war crimes and crimes against humanity in Darfur (where more than 250,000 people have died in the past 10 years) and he continues to kill civilians, steal money and sabotage any attempt to create a peaceful or working economy between the north and south entities created in 2011...Sudan military forces bomb refugee camps, block aid from the UN, indiscriminately kill and rape women and children, block access routes to prevent any escape from these atrocities and prevent journalists and aid workers from entering the country...

In March 2013, Sudan and South Sudan agreed to implement the 2012 Comprehensive Peace Agreements to resume the production of oil while the violence and human rights abuse continued on both sides of the border...In late 2013, violence in South Sudan killed over a thousand people with 20,000 displaced due to a "disagreement" between the president (the majority Dinka tribe) and the former vice president (the Nuer tribe) after President Salva Kiir dismissed his entire cabinet, including the vice president, Reik Machar...This is allowed by the new constitution but many believed it was a move to increase presidential power and instead of seeking a political solution, the humans who control and influence these militias, military's and ethnic populations have killed innocents determined by the language they speak, raped women and young girls because they can and destroyed infrastructure, homes and villages without purpose...There is no cohesive political structure in this country and the "army' is a mixed group of warring militias in a coalition who are paid to pretend they're a consistent, single unit...When there is violence, these groups within the army often split along ethnic divisions with loyalty to their commander...The "civilized" world is desperately pleading for reconciliation as thousands hide in UN compounds without a future and without hope...2014

In the Sudanese state of North Kordofan where this violence goes unchecked, the governor is Ahmed Haroun, wanted by the ICC for crimes against humanity in Darfur...South Sudan is landlocked and can only ship their vast oil wealth through Sudan whose negotiating stance is violence...Both sides trading accusations and fighting continues everywhere in the two countries...The UN Security Council in May 2012 called on Sudan and South Sudan to reach an agreement on the status of the disputed, oil rich Abyei border region (still disputed in 2015) to extend the UN security mission in the area by 6 months and create a jointly administered police force in Abyei...Median age is 19 and life expectancy is 61 years...Maternal mortality rate is ranked 9th in the world...Sanitation and potable water supplies are unreliable...Over 30% of children under age 5 are malnourished and school life is just 4 years...Over 80% of the population are in subsistence farming with nearly 50% of the population living below the poverty line in a nation with 21 billion cubic meters of natural gas reserves and more than a billion barrels of crude oil reserves...Media is highly restrictive with the state controlling the flow and content of information...Over two million humans have been killed with millions displaced internally over the last thirty years as millions of refugees flow between the borders of Chad, Ethiopia, Kenya, Central African Republic, Congo, Uganda and Sudan...Ongoing border disputes with Chad, Egypt and the new government in the south...

Civil war displaced over 2.2 million humans and thousands died between 2013 and 2015...

Woman and daughters in Juba (South Sudan) are often given away for cattle after age 12 and when asked what is the choice of the woman, the chief says "Women have no say"...A country where many children still die from malaria and malnutrition while war takes the lives of young men...Child soldiers number 15,000-16,000 (UNICEF)

2013...Documants leaked by someone close to Sudan's regime show officials in government planning to 'starve' the civilian population in the south and "a strategic relationship" with Iran to promote terrorism (Washington Post)

SOUTH SUDAN
(11 million humans)
Military Forces = 210,000 active duty / $550 million per year

Elections scheduled in June 2015 were canceled due to instability and ongoing conflict (the next election scheduled in 2018)...Presidential republic with the president elected by majority popular vote to serve a four year term...Median age is 17 years with birth rate 15th highest in the world...Maternal mortality rate is ranked 9th in the world...30% of children under age 5 are malnourished, a number that changes dramatically with an estimated three million humans hungry due to weather and the continued violence that disrupts the delivery of food aid and planting...Sanitation and potable water supplies are inadequate and unreliable...A country with a lack of basic infrastructure, six UN camps are trying to cope with the human crisis asking for $1.6 billion to help 4.5 million people in 2015...Nearly 50% of the population exist below the poverty line in a nation with over 63 billion cubic meters of natural gas reserves and 3.7 billion barrels of crude oil reserves...Both the armies of South Sudan and Sudan are preparing for more war and conflict while both nations deal with internal inflation, terrorists, protests, sex trafficking and violence...Primitive men with ruthless armies who dominate and terrorize anyone and everyone while the world watches...

Sudan continues to be one of the most primitive nations on Earth virtually ignored by the civilized world as government soldiers kill, rape, pillage and torture with no opposition...America help set up the new South Sudan in 2011 and left...UN says only 3% of it's humanitarian appeal for South Sudan is funded (NYT's) as more and more aid groups leave the country...No government services and mass starvation in a nation run by ignorance...

SWAZILAND
(1.4 million humans)
Military Forces = 3,000 active duty / $118 million per year

Bordered by South Africa and Mozambique, this nation is ruled by King Mswati III who allows no opposition parties and no challenges to his rule by decree signing a new constitution in 2005 that guarantee his power over 1.4 million people who live in poverty and are increasingly dependent on aid to survive...Highest HIV/AIDS prevalence in the world (over 26% of the population is infected) with a life expectancy of 51 years...70% of the king's "subjects" live below the poverty line while the king builds new palaces for his many wives...The nation survives economically on trade and aid from South Africa, sugar exports and mining...Unemployment rate is above 40%...Freedom of the press is severely curtailed with threats and arrest for any criticism of the state or the monarchy...Almost 30,000 children in the labor force with 6% of children under age 5 malnourished...No elections, a hereditary monarchy and a dying population is a recipe for unrest and upheaval in a country with no energy reserves, little industry and fading hope...Over 200,000 orphans in country, victims of the AIDS epidemic...

TANZANIA
(49 million humans)
Military Forces = 27,500 active duty / $220 million per year

Home of the Serengeti National Park, 14,763 square km of land containing millions of Africa's indigenous wildlife that is a World Heritage Site crucial to the survival and vitality of multiple species including lions, elephant, giraffe, wildebeest, buffalo, crocodile, zebra, gazelle, leopard, jackals and hundreds of bird species including the ostrich...This park is one of 15 in this nation that has dedicated over 30% of it's land resources to the protection of ecosystems and various species in a classically beautiful African landscape that includes the highest mountain on the continent, Kilimanjaro...Bordered on the East by the Indian Ocean with Lake Victoria to the north, Tanzania is a presidential republic with the president and vice president elected on the same ballot by a majority to serve 5 years...President is both chief of state and head of government...Medium age is just 18 years with a life expectancy of 60 years...Sanitation facilities are challenging for more than 80% of the population with a high prevalence of HIV/AIDS (humans with HIV/AIDS ranked 6th in the world)...14% of children under age five are malnourished and almost three million children are in the labor force...Economy is agriculture, gold mining, tourism and foreign aid...All land in Tanzania is owned by the government and controlled though leases...The country has over half a million refugees from neighboring nations and disputes with Malawi over boundaries...Recent discoveries of offshore energy reserves (6.5 billion cubic meters of natural gas) could boost an economy where over 65% of the population live below the poverty line...

TOGO
(7 million humans)
Military Forces = 9,600 active duty / $64 million per year

A history of military rule by a single leader who was replaced by his son in 2005 with a legacy of human rights abuses and staged elections...Togo is a vertical country on the west coast of Africa stretching 800 km north to south and less than 100 km east to west...To the west is Ghana, to the east is Benin and to the north is Burkina Faso...Current President Faure Gnassingbe Eyadema won a criticized election in 2010 to quiet dissent that arose after taking control of the nation after his father's death...The UNIR won a majority in the parliamentary elections of 2013 and the president was elected again in 2015 with no presidential term limits...A poor nation with more than 30% of the population below the poverty line, Togo's economy is agricultural exports and phosphate...Median age is 19 years with over 17% of children under age 5 malnourished...Drinking water and sanitation facilities are unreliable to most of the population...Child labor force is more than 700,000 and the literacy rate is under 70%...

TUNISIA
(10.8 million humans)
Military Forces = 40,000 active duty / $550 million per year

In north Africa on the Mediterranean Sea with Algeria to the west and Libya to the east, Tunisia is in a geographically strategic location for exporting textiles, food, chemicals and petroleum products to the European Union to support an economy strengthened by foreign investment and tourism...After independence from France in 1956, Tunisia was ruled by an enigmatic leader for over 30 years, Habib Bourguiba, advancing female rights in ways unknown in the Arab world while consolidating a single party state into a virtual dictatorship until dismissed in 1987...Parliament is multiparty with the prime minister holding much of the power...Elections in 2014 ushered in new president Beji Caid Essebsi and a new parliament operating under a new constitution...A relatively stable society with a median age of 31 and just under 3% of children below age 5 malnourished...425 million barrels of crude oil reserves and more than 65 billion cubic meters of natural gas reserves...Tunisia has the potential to become a major player in Mediterranean shipping if it can improve coastal structures and interior infrastructure to support future growth and meet the challenge of Islamic State militants (3 major incidents in 2015)...

In February 2014, the president of Uganda signed a draconian bill imprisoning humans for being gay, for not reporting gays and for talking about gays...Museveni suggested that females become lesbians because they are 'sexually starved" without marriage and a male...Africa remains one of most homophobic continents on Earth...An ignorance and lack of intelligence and awareness in a world where 95% of the population is and will always be heterosexual...36 of Africa's 54 countries have laws against same-sex 'acts' in their statutes...

UGANDA
(36 million humans)
Military Forces = 45,000 active duty / $275 million per year

North of Lake Victoria in central Africa with the Democratic Republic of Congo to the west, Rwanda and Tanzania to the south, Kenya to the east and South Sudan on the northern border...Uganda is still recovering from the disastrous rule of Idi Amin in the 1970's and Obote in the 1980's that killed almost half a million people and the legacy of the Lords Resistance Army (LRA) who were forced out of Uganda just seven years ago but still operate in the shadow of borders and lawless regions...Primary export is coffee but recent discoveries of oil and gas have the potential to boost an economy centered on agriculture...Over 20% of the population live below the poverty line in a nation of multiple ethnic groups and multiple languages...Medium age of 15 is one of the lowest on the continent with the 3rd highest birth rate on the planet...Sanitation facilities are inadequate, a high prevalence of HIV/AIDS and over 15% of children under age 5 malnourished...Life expectancy is 55 years...Current president Yoweri Museveni has been in office over 25 years having amended the constitution to remove term limits with accusations of election fraud...The government continues to ignore corruption at the highest levels and is under scrutiny for it's involvement in the Democratic Republic of Congo with incursions and aid to rebel groups...The government denies any illegal activities and the world community looks the other way if there is enough stability and profit to encourage this institutional blindness...

In 2016, Museveni won another term while his opposition opponent remains under house arrest...

ZAMBIA
(14.4 million humans)
Military Forces = 15,300 active duty / $245 million per year

Zambia is landlocked in south central Africa sharing a border with Zimbabwe on the Zambezi River, home to Victoria Falls, 354 ft high and over a mile wide...Heavily dependent on copper exports (the largest producer of copper in Africa) and agriculture to support an economy where over 60% of the population live below the poverty line...Over one million children in the labor force and a literacy rate of just 62% spending less than 2% of GDP on education...Birth rate is 5th in the world, life expectancy is 52 years and medium age is under 17 years of age...15% of children under age five are malnourished and prevalence of HIV/AIDS is 7th in the world...Less than half the population has access to improved sanitation and clean drinking water...Close relationships with China who signed a mining agreement in 2010 and is active in financing and building new power plants...As in most of Africa, corruption and cronyism is widespread and problematic as Zambia's citizens grow restless over worldwide increases in copper prices and revenues that never seem to trickle down to the general population...

ZIMBABWE
(13.5 million humans)
Military Forces = 30,000 active duty / $95 million per year

Robert Mugabe has been the leader of Zimbabwe for over 30 years and is responsible for the decline of this nation with mineral resources that include gold, diamonds, coal, copper, nickel and tin, a labor force of 5 million and an unemployment rate over 90%...Over 70% of the population exist below the poverty line as Mugabe, who is seemingly ignorant in almost all aspects of governance, destroyed land and commercial production farming with chaotic land redistribution programs, froze commodity prices, fought in conflicts out of country costing millions of dollars and printed money at will causing hyperinflation devaluing the currency to essentially nothing...More than half the population have no access to working sanitation facilities, deaths from HIV/AIDS is ranked 6th in the world and over 11% of children under age five are malnourished...Mugabe won elections in 2002 widely condemned as "seriously flawed" by international observers with an increased level of violence and in 2008 he was accused of attacking opposition supporters to win an election runoff that saw his opponent drop out of the race with threats against his life (Mugabe is scheduled to run for election again in 2018)...A small-minded African strongman who rules with intimidation and violence but he's old and tired and will soon be gone...Zimbabwe shares Victoria Falls (2 million year old path to the edge) with Zambia, is the home of the stone enclosures of Great Zimbabwe, ancient remnants of a past empire and has herds of elephant and other game roaming large open stretches of remote wilderness...A beautiful African wonderland with multiple national parks and wildlife preserves marred only by human habitation...

WORLD HEALTH

Unlimited access to healthcare should be a standard for the human species with the focus and emphasis always on preventive care, nutrition and exercise...Humans are animals who have run and played for thousands of years to survive and to explore an unknown world and our physiology is uniquely connected to consistent levels of activity and exercise...Few humans understand that just a single hour a day of aerobic exercise would make a huge difference in the world's health but there are other factors...Current industry and commerce introduce thousands of chemicals into Earth's environment every year having negative consequences of varying degrees and in this century political pressure on elected individuals by industry and corporations to enact laws benefiting an unregulated waste culture coupled with opaque industrial standards (trade secrets?) leave millions unaware of destructive practices and restrict concerned agencies from effecting change in a growing culture of greed and profit that allow this to continue with a growing measurable negative impact on humans and the multiple species that support human existence...

The dominant health organization on Earth is the World Health Organization (WHO), an agency of the United Nations established in 1948 to study the impact of global health policies, help establish priorities for intervention, combat disease worldwide and improve maternal care and child nutrition...In 2016 WHO constantly monitors the world with analysis and data in an attempt to create an accurate picture of a population of over seven billion humans and health threats current and future with a budget of $4.4 billion dollars in 2016-2017...It remains one of the most important organizations on the planet...

Over 800 females die every day of complications from pregnancy and childbirth, almost all of them poor...WHO

WHO Definition of Health
Health is a state of complete physical, mental and social well-being, not merely the absence of disease or infirmity...

In 2014, the WHO issued a report on the effects of the widespread use of antibiotics with a stern warning that microbes resistant to the most effective antibiotics in the human arsenal are now found all across the globe and are developing a high resistance to treatment...These include common microbes responsible for pneumonia, tuberculosis, diarrhea, E.coli, strep infections, salmonella, gonorrhea and infections of the blood, wounds and urinary tract...

Child marriage make up over 30% of unions worldwide...Women who have babies before age 15 are 5x's more likely to die in childbirth as women in their 20's...

HIV
Each year over a million humans worldwide die of AIDS, mostly in Africa and over 2 million become newly infected with the HIV virus...There are currently more than 36 million infected people on the planet and there are finally discussions on a global level asking if it is possible to give antiviral drugs to everyone as soon as the virus is discovered to prevent transmission from mother to children and between sexual partners...

If the human species has the knowledge, the resources and the technology to create a vaccine or drug that will prevent or cure a disease why is it not given to every human who has that condition? Money should be a non-factor in the control and treatment of human health...In 2016 there are unbelievable prices on specific drugs that should be available to whomever needs it but all activities on 21st century Earth are determined by profit no matter what the human cost (more on this at the end of the chapter)...

Chagas disease is believed to have infected up to 8 million people in Mexico, Central America and South America...Long incubation, hard or impossible to cure, transmitted by insects, blood and mother to child...25% of victims will develop enlarged hearts or intestines which can burst and cause sudden death...A disease of the poor with little research or money spent on finding treatments...

EBOLA

A virus discovered in 1976, Ebola is a group of five identified virus's and maybe the world's deadliest virus with a high mortality rate (90%), no approved vaccine to prevent onset and the only effective treatment after contracting the disease being "supportive intensive care" replacing fluids to keep patients hydrated...All five strains of the virus originated in Africa with outbreaks being relatively short and swiftly contained due to the rapid onset of symptoms after infection and the rapid death of victims limiting the spread of the virus to others...This changed with a new outbreak in West Africa in early 2014 that spread rapidly in three nations, Guinea, Liberia, Sierra Leone with scattered cases in several neighboring countries and isolated cases in the U.S., Spain and Great Britain...The international response to this outbreak was criticized by many international aid groups and healthcare professionals who said the response was not focused enough but in 2016 the WHO declared all three nations 'Ebola-free'...WHO reported 28,603 people had been infected with a death toll of 11,301...

A study published in late 2012 in Scientific Reports suggest laboratory studies have shown Ebola may be transmittable through the air although no airborne cases have ever been identified or reported and it's still believed the virus can only spread by direct contact with an infected carrier...

RABIES

Rabies is a viral infection of the nervous system that kills more than 53,000 humans every year with many victims poor children in Africa and Asia bitten by dogs, an infection that's 100% fatal unless treated before the onset of symptoms, usually a few weeks after being infected...Rabies is transmitted by the saliva of infected animals (often bats and dogs) but any mammal can be a carrier...In a real world civilization there would be an international hotline with a team of physicians and experts to administer shots to anyone infected, tracking and destroying the infected animal...53,000 humans dying every year from a treatable condition...In 2014 the head of the World Animal Health Organization (OIE) said this virus could be eliminated for 10% of what it costs to treat victims adding that inexpensive, ineffective vaccines and a lack of money with a focused effort are responsible for the painful deaths of thousands of children every year...Dog bites are the source of almost all rabies infections...

The World Health Organization estimates smoking killed over a 100 million humans in the 20th century and will kill over a billion in the 21st century as developing countries with huge populations consume a product that makes little sense...Nicotine is a powerful drug but beyond that there is nothing beneficial to biological creatures from inhaling fire and smoke from a burning plant...No substances contained in tobacco smoke are beneficial to life...

33% of the world's population are infected with TB, the leading killer among infectious diseases...

TUBERCULOSIS

In 2012, there were over 8 million new cases of this bacterial infection which is growing more and more resistant to antibiotics and is responsible for more than 1.8 million deaths a year...The World Health Organization reported over 40 million humans have been saved over the past 17 years through international cooperation in a major effort to break the disease and eliminate it from the long list of humanity's deadly afflictions but as is often the case in a world order motivated and driven by corporate profit, pledged funds are often not delivered and treatments are ineffective, antiquated and inadequate because of this lack of funding...After more than a dozen years with no new vaccines there are currently more than 10 being tested and there is hope a newly developed vaccine (MVA85A) given to thousand's of babies in Africa will be effective in preventing this treatable and devastating disease from becoming widespread...

WHO...The Millennium Development Goal target of halting and reversing the TB epidemic by 2015 has been met globally...TB incidence has fallen an average of 1.5% per year since 2000 and is 18% lower than the level in 2000...

Over 6.5 million children under age 5 died in 2012...Over 280,000 women died of complications in pregnancy and childbirth in 2010...44% of deaths under age 5 occur in the first 28 days...WHO (World Health Organization)

MALARIA

World Health Organization reports malaria death rates have dropped some 60% globally since 2000 although the majority of deaths are in sub-Saharan Africa (90%) with a majority being children under age five, thousands every day...Malaria is endemic in over ninety countries and believed to be responsible for a million deaths each year even though methods of treatment and prevention are well known...In Africa reductions are continuing with programs using drugs, intervention and mosquito nets but there are disturbing criminal elements with ineffective drugs repackaged to look genuine and with little oversight in China and India where a majority of the malarial drugs are manufactured and sold to distributors...Malaria is the deadliest vector borne disease in the world infecting 200 million people a year and yet it's easy to treat with drugs if identified early...Another disease that could be controlled or eliminated on a world with consensual awareness and reality intelligence but on this world it still kills thousands of children every year (over 300,000 children under age five in 2015)...The best treatment is prevention efforts which are ongoing in high risk countries with the help of aid organizations...

Diarrhea is the leading cause of malnutrition in children under five...

DIARRHOEAL DISEASE

Kills over 2 million humans every year (about 800,000 children) in developing nations through dehydration and fluid loss...*This condition is **preventable** by providing clean drinking water with adequate sanitation and good hygiene and this condition is easily **treatable** with a solution of sugar, salt, clean water and zinc tablets...*Another glaring example of a primitive civilization where millions of humans accumulate over a million dollars every year (many make much more) while almost a billion humans have no access to clean water and 2 billion humans have no access to improved sanitation...In an intelligent, connected civilization no one would die from this condition...

In the past 200 years six pandemics of cholera have killed millions of people on six continents with many millions getting desperately sick before recovery...A 7th pandemic started just 50 years ago in South Asia spreading to Africa in the 1970's and to the Americas in the 1990's...WHO

CHOLERA

Cholera is a diarrhoeal disease caused by a specific bacterium *Vibro cholerae*, a serious problem in areas of disaster where clean water and food are scarce in crowded unsanitary conditions allowing the microbe to spread rapidly through desperate populations...Another disease easily treatable with oral re-hydration, salts and intravenous fluids...Very aggressive disease with the capacity to kill within hours of infection...Recent statistics have shown an increase in cholera cases of more than 300% over the past five years with a continuing crisis in Haiti that began after a deadly earthquake destroyed infrastructure in 2010...Global estimates of up to 4 million cases every year causing 100,000 deaths...

INFANTILE COLIC

Infantile colic is a condition in babies (birth to around five months) defined by excessive crying for hours at a time day after day with no identifiable cause...This condition is enormously stressful on the child and on parents, a continuous crying that inhibits the growing of neurons in the first stages of life and many times leads to abuse in households by humans who are unable to cope with something not understood...Estimates are that infantile colic affects 10% to 25% of babies worldwide but there is hope with new research that is focusing on the emerging field of *probiotics,* live bacteria living in the human intestinal tract that are essential to nutrition...In a study released in 2014 (Flavia Indrio at the University of Bari in Italy) introducing friendly bacteria into a baby's digestive system very soon after birth (just a few days) had a dramatic effect in improving symptoms and may even prevent colic from developing at all...More research is being done and the biggest challenge is getting this information to those who are in need but it shows a definite learning curve in our understanding of how all life on Earth is interconnected and woven through a single ecosystem that supports human survival...

Any and all actions that help the youngest humans on this planet engage the world without distress or pain should be the highest priority in the development of effective treatments and medicinal science...

Baby Health

A study in India of 700 babies under 4 months of age has shown giving zinc to newborns being treated with antibiotics for serious life-threatening infections helps with recovery increasing survival rates dramatically...This demonstrates in another way how connected humans are to the chemistry of this planet and the need for vigilance in understanding the nutritional needs required by every human baby to develop in a healthy way...*Thousands of malnourished babies across the world do not get the minimum of their nutritional requirements and it must be the priority of every aid group to educate and assist caregivers in creating a stable, predictable and adequate diet for every baby born...*

Over one million infants die every year from the complications of preterm birth (born alive before 37 weeks)...

A study in Sweden randomly assigned 334 infants so that half had their umbilical cords clamped within 10 seconds after birth and the other half were clamped after 3 minutes or longer...Tested 2 days later, there was no difference in iron status...Tests at 4 months found concentrations of iron 45% higher in the group that were clamped after 3 minutes and iron deficiency more prevalent in the babies who were clamped at 10 seconds...

Conclusion: Delayed cord clamping compared with early clamping resulted in improved iron status and reduced prevalence of iron deficiency at 4 months of age and reduced prevalence of neonatal anemia without demonstrable adverse effects. Since iron deficiency in infants, even those without anemia, has been associated with impaired development, delayed cord clamping seems to benefit full term infants even in regions with a relatively low prevalence of iron deficiency anemia...BMJ 2011...*This study of iron in the human body illustrates how little is still known about the requirements of newborns and the effect of seemingly negligible actions...*

WHOOPING COUGH

Estimates are 16 million people develop this affliction every year with nearly 200,000 children dying from another preventable disease...Whooping cough *(pertussis)* is a highly contagious respiratory infection caused by a bacterium *Bordetella pertussis* and is spread through the air...Whooping cough is preventable with vaccines, DtaP for children under seven and Tdap for older children and adults...A preventable and treatable disease with 95% of all cases in developing nations...

Whooping cough is one of the leading causes of vaccine preventable death in the world, mostly infants...

Mortality rates for all illness is more than 5x's higher among severely underweight children...

Almost 10% of Earth's adult population has diabetes...1.4 million humans died of this disease in 2011...WHO

SYPHILIS

In a study published in *PLOS Medicine* (26 February 2013) researchers estimated 1.4 million pregnant women were infected with syphilis worldwide in 2008 with approximately 520,000 adverse outcomes including 215,000 stillbirths, 90,000 neonatal deaths, 65,000 low weight babies and 150,000 babies with congenital infections...The researchers found that almost 33% of the women with negative outcomes attended antenatal care but were not tested or treated for syphilis...Caused by the spirochete *Treponema pallidum,* syphilis develops through 3 stages with few or no symptoms which aids in it's spread through populations...As many as 500 million new cases of curable STD's (syphilis, chlamydia, gonorrhea and tricomoniasis) occur worldwide each year...

The leading cause of premature death on Earth is cardiovascular disease, preventable through lifestyle changes...

Alcohol is responsible for over 2.5 million deaths a year...

CANCER

A report released in 2014 by the World Health Organization (WHO) estimates by year 2030 there will be over 20 million new cases of cancer each year with an average of 11 million cancer deaths worldwide...This ancient human disease is an unpredictable variation of cellular reproduction that goes out of control caused by a variety of factors having both internal triggers, genetic predisposition and various external triggers including natural phenomenon (UV rays, radiation emissions) and man-made elements (chemicals, pollutants, smoking) of a variety more abundant in 2016 than at any other time in human history...The report was compiled by over 250 scientists globally who noted that worldwide the most prevalent cancer is lung cancer (smoking and an increase in environmental atmospheric pollution) followed by liver cancer and stomach cancer...Over half of new cases diagnosed in 2012 were in Asia, mostly China where air pollution is progressively worsening in large urban areas with negative effects on adults and children...Developing countries are the most vulnerable with poverty associated cancers caused by infection, disease, tobacco use and the lack of progressive detection and treatment...Alcohol, smoking, processed food sources and a lack of exercise are listed as primary causes in developed nations with the costs of treatment and care rising in all the healthcare sectors...The global activities of tobacco corporations are cited as a dominant factor in the rise of tobacco use in "poor" countries with the report concluding that over half of all cancers could be avoided through prevention, early detection and treatment...On a planet where regulatory structure governing the release of harmful substances in the environment grows weaker, pollutants grow stronger and a lack of healthcare for Earth's exploding populations give little hope of changing the numbers predicted by this report...

The highest child mortality rates are in Sub-Saharan Africa and Southern Asia...1 in 8 children die before age 5 in Sub-Saharan Africa with 70% of the deaths occurring in the first year...50% of under-five deaths occur in just five nations, India, Nigeria, China, Pakistan and Democratic Republic of Congo...UNICEF

The majority of under-five deaths are caused by pneumonia, diarrhoeal disease, preterm birth complications and birth asphyxia...Malnutrition is an underlying cause in over one-third of these deaths...UNICEF

Approximately 60 million humans die each year...

Approximately 130 million humans are born each year...

There are currently more than 7.4 billion humans alive on Earth...

Pharmaceuticals

The current market for pharmaceuticals globally is estimated to be between $300 billion and $400 billion in 2015 with over 30% of the sales controlled by just 10 corporations...An industry with a long list of interest conflicts and a high profit margin, drug production is now focused on medication requiring long-term use with companies often spending up to a third of all sales revenues on promotion and advertising, more than twice the amount spent on drug research and development...With political donations of millions of dollars the pharmaceutical industry is opaque with corporations controlling trials and testing, releasing information favorable for sales potential while withholding information that could negatively affect the profit margin....

A medication to treat seizures in infants (Acthar) can cost almost $300,000 a year...

A medication (Naglazyme) used to treat a connective tissue disorder in children costs $350,000 a year...

Soliris, used to treat a rare stem cell disorder, had global sales of over $1.5 billion in 2013...A single year of treatment on Soliris costs over $400,000...

Myozyme, a drug that treats a rare genetic disorder by replacing a deficient enzyme costs over $100,000 in the treatment of infants and as much as $300,000 to treat an adult...

Kalydeco, a drug that treats the underlying cause of cystic fibrosis costs around $300,000 per year...

It's an intensive process in most cases to develop effective medication with testing, animal trials, human trials, distribution, regulation and the implementation of treatment but it's also true that market forces and the influence of corporations on political and regulatory structures has created an exclusive, captive market beholden to pharmaceutical companies in a global climate of greed and profit which has created one of the most profitable industries on Earth...There are better, more efficient, more humane, more effective and more empathic ways to deliver necessary medicine to the humans who suffer from these rare disorders affecting thousands instead of millions and new research into genetic components that are the root cause of these conditions will one day 'cure' the problem before it happens...Treatment for widespread diseases affecting millions should be available at no cost to every human on Earth with an aggressive campaign to eliminate the source...(Rabies could be eliminated for 10% of what it costs to treat those who are infected?)...There's a balance but many pharmaceutical corporations continue to charge these exorbitant prices simply because they can with a blind eye to the suffering of the poor...

WHO reports over 5 million humans die in pain every year because of the unavailability of morphine in poorer nations due to ignorance, government restrictions and inconsistencies on a primitive planet...2012

"Evergreen" is a tactic used by drug companies to make a very slight change in a drug so they can continue to charge exorbitant prices for the drug and claim it is 'new'...Where is the empathy and reality intelligence in this practice that keeps an effective medication from the humans who need it the most and cannot get it if poor...With this kind of manipulation a specific drug can continue to cost $70,000 a year instead of $2,500 a year...

Another popular tactic is to "Pay for Delay" where drug companies pay generic manufacturers to delay production of a cheaper alternative so they can continue to make the brand name drug and charge very high prices...

A crisis unfolding in real time is the unavailability of effective drugs for entire populations as these corporate structures focus on shareholder profit stopping production of necessary drugs that are less profitable...

It's obvious the activities that create this market of inequality are not the result of incompetency or a lack of knowledge or a 'learning curve' but rather conscious, directed strategies to create as much profit for the corporation and/or individuals without awareness or empathy for millions of humans who suffer from the inability of billions of humans to understand human reality and human singularity...

America's healthcare industry spent $14 billion on advertising in 2014 with pharmaceutical corporations spending as much $4.8 billion a year (NYT's)...Direct to consumer advertising in one of the few nations that allow it...

What's clear is widespread disease, regional and global epidemics and the unacceptable suffering and deaths of million of children will continue until the intelligence and commitment of this species increases to a point where borders begin to disappear and a focus on infrastructure, prevention and global cooperation becomes a priority...In the ignorant primitive climate that dominates human reality in this century that future will take hundreds of years...

WORLD FOOD

Over 70 billion animals are killed every year for food...

The United Nations Food and Agriculture Organization estimates around 795 million people suffered from chronic undernourishment in 2014-2016...780 million live in developing nations, 11 million in developed nations...

Over one billion humans are continually hungry on Earth defined as consuming less than 1,800 calories a day...soaring food prices, mismanagement, violence and instability are among the reasons...UN

Over seven million people die every year from hunger...UN

It's estimated for every one degree Celsius increase in average world temperature crop yields will decrease by an average of 5% on a planet that will need to more than double current food production to feed an expected population of over 9 billion by 2050...One solution is 'climate-smart agriculture' using crop rotation, better water management, genetic variations of drought tolerant crops, zero tilling techniques and research into ways to make human agriculture an active carbon sink to restore balance to a continually changing world...

The UN reported in 2011 that acute hunger was afflicting 4.5 million humans in Ethiopia, 3.75 million humans in Kenya and over four million humans in Somalia...Yet another crisis in this desperate region of Earth experiencing the worst drought in 60 years, this possibility was recognized in 2010 by The Famine Early Warning Systems Network but serious action was not taken until almost a year later and it was too little, too late...The sad aspect is that food is available but these people live in abject poverty with no relief at any level of civilization...The Group of Eight, powerful countries with vast resources, promised to fund over $20 billion in aid for food security and agriculture development in 2009 but of the money that was delivered little was spent to help these populations...It costs far less to develop agriculture than it does to import food aid but local governments and political factions are clumsy and corrupt with inadequate transport and no protections for the populations...Aid workers who know concede that many things could be done to establish self-supporting farm communities, improve roads, transportation, sanitation, water collection and storage facilities if the support and the supplies were made available, something very possible in this world if people with technology and intelligence were managing and controlling resources and security...But these are not priorities on this continent and the tragedy of this chronic inaction is another generation of humans who will suffer distress and hunger from birth and will grow up without the dense brain growth necessary to understand reality...

Children who are malnourished can suffer up to 160 days of illness each year...

In Sudan (2014) the top UN humanitarian official (Toby Lanzer) reported almost 4 million people at severe risk of food insecurity after violence disrupted the planting season...Thousands have died from unchecked violence as the government grows weak and ineffective increasing the prospect of famine that will effect millions of humans and kill more than 50,000 children...Almost a million people have been displaced by fighting since December 2013 and almost 80,000 humans have sought UN protection...An estimated 146,000 tons of food was needed and as of January 2014 just $300 million have been received from donor nations who pledged $1.27 billion...

More than 2 million children under age 5 die from hunger every year...

Freshwater fisheries, marine fisheries and aquaculture (fish farming) supply roughly 15% of the world's calorie intake and it's growing clear this percentage will decline in the future as marine fisheries collapse from overfishing in an industry that's declined consistently for two decades...Currently, marine capture fisheries yield up to 150 million tonnes of seafood annually with around 70% going to human consumption and the rest converted to fish-meal or discarded while ocean habitat damage from decades of trawl nets, drift gill nets, careless pollution, sewage and agricultural runoff has created dead-zones where fish population's have declined dramatically...Ironic is a growing insustainability of aquaculture operations due to a decline in wild fish that are used as feed in fish farms...The continuing decline in marine fisheries globally is a serious problem with few serious solutions and it will continue to impact the lives of hundreds of millions of humans as Earth's population increases...

Discarded fish from marine fisheries, as much as 30 million tonnes annually, is the largest amount lost of any single food source produced or harvested from the wild...

The amount of fish discarded at sea could sustain a 50% increase in aquaculture production...

About one-third of arable land is now used to grow the animal feed necessary to produce meat demanded by world populations and it's fueled an increase in global deforestation with almost 70% of Amazon land turned into pasture, grazing land degraded due to soil erosion, overuse and the growing abundance of feed animals who are responsible for almost 20% of greenhouse gas emissions...It's increasingly clear that current methods of meat production at industrial scales developed to maximize industry profit are energy inefficient, environmentally harmful and unsustainable over time...

By 2050, 1,573 million tonnes of cereals will be used annually for non-food (FAO, 2006a) of which at least 1.45 million tonnes can be estimated to be used as animal feed...Each tonne of cereal can be modestly estimated to contain 3 million kcal which means the yearly use of cereals for non-food will represent 4,350 billion kcal...If we assume a human daily caloric need of 3,000 kcal, then about 1 million kcal/year is needed per person...Taking the energy value of the meat produced into consideration and the loss of calories by feeding the cereals to animals instead of using these cereals directly as human food, represents the annual caloric need for over 3.5 billion people...
United Nations Environment Programme (UNEP)

In 2013, Americans raised and slaughtered 9 billion animals for consumption or export...8.6 billion chickens, 240 million turkeys, 112 million pigs, 33 million cows, 24 million ducks and 2 million sheep...

Food contamination sickens almost 50 million humans in America each year with about 3,000 food-related deaths...

United States
The U.S. is the world's largest exporter of agricultural products...
The use of antibiotics in feed animals to accelerate growth and help protect the animals from pathogens while keeping them confined in crowded, filthy conditions is over 80% of the U.S. antibiotic market with accumulating negative effects on consumers as new drug resistant pathogens are created with a lack of regulation, no awareness of consequence and maximum corporate profit still the priority in free market America...The FDA is weak on enforcement due to continued political pressure from the lawmakers given political money by this industry and in 2012 the new proposed rules and regulations are *voluntary* for the meat industry *and* for the drug industry concerning the use of these antibiotics which in a free market society with little regulation means little will change...In the growing and killing of chickens for human consumption there are numerous antibiotics being used to promote rapid growth and control the process to produce the biggest chickens in the shortest time to maximize profit...CAFO (Concentrated Animal Feeding Operations) is a process of feeding enhanced product and administering multiple antibiotics to captive animals to accelerate growth and shorten the life of the animal...The USDA is formulating new ways to save money by removing inspectors from factory floors and shifting to a random testing procedure that will allow more substandard product to reach the consumers...New laws introduced and passed by state legislatures now punish the people that expose animal cruelty and regulatory violations in processing plants while continuing market pressure from large grocery chains and worldwide fast food outlets are instrumental in changing the practices in meat, pork and poultry industries who use cheap labor and a lack of regulation to produce their product as cheaply as possible to maximize profits and meet a growing demand for unhealthy food produced in an unhealthy way...

In the United States 30% of all food ($48.3 billion) is thrown away each year...It's estimated about half of the water used to produce this food also goes to waste since agriculture is the largest human use of water...

Chemicals and antibiotics used regularly in the processing of American food are banned in many nations around the world...These include growth hormones rBGH and rBST, food preservatives BHA and BHT, the additive Potassium bromate (a know carcinogen), arsenic given to chickens to increase body weight and various colors and dyes that have been used for years (also known carcinogens with some implications in birth defects)...Scientists have linked rBST to prostate, breast, and colon cancers...Banned overseas, it is still legal in the U.S...

A California slaughterhouse where the videotaped mistreatment of cows prompted the largest meat recall in U.S. history reached a $500 million settlement with an animal welfare group and the federal government which is not expected to be paid because Hallmark Meat Co. is bankrupt...It was the first time federal fraud statutes were used in an animal abuse case because this company supplied meat for national school lunch programs...Another abuse investigation in central California of the Central Valley Meat Company was initiated after videos showed animals who had fallen down being repeatedly kicked and shocked...

Taking videos of animals being abused and mistreated in slaughterhouses is now illegal in multiple states...

Over 32 million children in the United States eat lunch at schools that participate in the National School Lunch Program using agricultural surplus to feed the children with over twenty million of these children eating at reduced prices or for free at a government cost of over $13 billion per year...Feeding children is critical but as with many things in the U.S. this program has little oversight, widespread mismanagement, exploitation and privatization for profit by large corporations whose singular priority is profit...The result is an Agricultural Department that pays over $1 billion for commodities with over half being processed by corporate food manufacturers who turn healthy food into fast food with a lower nutritional value at a higher cost...Money mismanaged by the government turns into large profits for the corporations and poor health for the children who eat the food...Management companies contracted by schools to distribute taxpayer money in an efficient and equitable way charge the schools full price for food processor contracts and receive rebates from the corporations who are awarded these large contracts...Bribery and corruption with neither management or processors admitting to any impropriety or any responsibility...Privately managed school cafeterias are consistently shown to offer food higher in sugar and fat with unhealthy foods available to students while lobbyists for the food processing companies have delayed and prevented the passage of laws that would put limits on this waste, limits on unhealthy food and require more accountability at every step of the process...

GRAS

Researchers for the Pew Charitable Trust released a report in 2013 showing financial conflicts of interest are almost ubiquitous when food company employees, paid consultants or professional experts conclude that chemicals are "generally recognized as safe" (GRAS) and can be added to almost any food... GRAS determinations require no input from or even notification to the U.S. Food and Drug Administration (FDA) and almost all the determinations studied were made by panels or individuals selected by the additive manufacturer or by a consulting firm representing the additive manufacturer with the same individual(s) serving on multiple panels making the GRAS determinations...Corruption, greed, malfeasance and collusion...In 1958, Congress created the GRAS exemption to the more formal food additive petition process but this exception was intended for common food ingredients such as oil and vinegar...Since that time companies have expanded the use of the GRAS loophole to include a wide variety of chemicals whose safety is far from "generally accepted." and over the past ten years the loophole has now *become the law* with thousands of chemicals being cleared via GRAS determinations rather than the food additive petition process...A 2010 report by the GAO (Government Accountability Office) revealed the same "conflicts of interest" and has recommended action...Nothing has changed as corporate and lobbyist money continue to fill the pockets of America's public servants...

Out of 450 GRAS determinations made between 1997 and 2012, 22% were made by an employee of the additive manufacturer, 13% were made by an employee of the consulting firm selected by the additive manufacturer and 64% were decided by an "expert panel" selected by the additive manufacturer's consulting firm...

The U.S. government does not have the power to recall food off the shelves...

In the United States pigs produced for the pork industry are confined in 'gestation crates', steel cages just 24" wide for up to three years until slaughtered...Nine U.S. states and most of Europe have outlawed the crates...

Kosher Meat

Kashrut is the set of dietary laws primarily based on religious beliefs that dictate everything from what is permitted to be eaten to how animals are killed in the preparation of food...Animals are killed by rabbis who specialize in *shechita* or ritualistic slaughter (*how does one specialize in ritualistic slaughter?*) who make one long cut across the throat of the animal in an act that is almost always painful and inhumane to the animal both in the preparation of the animal (*flipped upside down to expose the throat*) and the killing itself, often requiring second cuts while the animal is still conscious to finish this process...The animal must be kosher and this is restricted to ruminants that have split hooves...Ritualistic slaughter is banned in Switzerland, Norway, Sweden, Poland and Iceland...All birds not specifically excluded in Deuteronomy 14:12–18 (biblical reference) are permitted as food but according to strict rabbinical law only the birds with a tradition of being eaten are allowed...Before slaughtering every animal must be healthy, uninjured, and viable...The animal cannot be stunned by electronarcosis, a captive-bolt shot to the brain, or gas, (both common practices in modern animal slaughter) as this would injure/kill the animal rendering the ritualistic slaughter invalid...This refusal to kill the animal in an instantaneous way promises a painful death for the animal and this entire primitive process is done thousand's of times each day based on questionable religious beliefs...

In February 2014, Denmark outlawed ritualistic slaughter saying the "animal's rights come first" and all animals must be "stunned' before killing...Jewish organizations reacted negatively to the ban saying it is an "attack on Jewish religious practice" and the Conference of European Rabbis is looking for options to "stop these continuous attacks against religious minorities in Europe"...

Over 50,000 cows are killed every 90 minutes to feed humans...

Industrial Food

Tyson Foods in Arkansas is currently the largest meat production company in America, a fully integrated operation from farm to store that supplies food to over 130 countries worldwide including beef, poultry, pork and prepared foods...In 2016 there were over 6,200 slaughterhouses in the U.S. but just four large corporations supply 80% of all the meat in the U.S....In a new normal of corruption (legal), political donations have altered and moderated the state and federal laws designed protect the consumer in favor of corporate profit, new laws that took away the USDA's authority to shut down a processing plant even when it repeatedly produces contaminated products...Regulation by oversight is weak in this industry where close ties between regulators and the industry are obvious and corrupt with corporations allowed to choose their own regulators and inspectors....

Food has changed more in the past 50 years than in the previous 10,000 years...

The FDA made over 50,000 safety inspections in 1972 when there were 3 billion less humans in the world and only 9,000 safety inspections in 2006 in an industry that's bought favor with politicians for years to create maximum profit without serious oversight...Some of these companies have a long history of opposing regulation and safety rules with key political players who introduce amendments written by corporation lobbyists to favor any legislation toward the companies, not legislation written to protect the consumers...Monsanto, a large multinational agricultural corporation based in Missouri, controls over 90% of all soybean farming in the U.S where they can require farmers to buy only their seeds through 'patent laws' passed by Congress...Monsanto can then control and police the seeds the farmers plant with their own investigators who travel the country to insure their policies are strictly followed...They've used their own funding to pay for "expert" groups to approve their products and it's not surprising many Monsanto executives have been political appointees in multiple administrations in the White House...The food industry also has laws enacted by state legislations to put people in jail if they say anything negative about their product, ridiculous laws called "veggie libel laws" that prohibit consumers from saying anything about food quality and production methods, anything critical whether true or not...Products that humans buy and consume...These companies fight food labeling by arguing the consumer would be confused, not understand the new labels and consumers do not need to know where food comes from or what exactly is in a product they're buying...An obvious and glaring lack of transparency in a profitable industry controlling the methodology of food production and consumption that is vital to maintaining human health at every level of existence...

The U.S. raises and kills more than 8 billion chickens a year...

It's estimated over a billion tons of food is lost or wasted annually during production in low-income countries and during consumption in high-income countries...

How can financially and politically connected corporations be allowed to control "seeds", a commodity that has been used by farmers worldwide for thousands of years to grow crops? Monsanto, DuPont, Syngenta, Dow and Bayer control "patents" on seeds and many pesticides that adversely affect the freedom of agriculture and farmers worldwide with a singular motive of profit...All modified seeds were and are variations of seeds that have supported the human species since the beginning of agricultural societies and the control of this product for profit is a reflection of Earth's new paradigm in this century of greed, control and profit by any means necessary...It's estimated over 80% of corn seeds are coated with pesticides (neonicotinoids) made by Bayer that have been implicated in the destruction of bee colonies...A proposed ban on these chemicals by the EU brought threats of lawsuits and intense lobbying by Bayer and Syngenta to prevent any changes in regulation...Thousand's of humans in Europe and in over 50 nations have protested Monsanto's control of genetically modified seeds and their continued resistance to food labeling that would give consumers knowledge and a choice...The control of the agricultural market by five or six corporate giants is abusive...These companies spend millions of dollars lobbying to influence policy and prevent legislation that would affect their operations or affect their profit margins and many lobbyists employed by these corporations are former government employees with inside connections to promote corporation policies...In addition to excessive control over agricultural policies worldwide, many of these companies have a reputation and a record of environmental pollution with cleanups, fines and financial settlements being an accepted expense of doing business...

The Rest of the World

The famines of the past 50 years have created millions of humans who are unable to understand the nature of human existence...A baby's brain does not grow dense neural pathways if there is chronic hunger...

United Kingdom households waste an estimated 6.7 million tonnes of food every year, around one third of the 21.7 million tonnes purchased...Approximately 32% of all food purchased every year is not eaten...

In developing countries the food producer and the consumer have a closer connection than in developed countries...With fewer processed and packaged foods, more fresh food is traded at market by street vendors and perishable food is most often prepared and consumed immediately with minimal storage...Problems in developing countries include inappropriate use of agricultural chemicals, the use of untreated wastewater, the use of sewage and animal manure on crops, the absence of food inspection including meat inspection and a lack of infrastructure throughout the food processing chain including a lack of refrigeration, poor hygiene and the scarcity of clean water...

India

A estimate by the Ministry of Food Processing reported that agricultural produce worth 580 billion Rupees is wasted in India every year...

Over 250 million Indians (21% of the population) do not get enough to eat despite a 50% increase in food production over the last two decades...40% of the 250 million are children...

India ranks first in world buffalo population with more than 115 million buffalo producing over 1.5 million tonnes of buffalo meat every year with exports to other countries comprising about 25% of the total...The country also processes sheep, goats, pigs and poultry with a production of more than 33 billion eggs annually, nearly 4,000 slaughter houses and 200 meat processing units with license...India is the largest milk producer in the world (75 million dairy farms) with over 132 million tons of milk produced annually using a system of millions of cooperative milk producers who collectively can earn almost $5 billion a year...India's food policy is focused on providing farmers with higher and more consistent prices for their crops than they would get from the open market and then selling food grains to the poor at lower prices than they would pay at private stores...

In 2011-2012 India produced a record 250 million tons of food grains...

In a country with the world's highest incidence of child malnutrition, children of middle class families eat up to four times the recommended quantity of food resulting in an increasing rate of obesity in India's urban areas...

The enormous increase in India's food production that began with the "Green Revolution" in the 1960's and 70's has not happened without a price...In many agricultural areas groundwater is depleted forcing the use of high-horsepower pumps to go deep underground for needed water and the current system with the state buying grain for distribution is riddled with corruption from the top down causing many of the poor and hungry to continue without adequate food on a daily basis...The Supreme Court even weighed in urging the government to "give the food away for free" rather than letting it rot in warehouses because of a broken system...A National Food Security Act designed to give over 70% of India's population a "legal right to food" was passed into law in September 2013...

Agriculture in India uses more than 80% of all freshwater supplies...

China

12% of China's population (162 million humans) are chronically hungry...

The most populous nation on Earth has undergone a dramatic change in food production over the last 20 years that today is feeding nearly 25% of the world's population on less than 10% of arable land but the environmental impact of this achievement is becoming apparent in dramatic ways...China has just 0.09 hectares of arable land per capita and more than 20% of that arable land is polluted with industrial effluent, agricultural chemicals, sewage and even mine runoff...The soil in China is becoming increasingly toxic and China uses more than one third of the world's nitrogen output for fertilizer to aggressively sustain growth in the fields...This enormous use of fertilizer has created "dead zones" in five of China's largest fresh water lakes and accelerated the acidification of the soil decreasing crop production by as much as 50% in some areas...

China's annual "number one document", released at the start of each year, was focused in 2014 on curbing rural pollution and stating the country "would not relax domestic food production" seeking grain self-sufficiency with limited imports...Approximately 3.33 million hectares of farmland are too polluted to grow crops...

China exports tilapia, cod, apple juice, mushrooms, and garlic into the U.S. with less than 3% of these imports inspected by the Food and Drug Administration...This from a country with many food related problems including multiple "tainted milk" scandals and an incident involving the deaths of thousand's of pigs (11,000) found floating in a major river...In a decision made in 2013 the U.S. Department of Agriculture has decided to allow chickens killed in the U.S. to be sent to China for processing and then returned to the U.S. for sale (an enormous waste of resources)...In 2012, the U.S. exported over $350 million in poultry to China and imported over $1.9 billion in seafood from China...

In 2012, the volume of beef processed in China was over 6 million tons, 10% of the global beef output...The United Nations Food and Agriculture Organization estimates that a 70% increase in global food production and a 50% rise in investment in agriculture will be needed to feed a population of 9.5 billion expected by 2050...

The total volume of demand for meat is expected to increase by more than 200 million tons over the next three decades with the majority of demand coming from the United States...

Food grains including rice, wheat and coarse grains (cereal grains used primarily as animal feed) account for most of the world's crop area...Over two billion tons in 2012, mostly corn, wheat and rice...

Asia

Experts are convinced the future of agriculture and food security in Asia will be dominated by two major forces: China and Biotechnology...With one of the largest footprints globally, decisions by China involving trade, agricultural infrastructure, genetic crop engineering, commodity exchange and population growth will be influential in re-defining global agribusiness and equitable trade across the entire region...Bio-tech and genetically modified food (GMO) will increase with acceptance in a world challenged by changing climate and growing populations that will increase demand for food produced and processed within smaller regions...Advances in bio-science will aid in growing food in challenging inhospitable environments and augment a growing need for water management and conservation...

There is a steady growth in Asia and around the world in prepared food and packaged food gradually eliminating the need for full service kitchens in the home...Almost 40% of consumer food expenditure's in Japan in 2010 were for food prepared outside the home...In America food prepared outside the home accounted for over 50% of food costs...

The World Health Organization (WHO) classified 45% of men in China as overweight in 2010, an increase of almost 18% in 8 years...Primary causes are an increased access to processed foods and beverages...

There is a growing, increased focus in all of Asia on food safety and the dependability of supply systems in the management of ingredients and preparation...In China (2013) approximately 900 people were arrested in a "crackdown" on selling mislabeled meat products and meat-related crimes...In India it's estimated almost 90% of food production is "informal" and primarily based in poor urban areas...To this end China has consolidated the responsibility for non-agricultural food and drug safety under the new China Food and Drug Administration (CFDA) to improve oversight and enforcement in a nation beginning to understand the economic cost of unsafe food as the demand for imported food rises...Also problematic is emerging markets Philippines, Taiwan, Indonesia, India and China who continue to experience exponential growth in chemically intensive agriculture with farmers ignorant of application procedures, re-harvest preparation and post-harvest treatments to extend a product's shelf life...Large global chemical corporations aggressively market these substances and the long-term consequences are ignored in the effort to maximize profit and feed increasing populations...

A growing movement in developing countries is focusing on the reorientation of family farms designed to produce food for domestic populations, a new strategy with the G8 changing its focus from severe hunger supply scenarios to funding support programs to develop sustainable local and national food systems...Support and resources for food sustainability in Africa should have been prioritized a long time ago instead of short-sighted reactions to famine that produced enormous waste and fraud...

Africa

18 of 20 countries with the highest fertility rate are located in Sub-Saharan Africa and with a population expected to double by 2050 Africa is very challenged by food demand and food security...A 2012 estimate reported over 240 million undernourished humans living on the continent, an increase of over 30% in just 20 years...Modern agriculture is mostly dependent on water, fossil fuels and fertilizer phosphate and in sub-Saharan Africa there are challenges in the acquisition and total resources of all three products...A report by the U.S. Department of Agriculture concluded "50% of the land in Africa is unsuitable for any kind of agriculture" and when deforestation, drought and a changing climate are fed into the equation the International Panel on Climate Change (IPCC) estimated a reduction in yields of over 50% by 2020...Africa's focus this decade will be less on world exports and an increase of domestic production but there are no easy solutions with the expected exponential growth of people...

Sub-Saharan Africa has the highest prevalence of hunger in the world, 25% of the continent's population...

A majority of the poor and malnourished people in Africa live in remote, rural areas dependent on subsistence farming and traditional rural-based activities but many African nations decimated by continuous war and corrupt leaders have no infrastructure or any adequate systems to collect, store and distribute food to populations or even a capacity to know what food they have and where that food is going...New tracking requirements initiated just last decade by the Comprehensive Africa Agricultural Development Program are still weak, uncoordinated and unsustainable...There is so little stability in many of these countries that a comprehensive program relying on statistical data to analyze human needs and current conditions will not alleviate the hunger of more than 25% of the population...One solution is to give the people, every individual on the continent, $1,000 per year for five years and let the marketplace work as it should...At the end of five years there would be statistics, stability and a way forward to build agricultural resources to help the entire continent...This will not happen...

Russia

Russia's goal by 2020 is to be self-sufficient in supplying all the major food groups within the country through modernization and efficiency concerning the production of multiple animal products, an increase in agriculture and a growing fish industry...Poultry and pork are the fastest growing sectors and farming is exploding with an estimated 45 million arable hectares unused in 2014...

Futurists expect Russia's northern territories to become vast food production fields in a future of climate change...

With a focus on in-country production and a restriction on imports, Russia is upgrading fishing fleets and coastal facilities to meet new demand...Wheat, potatoes, cereals and sugar beet are important crops and global warming is already increasing a limited growing season by five to ten days a year...In the near future many believe whole new varieties of crops will be sustainable in a warming climate but along with this change come pests, disease and an increased need for chemical pesticides to protect yields...Russia exported over $7 billion worth of food products in 2012 with almost two-thirds going to former Soviet satellite countries and former Soviet republics and there is a huge push by Russia's food producers to bring their cuisine and traditions to the rest of the world citing innovative food products not found anywhere else on Earth...One of the main challenges are the restrictions and regulations enforced by the EU on imports that are not understood due to a variety of ingredients or the techniques incorporated into the manufacturing of Russian food products...

The parliamentary majority United Russia has put forth a bill severely restricting the imports of genetically modified products (GMO) into Russia although the production of GMO food in country is still allowed...Producers must label the product and warn consumers if the the percentage of GMO additives exceeds 9%...

In 2011 Russia produced 9 million tons of grain, 4% of the world's grain and rice production...

The poverty rate in South America is estimated at more than 66% of the population...

Latin America & South America

A region that is home to over 25% of new arable land and a substantial quantity of renewable water sources, Latin America has increased agricultural exports at a rate of 8% annually since the mid 1990's and is becoming a major player in a growing world food market..Still needed are upgrades in infrastructure, a review of regulations and a revamping of trade policies to reach a potential that experts say could supply one-third of the meat and one-third of the fruits and vegetables traded on the world markets by 2050...Future free-trade agreements and local politics will determine an increase in food production that must grow 70% to 80% by 2050 to meets the needs of an expected 35% increase in population...The poor in this region of the world spend up to 70% of available income on food...

Around 45 million humans in Latin America and the Caribbean are chronically malnourished...

Brazil is the world's largest exporter of beef sending over 1.35 million tonnes to 130 nations in 2013 with a value of more than $6 billion...Brazil has the largest concentration of cattle in the world with over 190 million animals...

South America is a major exporter of food with corn being the most widely cultivated crop on the continent, beans and potatoes a staple in many European nations, tomatoes in the west, cashews in the tropics and coffee an essential product in almost every country on Earth...Soybean production hit new records in 2013 with Brazil marketing 82 million tons, Argentina with almost 50 million acres devoted to soybeans and Paraguay, Uruguay and Bolivia maximizing production to take advantage of America's fixation on corn and bio-fuels...In 2014 these five South American nations are expected to plant almost 134 million acres of soybeans with the majority of the crop going to China but there are negative consequences from this surge in production including continued deforestation in Brazil (35% more in 2013 than in 2012) and a growing widespread use of pesticides made possible by genetically-modified soybeans more resistant to the effect of heavy pesticide use...

Argentina crop production is focused on sugar cane, wheat, soybeans and corn...Livestock (over 55 million cattle) and dairy are strong exports with an active, vibrant fishing industry that routinely exceeds a half a million tons annually...Many exports are limited to countries on the continent, Latin America and the Caribbean...Columbia is the world's fourth largest coffee producer with cattle farms taking almost 75% of agricultural land (most beef is consumed domestically)...Banana's are the 3rd largest agricultural export with the majority going to the European Union and the U.S...

25% of the humans in Columbia and Peru are "hungry" and in Brazil 21 million humans suffer a lack of food...

Venezuela imports an enormous amount of food stocks (as much as 60%) with price controls that effect the acquisition and availability of nearly everything and a president (Nicolas Maduro) who blames "economic war" from the private sector and America's CIA interference for the food shortages that have grown routine over the past two years...At the end of 2013 the Ministry of Agriculture and Lands in Venezuela announced that domestic agriculture had grown by more than 10% compared with 2012 including an increased production of meat, milk, sugarcane, potatoes and corn maize...This met about 40% of the domestic needs requiring supplementation with imports and there are still shortages of chicken, milk, beef, pasta and sugar...Inconsistencies in the government are mostly to blame...

The United Nations reports that Venezuela has reduced the number of people suffering from malnutrition from more than 13% in 1991 to less than 5% in 2011...

"Our goal is to produce the food that we consume and transform Venezuela into an exporting powerhouse..."
President Nicolas Maduro 2013

20% of irrigated cropland use 70% of global freshwater supplies...

The Future of Food

No organization on Earth can accurately predict the cumulative effects of markets, money, supply, population, climate and politics that will determine food availability and resource sustainability in the 21st century...Critical in the equation is the preservation of crop biodiversity and creating regional footprints of local crop growth and food production that promise a safer product and ease demands on transport in a world where energy use will be the determinant of any and all activity...Also critical is the environmental cost of food production which today sees a majority of cereal grain fed to animals in a global society increasingly eating meat...New technology, seed manipulation and innovative strategies for water use will challenge entrenched farming techniques and these new ways of defining agricultural production will be fluid and very different from any predictions put forth today...More alarming than any land use conundrum is the ongoing collapse of marine fisheries worldwide that if continued will have real-time consequences for populations everywhere on Earth...Predicted by 2050 is a world where 80% of seafood consumed for food will come from aqua-farms who continue to crowd fish together in unimaginable ways (50,000 farm salmon per enclosure, a creature about two feet in length living life in a space slightly bigger than a kitchen sink) with the farm fish being fed ocean fish and antibiotics to counteract the contaminants found at high levels in these facilities...Around 25% of all farm fish die before slaughter from stress, disease and other factors...The majority of aquaculture is in Asia (65% in China) this century with facilities run by multinational corporations with little thought to environmental impact and sustainability...This is slowly changing...

Defined by the United Nations Food and Agriculture Organization (FAO), aquaculture is the "farming of aquatic organisms including fish, mollusks, crustaceans and aquatic plants implicating some sort of intervention in the rearing process to enhance production such as regular stocking, feeding, protection from predators, etc...Farming also implies individual or corporate ownership of the stock being cultivated..."

Aquaculture is the fastest growing sector of the world's food economy increasing by more than 10% per year...

Around three billion people worldwide depend on fish as their main source of animal protein...

A report in the journal Marine Policy estimated that 20% to 30% of wild seafood imported to the United States every year is from "pirate" fishing, the illegal fishing of tuna, lobster, crab and snapper worldwide believed to represent between 13% to 31% of all seafood catches...Many processing plants in foreign nations have few controls to regulate the source of fish and the open ocean is not yet manageable in the 21st century...

The U.S. kills about 6 billion fish each year for consumption with over 100 billion killed globally...

Aquaculture

Aquaculture in today's economy provides more seafood to consumer markets worldwide than wild fish demonstrating the rapid increase in this industry whose intent is to provide a constant supply of seafood (mostly salmon) to a hungry world...The many benefits of a steady predictable supply are obvious but the problems are many including pollution of farms, overcrowding of the fish, the spread of disease in a closed environment and a negative ratio of feed per fish with 2 to 4 pounds of wild bait fish needed to grow a pound of salmon...The Aquaculture Stewardship Council or ASC (an independent non-profit based in the Netherlands) is an organization created to address aquaculture sustainability and the environmental impact of this industry with a focus on providing a standard for the industry and it's having an impact with the recent formation of the Global Salmon Initiative (GSI), a consortium of the world's largest salmon farm operations all pledged to use salmon only from farms that meet ASC standards (by 2020)...These new standards include sustainability of wild fish used to feed the salmon, limits on antibiotics and a ban on genetic engineering...Active testing is ongoing at multiple test sites with Norway trying to raise salmon using alternative proteins without having to feed them fish and a DuPont subsidiary, Verlasso, feeding the salmon pellets of blended yeast, fish meal, plants and fish oil that reduced the use of wild fish to a one to one ratio *and* they keep the density of salmon in their pens at 12 kilograms of fish per cubic meter of water as opposed to accepted industry standards of 25 kilograms of salmon per cubic meter of water...Despite ASC guidelines there is still a high interest in developing and marketing genetically engineered fish who can grow to market size in less than half the time of normal fish...

The U.S. consumed more than 282,000 metric tons of farmed salmon in 2013, an increase of almost 15% in five years...Global demand for salmon will double by 2020...

Major concerns in this global industry is the environmental impact on hundreds of thousands of hectares of vital coastal wetlands and mangrove forests destroyed to construct aquaculture facilities and intense harvesting of small ocean fish used as fish meal that has an unintended impact on larger fish species who depend on small fish for food...The danger is an industry growing so rapidly that a focus on profit will override sensible policies necessary to regulate escalating activities with unknown long-term consequences before science and industry provide answers and solutions to environmental impacts and aqua-farms can focus on omnivorous species, some of the easiest fish to feed and herbivores like tilapia and catfish...The promise of farming seafood is it provides a healthy product with a high level of protein and essential nutrients for growing populations with a smaller environmental impact than the production of beef...The Earth Policy Institute reported that aquaculture produced 66 million tons of product compared to 63 million tons of beef product in 2012, the first time this has happened in modern history...On a crowded planet with unchecked population growth farmers will need all the agricultural knowledge acquired over the past 13,000 years to meet the challenge of sustainable food production for billions of future humans...

The Food & Agriculture Organization of the United Nations reported that global aquaculture produced 73.8 million metric tons of aquatic animals in 2014 worth around $160 billion...

The Svalbard Global Seed Vault (Earth's Doomsday Vault) is a facility located on an archipelago north of Norway where more than 100 nations have stored over 800,000 duplicates of seed samples from around the world in an environment intended to preserve these samples for centuries...It is owned by the Norwegian government and managed by the Global Crop Diversity Trust and the Nordic Genetics Resource Center...

Interesting is a new focus on 'vertical farming' conducted in large high-rise structures where all aspects of growth can be controlled and regulated (light, temperature, nutrients, etc)...It's an exciting area of development with the ability to produce food for regional areas year round reducing the wasteful transport of food worldwide...The two largest vertical farms in the U.S. are Green Sense Farms (hydroponic with roots under water) and AeroFarms (aeroponic) with the roots exposed to the air...Control over all environmental determinants is a huge advantage as well as reducing pollution, pesticides (unnecessary with no insects) and a known environmental footprint...

Alternatives

Algae...There is a movement in regions of the world to create commercial algae farms to supply populations with food and produce biofuel in much greater quantities than the ethanol biofuel produced today...Algae are simple organisms that can thrive in many areas unsuitable for crop production and the increased production of algae on a large scale will prevent the waste of billions of gallons of water and provide food for humans, feed for animals and fertilizers for crop growth...

Terra-forming...Norway is backing a new experimental large-scale project (The Sahara Forest Project) being built in Jordan to test the real world feasibility of seawater greenhouses that use a model of the planet's natural water cycle to feed hot desert air into a structure containing seawater where it is cooled, humidified and fed into an evaporator to condense and collect freshwater...The greenhouse operation is combined with a Concentrated Solar Power electrical generation facility with the goal of creating green jobs, sustainable vegetation and food production with clean electricity...

A large-scale attempt at terra-forming adopted in 2007 by African governments and still being studied and planned in 2014 is the Great Green Wall Initiative, a project that would develop a barrier to desertification 15 kilometers wide and 8,000 kilometers long passing through 11 nations that border the South Sahara desert...This "living wall" of trees, bushes, animals and birds would protect the waters of Lake Chad, help restore habitat and reduce erosion creating a bio-diverse environment for sustainable crop growth and development...Over 80% of rural African inhabitants depend on the land for survival and it's estimated more than 40% of the land in this region is degraded and growing more so with rising temperatures and increasing drought (Less than 5% of available cropland employs sustainable water and land management)...Recent efforts at land restoration have been successful in Senegal (planting 11 million trees over 27,000 hectares) and in Burkina Faso restoring over 50,000 hectares of agro-forestry using traditional land use practices merged with new available technology...It's been obvious for years to researchers, scientists and the global organizations who live and work on this continent that the emphasis must be on improving techniques, infrastructure and sustainability to create resiliency and a predictability that will allow the humans in this region to survive drought, floods and uncertainty with options that are now possible with new technology and a global connection...

As of May 2016 the Great Green Wall in Africa is about 15% complete...

Artificial Meat...This has the potential to make a real difference with cattle occupying over 20% of ranch and farmland and the production of feed to supply cattle taking up more than 25% of available agricultural land...An increased demand for meat worldwide is a major driver of deforestation and the enormous number of animals in captivity contribute methane to the atmosphere, a highly potent greenhouse gas critical in all climate change predictions...Still in an early stage of experimentation and development, artificial meat would solve many problems that are associated with growing live animals for slaughter and consumption...*With the growth of new technology, this will happen*...Artificial meat requires no growth hormones or antibiotics...

Genetically Modified Crops...In 2011, almost 400 million acres of farmland were planted with genetically modified crops (mostly for animal feed) using technology that grows daily and it will be a standard one day on a world that will have the knowledge and skill to safely create or modify food in response to the demands of a planet with billions of hungry humans...Our species has been modifying growing techniques, mixing and experimenting with different soil combinations and understanding the effects of temperate climates and growing seasons for 10,000 years to control what we grow and how we grow it...GMO crops are just an extension of a technological advanced species discovering better ways to master the large scale production of food and in a thousand years almost everything humans consume will be determined by a deep understanding of the science of plant and animal life...

What's alarming in this century is the misguided priorities of large corporations who modify crops to enable them to withstand the use of deadly pesticides proven harmful to the environment and to biological life...*This interconnected world of greed and profit is undeniable when the corporation (Monsanto) that creates GMO crops also controls the production and sale of a highly toxic pesticide the crop is designed to withstand*...The use of *glyphosate*, a broad-spectrum herbicide and a main ingredient in Roundup, has increased to over 200 million pounds a year in American agriculture and is used in many applications around the globe...There is a growing body of evidence demonstrating the characteristics of this herbicide which can enter animals fed on the corn and soybeans created to withstand the heavy use of *glyphosate* and it's rapidly becoming a serious environmental threat but with the biggest corporations making the biggest political donations in a systemic hierarchical structure of corruption and collusion, nothing is likely to change...In a future world with intellectual awareness the promise of GMO technology is growing food without the use of pesticides in a way that maximizes water and soil conservation...

Insects?...There are over 1,500 species of insects currently consumed in Asia, Africa, South America and Latin America including worms, crickets, millipedes, ants, spiders, wasps, grasshoppers, etc...Proponents say insects are low in fat, rich in protein, high in calcium and very easy to cultivate with little impact on the environment...Insects eat just about anything, emit almost no greenhouse gases and could solve food shortages in a hungry world...Currently the UN and the EU are investigating production at large scales...Many species on Earth survive on these creatures...

The Potsdam Institute for Climate Impact Research in Germany released a report in 2016 showing that food waste per person per day increased over 50% between 1965 and 2010 and the higher a nation's standard of living, the more food they wasted...In 2010, there was 20% more food available globally than was needed by Earth's populations...
Published in Environmental Science & Technology

EUROPE

Europe is located on the western peninsula of the Asian landmass and is by convention one of the seven continents of the world...Divided on the east from Asia by the Ural and Caucasus mountains, the Caspian Sea, the Black Sea and the waterways connecting the Black and Aegean Seas, Europe is the second smallest continent by surface area (3,930,000 sq miles) bordered on the west by the Atlantic Ocean, on the north by the Arctic Sea and on the south by the Mediterranean Sea...The third most populous continent on Earth with almost 800 million people, modern humans appeared in Europe about 40,000 years ago with fossil evidence and elaborate cave paintings dating back more than 30,000 years representing a human intelligence and awareness which is an important clue in the history of emerging cultures and human migration in this region...Europe is considered the birthplace of modern western civilization with the rise of Greek and Roman culture which spread from the Mediterranean Sea north to the Arctic Ocean and is home to the island nation England, one of the oldest western civilizations on Earth...The collapse of the Western Roman Empire in the first millennium began a thousand year history of war, religious crusades, plagues, lawlessness and dynastic rivalries as well as technological innovation, unparalleled progress in human rights, the establishment of educational centers, advances in agricultural production and the rise of modern philosophy and intellectualism that would lead to the establishment of permanent boundaries between nations...An increase in trade across the known world, revolution in France, an unparalleled explosion of science and art from the 14th century onward, an industrial revolution just 200 years old and the endless conflicts and wars fought on this continent over a thousand years have shaped boundaries, borders and political structures which still reverberate today, the foundation of modern Western Civilization...In the last two centuries of the second millennium European nations controlled the Americas and most of Africa establishing colonies in the majority of African nations with occupation and violence that is the root and causality of many problems impacting Africa today...The two great wars of the 20th century, the most violent destructive confrontations in the history of the human species, were staged primarily in Europe killing more than 50 million humans destroying entire societies, infrastructure and commerce that took decades to rebuild...Today most of Europe is united with 28 member states joining the European Union and 24 countries sharing a single currency, the *euro,* the world's No. 2 reserve currency on a continent undergoing continuing financial restructuring following the economic crisis of 2008...A crisis initiated by the corruption, malfeasance, fraud and wrongful conduct of governments, corporations and the biggest financial institutions (banks) in the world...At this writing no executive officer of any bank in the U.S. or Europe responsible for the theft of billions of dollars from millions of humans have faced any charges or any consequences for their crimes...An economic reality of the 1st World Civilization...

In August 2013, two traders who had worked for JP Morgan Chase were charged with multiple counts of wire fraud and other charges in an alleged scheme to cover billions of dollars in losses through record manipulation and falsification in the "London Whale" incident costing JP Morgan $6 billion...

The only criminal trial against any Wall Street executives for alleged wrongdoing related to the 2008 financial crisis involved two former Bear Sterns hedge-fund managers who were acquitted of all charges in 2009...

Money in this world is controlled by powerful entities including the Federal Reserve in America and the European Central Bank in Europe and this control of money buys and controls the governments of many nations, governments who make laws to protect themselves and the financial institutions from any persecution or liability no matter how large and grievous the crime with populations paying for this arrangement while struggling to survive economic hardship...The European Central Bank has lent over a trillion Euro to banks since December 2011 with some relief but no real plan going forward to deal with a monetary crisis of debt in the Euro-zone where more than 300 million people in 24 different countries use the euro as currency...This century large untouchable financial institutions are forced to use collusion, coercion, threats, promises and sleight of hand to keep the millions of humans on the bottom of the financial ladder believing the banks are legitimate and on the side of the people with their best interests in mind...This is far from the truth but in the marriage between finance and government that exists today there is little the public can do to change the economic distribution of wealth...Over many years with eyes closed and ears that listened to hear what they wished, modern western society has given control of all the finance in their world to a few powerful institutions and governments which structure themselves to work in concert with favorable laws and practices created by the politicians beholden to these institutions to acquire great wealth at the expense of many...The result is a growing income inequality increasingly visible in every nation on Earth...

In December 2013, the European Commission fined eight banks a total of $2.3 billion dollars for forming illegal cartels in concert with one another to manipulate interest rates and rob taxpayers of billions of dollars...UBS and Barclays were excused from any financial penalties (the largest fines) for assisting the investigation...RBS, Deutsche Bank, Societe Generate, JP Morgan, Citibank and RP Martin were all implicated in fraudulent activity with more investigations to follow...JP Morgan is paying billions in fines for it's activity on both sides of the Atlantic Ocean but no top executive involved has ever been charged with criminal activity for clearly fraudulent activity...

The other major challenge facing European nations and the European economy in this century is the aggregated effect of widespread immigration, mostly from Muslim communities, that has polarized citizens and nations as the cost of goods and services rise, crime and violence increase and many of these insular communities become isolated and angry...Islamists who migrate to other societies often do not 'merge' with local populations accepting customs and laws of the host nation as their own...There are many isolated enclaves and insular communities in Europe where immigrant populations practice tradition, communal laws and religious structure many centuries old that does not merge well with the growing awareness and intelligence of 21ˢᵗ century Earth...Recent statistics suggest rising crime rates and violence against women across all of Europe and many of the suspected are immigrants angry at western society, angry at the world, ignorant men who protect their "honor" with revenge killings of women and girls who "shamed" the family or have just "glanced at a man" breaking strict misogynous Islamic laws...The European Union has been accepting almost 2 million immigrants yearly with the United Kingdom, France and Italy allowing immigration from former colonies for decades...Add to that the increase in illegal immigration, the influx of millions of refugees from Syria, Africa, Afghanistan and Iraq, a recession making austerity a priority in many European nations and you have a situational powder keg with nations and governments passing new anti-immigration measures and new restrictions where there have been none before...Violence and chaos in the regions surrounding Europe will only send more humans in search of safety and security and it is unknown how sovereign nations and tired populations will respond...A real solution would be to raise the standard of living and reduce violence in the immigrants countries of origin but the world grows more chaotic and fractured, not less, and the planet's changing climate, endless wars, unpredictable drought, famine and the pressures of growing world populations will combine to relocate humans globally in unpredictable ways...

Europe's Muslim population is projected to grow to 30 million over the next 20 years...

The U.N. Refugee Agency estimates there are 65 million displaced humans worldwide...2016

In 2012, conflict, persecution and violence forced an average of 23,000 humans per day to leave their homes and seek shelter and protection in another country or in an alternative region of their native country...

Over 50% of refugees in 2016 are from Syria, Iraq and Afghanistan...Desperation, instability, chaos and war...

ALBANIA
(3.1 million humans)
Military Forces = 15,500 active duty / $120 million per year

Located on the east coast of the Adriatic Sea across the water from Italy, Albania has been a multiparty democracy for just more than two decades, a member of NATO for just five years and one of the most corrupt nations in Europe...Challenges include a formula for acceptance to the European Union, a continuing fight against organized crime, corruption and rebuilding Albania's infrastructure, railways, highways and neglected public facilities which are old and unreliable...Over 14% of the population exist below the poverty line with high unemployment (17%) and over 70,000 children in the labor force...Some energy wealth with almost 170 million barrels of crude oil reserves and 850 million cubic meters of natural gas reserves...Albania is attempting to stabilize and catch up with a global economy by strengthening financial institutions and modernizing their country to attract foreign investment while addressing the trafficking of both people & drugs, high unemployment, crime and a fractured economy in need of a cohesive direction...

ANDORRA
(86,000 humans)
Military Forces = No Standing Army

A small nation in the Pyrenees between France and Spain with an odd parliamentary democracy that has two heads of state, a Bishop from Spain and the President of France...Primary economy is tourism and a financial sector used as a secure tax haven for years...Very few humans live in poverty, unemployment rate is low, median age is 40+ years and a life expectancy of 82 years (6th highest in the world)...Literacy is 100%...A strategic location in the mountains between two of the largest nations in Europe account for a tourism industry with 10 million visitors a year that is the majority of Andorra's GDP...With no military Andorra relies on treaties with France and Spain to protect it from invasion...

ARMENIA
(3.1 million humans)
Military Forces = 69,000 active duty / $250 million per year

This nation located between Turkey and Azerbaijan with Georgia to the north and Iran to the south has been defined by border disputes and ethnic tension in a region still bitter over the political and military decisions made a century ago that reverberate through the politics and policies of the 21st century...This inability of the current governments in Azerbaijan, Armenia and Turkey to move forward with a resolution causes suffering among the general population, rarely among policymakers or their cronies...The long dispute with Azerbaijan over the Nagorno-Karabakh region and with Turkey over the deaths of more than one million Armenians in World War One are complex and detailed in other books but it's safe to say all economies suffer from uncertainty and hostility, unpredictable and unending...The economy is supported by industry, small-scale agricultural farming, foreign investment, foreign aid and remittances from abroad...Russia is a huge player in this region and supplies natural gas and power to the country...More than 30% of the population live below the poverty line with an unemployment rate around 17%...Life expectancy is around 74 years of age and over 6% of children under age five suffer malnutrition...Media censorship is prohibited by law but the interpetation of state laws by government officials can result in imprisonment for journalists who expose a truth challenging the government...

It's estimated that as many as 60% of all Armenians live outside the boundaries of the nation...What is clear is all the conflicts and disputes in this area of the world are linked to the influence of Russia today and the impact of the Soviet Union's disintegration in 1991...The inability of leaders to resolve differences is decades and even centuries old...

AUSTRIA
(8.6 million humans)
Military Forces = 28,000 active duty / $3.3 billion per year

After almost two decades of annexation and occupation during the chaos that was World War II and it's aftermath, Austria was declared independent and neutral in 1955 to remove Soviet occupation and maintain national unity under allied powers...Austria joined the European Union in 1995 and is a market economy with a high standard of living located in a stunningly beautiful region of the Alps with over thirty summits above 11,000 feet...Austria maintains it's postwar "neutrality" stance even with a ready military and is a major player in the global arena with IAEA (International Atomic Energy Agency), OPEC (Organization of Petroleum Exporting Countries) and OSCE (Organization for Security and Cooperation in Europe) all headquartered in Vienna, a city with a long storied history of culture, art and philosophy...Austria is a federal republic with a strong chancellor as head of government and a less powerful president as head of state...A welfare state with benefits and subsidies to protect business and provide for the citizens, Austria's unemployment rate is around 3.7%and about 6% of the population exist below the poverty line...Median age is 43 with a life expectancy of 83 years...Energy wealth with 61 million barrels of crude oil reserves and 11 billion cubic meters of natural gas reserves...Service sector, financial and tourism account for more than 60% of Austria's GDP, all challenged by Europe's exploding migrant crisis...

AZERBAIJAN
(9.6 million humans)
Military Forces = 69,500 active duty / $3.1 billion per year

One of six former Soviet states with a majority Muslim population, Azerbaijan is an autocracy disguised as a republic that accommodates the United State's military as a transfer point for supplies going to Afghanistan...Located on the coast of the Caspian Sea, this country has large supplies of oil and gas reserves, pipelines to deliver oil to Europe and a heavy foreign investment in energy supply and transport...Seven billion barrels of crude oil reserves and almost a trillion cubic meters of natural gas reserves has energized foreign investment with BP revealing plans in 2010 to build a new pipeline to Europe which would bypass Russia...Recent presidential elections have not been recognized as fair by OSCE observers after the two-term limit was eliminated and freedom of speech is stifled by a reality of government watchdogs and state run media ...Median age is almost 30 with a life expectancy of 70 years...Nearly 10% of children under age five are malnourished and over 140,000 children are in the labor force...The economy is energy with multiple pipelines for oil and gas transport and more being built...Challenges include corruption and border disputes with Iran over the Caspian Sea, Turkmenistan over the Caspian seabed and Armenia over the Nagorno-Karabakh region...15% of the country in the north is occupied by Armenia after a 1994 ceasefire left half a million humans displaced...

BELARUS
(9.5 million humans)
Military Forces = 69,000 active duty / $722 million per year

A stagnate country ruled by a small-minded dictator who remains in power (since 1994) through violence and intimidation of all opposition with no in country election since 1994 seen as free or fair by international observers...All sectors of economy depend on subsidies from a dysfunctional government which allows little foreign investment, controls financial institutions and continues to suppress free speech with imprisonment for critical reporting of the government...Belarus is heavily dependent on Russia for loans, energy and trade while poor governance has resulted in over 30% of the population living below the poverty line with 50,000 children in the labor force...Death rate is one of the highest in the world with a life expectancy of around 72 years...Energy wealth is estimated at 198 million barrels of crude oil reserves and 2.8 billion cubic meters of natural gas reserves...The nation is landlocked with

Russia to the east, Ukraine in the south, Poland to the west, a small border with Latvia to the north and Lithuania northwest...In a society with corruption and a government who enrich themselves, trafficking drugs and humans for sex and forced labor is common...Poor regulation of financial institutions make money laundering profitable with little risk...

BELGUIM
(11 million humans)
Military Forces = 35,500 active duty / $5 billion per year

Located in northwest Europe bordering France, Germany, Netherlands, Luxembourg and the North Sea, Belgium remains a focal point of European politics and war since independence...Another European power whose former colonial holdings Democratic Republic of Congo, Rwanda and Burundi are all struggling in the 21st century...Congo was 'acquired' by Belgium's King Leopold II as a personal wealth machine with forced labor and the deaths of close to 10 million humans which made Leopold II immensely wealthy during his 44 years as king...Today King Philippe is the head of a constitutional monarchy still struggling with language and ethnic divisions in three recognized regions...Dutch is the majority language but French and German are officially recognized as legal languages by the federal government...In late 2011, after 18 months of political deadlock, Belgium formed a governing coalition with six parties and a new prime minister...The nation's economy is recovering from the 2008 recession with a majority of the labor force employed in the service sector and industry...A multi-ethnic nation of various cultures and flavors that will struggle to maintain a cohesive federation as separatists make political gains and increase support for change amid political instability and uncertainty of the current political structure...These ethnic divisions were in full focus in 2015 after terrorists attacks in France were linked with multiple terrorist cells in Belgium and in 2016 with an IISS attack in Brussels that killed 35 and injured over 300 people...

The capital of Belgium is Brussels, home to NATO (North Atlantic Treaty Organization) and the European Union...

BULGARIA
(7.1 million)
Military Forces = 36,000 active duty / $690 million per year

A member of NATO and the European Union, Bulgaria is located on the Black Sea north of Turkey and Greece, east of Serbia and south of Romania...A former Soviet state, Bulgaria is one of the poorest countries in Europe with the economy supported by a labor force of 2.5 million working in the service sector and industry...22% of the population live below the poverty line with an unemployment rate of more than 11%...Energy reserves total 15 million barrels of crude oil reserves and 5.6 billion cubic meters of natural gas reserves with investments in multiple pipelines...Bulgaria's nuclear power plant (six reactors) has undergone major revisions after several reactors were declared among the "most dangerous in the world" and a referendum to build a 2nd nuclear power plant failed in 2013... Ruled in proxy by Alexander the Great and an ancient property of the Roman Empire, the Byzantine Empire and the Ottoman Empire, Bulgaria's history stretches back 2,500 years with traditions, legends, monasteries and treasures that are still very much a part of the nation today...The Festival of Roses origins go back to the Roman Empire and roses are still a staple of this economy...The Thracian Mask, made of solid gold, is thought to date back to 5th century BC and the Varna Chalcolithic Necropolis is from a past beyond 6,000 years...It's intriguing to imagine what still remains uncovered across the European and Asian continent waiting to be discovered someday on a future Earth when all humans share this planet with intelligence, curiosity and purpose...

Bulgaria was the site of one of the great treasure discoveries in all of Europe with thousands of golden artifacts dating back 6,000 years found in a burial site still not fully explored...

CROATIA
(4.4 million humans)
Military Forces = 21,000 active duty / $960 million per year

Located on the Adriatic Sea across from Italy, Croatia is a former republic of the Soviet state Yugoslavia (which ceased to exist in the 1990's) declaring independence in 1991 in a regional conflict which saw the dissolution of Yugoslavia and the creation of multiple states with varying degrees of autonomy...Croatia joined the European Union in 2013 and is attempting to rebuild an economy hard hit by the recession in 2008 after years of governmental corruption and the organized crime common in many former Soviet states...A familiar pattern of extreme ethnic violence in the 1990's with UN troops separating Croats and Serbs as humans who are essentially identical in every way killed thousands of each other and created refugees by the millions in a Bosnian war that was, as every war is, all about territory...In 1991 Serbs were over 12% of Croatia's population and in 2016 Serbs are maybe 5% of the population...Median age is 42 with a life expectancy of around 76 years, high unemployment (20%) with 21% of the population existing below the poverty line...Economy is strained with an unequal reliance on tourism and growing foreign debt...The nation is expected to adopt the Euro as its standard currency in the near future...Some energy wealth with 25 billion cubic meters of natural gas reserves and 71 million barrels of crude oil reserves...A parliamentary system with the president able to dissolve parliament and call elections...Challenges include widespread corruption, crime and ongoing disputes over borders and maritime access with Slovenia, Bosnia and Herzegovina...

CZECH REPUBLIC
(10.5 million humans)
Military Forces = 23,000 active duty / $2.3 billion per year

Part of the former Soviet state of Czechoslovakia, the Czech Republic is a landlocked nation bordering Germany, Austria, Poland and Slovakia (another part of former Czechoslovakia)...A member of the European Union, NATO and an ally with the west, this region has a storied history stretching back more than a thousand years involving the Roman Empire, Catholicism, the House of Hapsburg and the Austrian Empire resulting in a multicultural footprint that is embodied by the architecture and art in the capital Prague and the nation...After the creation of the Czech Republic and Slovakia in 1993 the country was led by Vaclav Havel, a former dissident under the communist regime until 2003 when he was succeeded by Vaclav Klaus...A multiparty parliamentary democratic structure with three major political parties (Christian Democrats, Social Democrats, ANO) joined in a fragile coalition framework to govern and create stability...Economy is supported by an automobile industry which produced over a million vehicles in 2010 for export to EU members...Unemployment is around 7%...Energy reserves are an estimated 4 billion cubic meters of natural gas and 15 million barrels of crude oil...

DENMARK
(5.5 million humans)
Military Forces = 24,000 active duty / $4.3 billion per year

Stretching north of Germany into the North Sea, Denmark is a Nordic kingdom from the 10th century with a history that includes the infamous Vikings from a long past and a striking influence over events on the European continent despite it's small size and geographical location...The first nation on Earth to recognize and legalize registered partnerships between same sex couples (1989), Denmark is a good example of what can be done given perspective, cooperation and long-range planning...Before the oil crisis in 1973 this country got 99% of their energy from the Middle East and in 2012 that number is zero...Over 20% of Denmark's electricity is from wind and the nation is a leading manufacturer of wind turbines and energy-efficient products worldwide achieved with conservation, CO2 taxes, building efficiency standards, gasoline taxes and a consensus by the people not to be dependent upon the dirty, expensive and temporary sources of energy which harm Earth's environment and contribute to health problems ignored by too many other countries...Economy is supported by an extensive service sector (70%) along with modern industry including pharmaceuticals and renewable energy...Unemployment is around 5% with 13% of the population below the poverty line...Median age is just over 40 years with a life expectancy of 79 years, spending on health is ranked 12th in the world and spending on education is ranked 7th in the world...Energy wealth includes an estimated 43 billion cubic meters of natural gas reserves and 610 million barrels of crude oil reserves...Denmark maintains territorial control of both Greenland and the Faroe Islands (both are self-governed) although the Faroe Islands are exploring transition scenarios to full independence...The nation is led by Queen Margrethe II and governed by the prime minister with the queen as head of the Council of State whose duty is to enact new legislation and provide input on matters of government...The nation is actively monitoring the Arctic Ocean as the disappearance of Arctic ice presents new challenges and opportunities...Under construction is a tunnel connecting Germany with an estimated cost of more than $5 billion...With its healthcare system and social services Denmark is a model of how nations should tax and support a social safety net that does not restrict entrepreneurial pursuits while ensuring the health and welfare of all citizens...

ESTONIA
(1.3 million humans)
Military Forces = 17,000 active duty / $485 million per year

A former member of the USSR located north of Latvia and west of Russia in the Baltic Sea, Estonia is a member of NATO and the European Union allied with the West in matters of defense and foreign affairs...After centuries of foreign rule Estonia gained independence in 1918 absorbed into the USSR in 1940 and regained independence in 1991...Over 30% of the population is Russian...Economy is recovering from the 2008 recession with growth in telecommunications and the electronic industry, unemployment is around 7% with 20% of the population below the poverty line...President is mostly a ceremonial role of foreign representation and the title of Supreme Commander of Armed Forces...The government is run by the prime minister in a coalition with Social Democrats and the Reform Party...A median age of 42 with a life expectancy of 76 years...Highly efficient internet system with government support and services...Adoption of the Euro in 2011...Challenges include a continuing tension with Russia over issues relating to Soviet occupation, border disputes with Russia, drug production, drug trafficking and a rise in the use of opiates as well as almost 100,000 stateless humans, mostly ethnic Russians with language barriers worried that travel between Russia and Estonia might be restricted if they were granted Estonian citizenship...

FINLAND
(5.4 million humans)
Military Forces = 35,700 active duty / $3.6 billion per year

Finland is a free market modern welfare state with a strong service sector, telecommunications industry, engineered metals products and a timber industry with almost 60% of this sparsely populated country covered by forests...A member of the EU since 1995, Finland was the property of Sweden for almost a thousand years until it was "given" to Russia in the early years of the 19th century...After independence in 1917, disputes with Russia continued through World War 2 until the signing of a peace treaty in 1947 that made Finland a neutral player in world politics through the Cold War era until the collapse of the USSR (1991)...Finland is a republic with a conservative majority in parliament...Median age is 42 with a life expectancy of over 80 years...Investment in education, research and healthcare are priorities with a focus on an intelligent workforce in a modern economy...North of Europe with Sweden to the west and Russia to the north, Finland is beautiful and cold with regular displays of the northern lights and an interior still unspoiled by commerce...Nuclear power in two locations generate 30% of the nation's power needs with coal and gas supplying the rest...To satisfy a state policy requiring all nuclear waste created in Finland be stored in Finland there's an underground storage facility under construction

(Onkalo) scheduled to be completed in 2020 at a depth of 1,710 ft (520 meters) in granite bedrock...The fuel is buried in canisters enclosed in capsules packed with clay and the facility will accept spent fuel for a hundred years before it's sealed...Current estimates are between 15,000 to 20,000 nuclear plants would be needed to supply all the world with energy in 2050...In 2016 there were less than 500 nuclear power plants operating in 31 countries with 68 plants under construction...On a planet with a wide variety of "green energy" options including solar, hydroelectric, geothermal and wind, it's clear the entirety of global energy demand will not be dependent on nuclear power but nuclear power could be a powerful technology to move a world off fossil fuels...Humans cannot and should not bury radioactive waste all over Earth as a short-term solution for a substance radioactive for thousands of years and other options will be possible from the development of new technology, nanotechnology to render the waste inert or storing waste on an inhospitable moon or asteroid...Holes in Earth filled with radioactive material whose long-term effects are unknown is not a solution...

FRANCE
(66 million humans)
Military Forces = 230,000 active duty / $40 billion per year

France dominates the European continent with coastal access to the North Sea, Atlantic Ocean and the Mediterranean Sea in a central geographical position which made this nation a key participant, in the many political and military conflicts that have shaped this region of the world...A permanent member of NATO, G-8, G-20, the EU and UN Security Council, France's sovereignty extends overseas to include Reunion (Indian Ocean), French Guiana (South America), Martinique (Caribbean Sea), Mayotte (Indian Ocean), Guadeloupe (Caribbean Sea) and Corsica in the Mediterranean Sea...A participant in the colonization of Africa in the 19th century, France still plays a role in rendering aid and military force in ongoing conflicts that continue to plague the African continent...With a long policy of open immigration from former colonies (more selective and reformed today) there are currently over 5 million inhabitants of African and Arab descent living in France which is becoming more problematic with isolated enclaves who refuse to merge into French society clashing with police and citizens over employment opportunities and benefits of the state...A rich cultural heritage in art, philosophy, music, architecture, food and wine make France and it's capital Paris one of the most vibrant centers of population on Earth...Government is a democratic multiparty coalition with a parliament of 348 seats in a stable republic with many voices...Median age is 41 with a life expectancy of 82 years...Ranked 10th in the world on health expenditures and recognized as one of the best healthcare systems in the world with around 75% of all health costs funded by the government...Unemployment is 10% with 8% of the population living below the poverty line...An economy supported by tourism (3rd largest income from tourism in the world) energy exports, defense industries, service industries and a social safety net that is one the best in Europe...A leader in nuclear technologies, France has over 50 nuclear reactors and is the planet's largest net exporter of electricity...Crude oil reserves of 85 million barrels and 10 billion cubic meters of natural gas reserves...With a strong military and strategic relationships with the other major powers on Earth, France will be pivotal in the 21st century in the stabilization of the world's economy and in addressing many humanitarian needs that will only increase on a planet growing more chaotic and more crowded with threats from home and abroad...

France is the world's top tourist destination with over 83 million visitors every year...

France is the largest nation on the European Continent at 643,801 square kilometers (248,573 square miles)

Many cave sites in France have been connected to Neanderthal's, an extinct species that cohabited with our species thousands of years ago until disappearing around 40,000 years ago...A new discovery in May 2016 indicates that the Neanderthals may have been in France more than 175,000 years ago predating the arrival of humans by 100,000 years...In this cave are structures that indicate planning and building with purpose, possibly ritualistic...

GEORGIA
(4.9 million humans)
Military Forces = 39,000 active duty / $400 million per year

On the Black Sea south of Russia, north of Turkey, north of Armenia and west of Azerbaijan at a crossroads between Europe, Middle East and Asia, Georgia has had a tumultuous decade of conflicts with Russia, the loss of two separatist regions (South Ossetia and Abkhazia) now occupied by Russia and political change with an uncertain agenda...The reality today is many differing political parties with conflicting views of the future, an economy of mining and agriculture with a large dependence on imports for nearly all gas and oil but a strategic location and a rail system (Kara-Akhalkalaki Railroad) that's a transit point for energy and manufactured goods flowing between Europe and Asia...10% of the population live below the poverty line with an unemployment rate over 16%...Median age is 38 with a life expectancy of 76 years...Around 120,000 children work in the labor force and a large population of refugees (200,000) exist, victims of the violence and conflict in the separatist regions...Some energy wealth with 35 million barrels of crude oil reserves and over eight billion cubic meters of natural gas reserves...Home to the Great Caucasus Mountains in the north and the Lesser Caucasus Mountains in the south, Georgia is hopeful a growing tourist industry will boost their economy...It's recognized that the previous president (Saakashvili) instituted many reforms over his ten-year reign that changed the power structure of the prime minister, the president and the parliament as well as addressing the culture of corruption in security and education but the future of this country hinges on a great many unknowns including Georgia's political direction, future economic policies and neighbor Russia...

GERMANY
(81 million humans)
Military Forces = 188,000 active duty / $39 billion per year

Located in north central Europe with access to the North Sea, Germany has a strong economy supported by automobile industry exports, chemical industries, quality machinery and a strong service sector while providing a substantial safety net and stable benefits for a highly skilled work force...A nation whose 20th century politics and military impacted almost every nation on Earth as the focal point of the two greatest conflicts ever fought on this planet...Wars which created the boundaries and borders seen today and the beginnings of a world civilization and a world unity (whose potential is not yet realized) creating the geopolitical blueprint of this century with an aggressive NATO alliance and a sprawling United Nations who created multiple global organizations to create a world unity, a unique awareness in the history of human existence (WHO, IMF, IAEA, WMO, IMO, FAO, etc)...In this century Germany celebrates a legacy of great art and music attracting more than 20 million tourists yearly to thousands of museums, libraries, theaters and musical venue with many of them subsidized by government...Today Germany's a powerhouse of economic stability after a reunification of East and West Germany in 1990 and is a major force in holding the European Union together while recovering from a recession felt around the world...Government is a federal republic, a coalition with a legislative body (Bundestag) of over 600 members elected every four years by popular vote...Unemployment is around 6% with 15% of the population living below the poverty line...Median age is 46 with a life expectancy of 80 years...Highly efficient healthcare structure with employees and employers paying less than 10% each to nonprofit insurance companies with a nationwide belief that all citizens are entitled to the same medical care no matter their status...An estimated 227 million barrels of crude oil reserves and 116 billion cubic meters of natural gas reserves, Germany plans to shut down all nuclear reactors by 2022 in a decision prompted by the recent nuclear accident at Fukushima replacing them with renewable energy...A strong western ally with over 65,000 US military personnel in country...With refugees and immigrants from Africa, the Middle East and elsewhere on the continent (Serbia, Syria, Turkey, Ukraine) Germany has been used by extremists and terrorists as a launching point for plots against the U.S. and neighboring European nations, a problem that will only get worse in this century...

GREECE
(11 million humans)
Military Forces = 180,000 active duty / $6.9 billion per year

Greece is a small country in the north central Mediterranean Sea with a mainland bordered on the north by Albania, Bulgaria and Macedonia and over 1,300 islands, most located in the Aegean Sea across from Turkey...Considered by scholars as the birthplace of western civilization with evidence of habitation in 7,000 BCE and the documented records and teachings of Socrates and Plato around 400 BCE, many of the ideals, philosophies, arts and culture of modern western civilization can be traced back to the civilizations that inhabited this region over 2,000 years ago...A member of NATO and the European Union, Greece settled into a kind of laisseze-faire existence after abolishing the monarchy in 1973 which came to a head in 2009 requiring a bailout from the EU to avoid default and a restructuring of labor policies with severe austerity measures highly unpopular with the Greek population...Today a new government (parliamentary republic) is challenging the EU over agreements made by the former government with the youngest prime minister in history and growing dissent from a population angry of the stranglehold the EU has on the nation's economy...Economy is dependent on foreign aid which supports a large public sector, exports in textiles, some agriculture and a tourism industry that accounts for over 18% of GDP...Challenges include pensions, healthcare costs and creating a tax system that can collect revenue from a population accustomed to weak enforcement...Over 35% of the population live below the poverty line with unemployment over 27%...Median age is 44 with a low population growth rate and a life expectancy of 80 years...Energy reserves include 990 million cubic meters of natural gas and 10 million barrels of crude oil...Expected given the complex geographical footprint of this region are the long-standing disputes with Turkey over ownership, boundaries and access to the Aegean Sea while organized crime, trafficking and corruption remain chronic barriers to a full economic recovery...

The Olympic Games originated in Greece and are a fascinating insight into a primitive ritualistic obsession by the planet's most intelligent creature (the only species on Earth to control fire) paying incredible homage and respect to the actual phenomenon of fire where it is a grave disappointment if the Cauldron of Fire (representing the beginning and ending of the Games) goes out for any reason...An interesting aspect of intellectual development and the challenge of human survival...A million years worshiping the power of combustible ignition...

HUNGARY
(9.9 million humans)
Military Forces = 26,000 active duty / $1.1 billion per year

Landlocked in East Europe with Austria to the west, Slovakia north, Ukraine and Romania to the east and Croatia in the south, Hungary was a region of Renaissance medieval life that centralized in Budapest, a storied city which has maintained popularity and prosperity through different time eras and multiple empires...Today the city is home to almost 2 million humans with a rich history in music, art and architecture...An independent republic after World War I, Hungary became an ally of Germany in World War II and a communist state after the war leading to years of revolts and uprisings against the Soviet state fueled by opposition groups that ended with the fall of the Berlin Wall and subsequently, the fall of the Soviet Union in 1991...A multiparty parliamentary democracy led by chief of state and prime minister, a member of the European Union and NATO and another EU member state who needed a bailout from the European Central Bank following the 2008 recession...The constitution has been revised multiple times and criticized by many for restricting freedom and voter rights amid aggressive policies regarding taxes and private investment...A market economy with billions in foreign investment responsible for over 80% of the GDP, exports in food products, machinery and pharmaceuticals, a large service sector and an unemployment rate of 8%...Over 15% of the population living under the poverty line (many of them in the Roma community virtual outcasts who live on the fringe of society)...Median age is 41 with a life expectancy over 75 years...Some energy wealth with 27 million barrels of crude oil reserves and eight billion cubic meters of natural gas reserves...Current challenges include an ongoing dispute with Slovakia over the Gabcikovo-Nagymaros Dam on the river Danube that is an unfinished project thirty years in dispute with an ICJ ruling that Hungary must complete construction amid concerns of impact on wildlife and nature...Common to all major cities (Budapest) in this region of the world is an entrenched criminal element using these population centers as transnational shipment and distribution points for illegal drugs with destinations in western Europe, Asia and South America...

ITALY
(62 million humans)
Military Forces = 320,000 active duty / $34 billion per year

The boot of Europe that includes the large islands Sicily and Sardinia, Italy has been recognized for years as one of the most corrupt nations on the European continent with politicians and corporations working hand in hand with Italy's notorious Mafia to pocket millions of dollars from the public through collusion, corruption and sleight of hand that enriched the well-connected with unlimited access to the public's money...These revelations in the 1990's led to elections in 2001 that brought Silvio Berlusconi to power for the majority of the 1st decade of this century, an individual who used the power of politics to build an personal empire enriching himself while ignoring the laws and the citizens of Italy...Wealthy and unapologetic, Berlusconi controlled a media empire, controlled the news and controlled elections keeping him in power and keeping him out of jail...Today the fourth largest economy in Europe is struggling with austerity policies and an unhappy population with more than $2.5 trillion in public debt, a vibrant underground economy believed to be more than 16% of GDP and an above-ground economy dependent on tourism, industries renown for high quality products and agriculture in the south of Italy with exports of wine, vehicles and quality machinery...Unemployment is around 12% with 30% of the population living below the poverty line...Median age is 44 years with a life expectancy of nearly 83 years...High youth unemployment and an aging population are concerns when the government's hands are tied by massive public debt and loan commitments...An estimated 544 million barrels of crude oil reserves and 60 billion cubic meters of natural gas reserves...As with every country in the EU, Italy is recovering from the 2008 recession with uncertainty in the financial and social institutions of the past, instability with forced immigration from the south (almost 180,000 refugees and migrants in 2015-16) and a changing focus on what's needed to invigorate this nation whose cultural heritage of architecture, music and art is unsurpassed by any other nation in the western world...

Italy is the only G-8 nation without any nuclear power plants and is the world's largest net importer of energy with around 15% of the nation's electricity coming from France, the world's largest exporter of power...

LATVIA
(2.1 million humans)
Military Forces = 13,300 active duty / $280 million per year

On the Baltic Sea west of Russia, north of Lithuania, Belarus and south of Estonia, Latvia is a member of the European Union and NATO with an economy based on exports in timber, agriculture, a growing electronics industry, a fishing culture in the capital (Riga) and in the many villages along the coastline (500 kilometers) of the Baltic Sea...Independent since the 1991 collapse of the Soviet Union, the nation is still recovering from the 2009 economic crisis with an unemployment rate of more than 10%...Current president oversees a political structure with a history of corruption in a ruling coalition that joined the Euro-zone in 2014...Challenges include a large minority Russian population, organized crime, boundary disputes with Russia and Lithuania and more than a quarter of a million "non-citizens" who cannot vote and must pass rigid language and history exams to acquire Latvian citizenship...

LIECHTENSTEIN
(37,000 humans)

An independent state since 1866, Liechtenstein is a tiny country nestled in mountains between Switzerland and Austria with an economy of low corporate taxes to attract business and a monetary framework that's made it a "tax haven" for years, a label Liechtenstein's government is trying to remove with economic reforms to increase financial transparency...More than half of the labor force commute daily to Germany, Austria or Switzerland with an unemployment rate of less than 3% and one of the highest per capita incomes on Earth...An absolute monarchy with a parliament of 25 seats...The nation has a modern Police Force but no military...Currency is the Swiss Franc...

LITHUANIA
(3.2 million humans)
Military Forces = 14,500 active duty / $428 million per year

Lithuania is a member of the EU and NATO with 90 kilometers of coastline on the Baltic Sea, Poland and Belarus to the south, Latvia north and the small Russian enclave of Kaliningrad located in the southwest corner on the Baltic Sea...Occupied by major powers (Germany, Russia, France, Sweden) for a large part of history, Lithuania declared independence after World War I only to become occupied by Germany in World War II and then annexed by the Soviet Union after the war...Independence came in 1991 after years of rebellion against Soviet rule and Lithuania became a parliamentary democracy with a coalition of parties in a parliament of 141 seats...A major trading partner with Russia, Lithuania survived the 2008 global economic crisis with the implementation of austerity measures and a focus on market reform and foreign investment...An aging country with a median age of 43 years and a negative population growth...Youth unemployment is high (20%) while the overall employment rate is around 10%...Life expectancy is 80 for women and 70 years for men, an almost ten year difference attributed to lifestyle and occupation...Lithuania shut down it's nuclear reactors in 2009 to join the EU...

LUXEMBOURG
(550,000 humans)
Military Forces = 1,000 active duty / $300 million per year

A nation more than a thousand years old located west of Germany, east of Belgium and north of France, Luxembourg is one of the central financial centers in Europe, home to the European Investment Bank and one of six founding members of the European Economic Community which later became the European Union...Industry is steel, chemicals, tourism and financial management with many foreign owned banks doing business globally...Luxembourg has been a member of NATO since 1949 and is a member of the Eurozone...Unemployment is less than 5% with one of the highest standards of living in Europe...Government is a constitutional monarchy with a prime minister, a council of ministers and a 60 seat legislative branch elected by popular vote...Life expectancy is 84 for females and 79 for males...

In 2016, Luxembourg announced they will invest in the industry of asteroid mining, a very long term commitment...

MACEDONIA
(2 million humans)
Military Forces = 8,000 active duty / $140 million per year

One of the Balkan states which escaped a lot of the violence in the 1990's, this former republic of Yugoslavia is a parliamentary democracy with the president elected by popular vote and the prime minister elected by an assembly of 123 seats...Macedonia is surrounded by Serbia, Albania, Bulgaria and Greece with a large ethnic Albanian population and an endless dispute with Greece over the name Macedonia...Median age of almost 37 with a life expectancy of 76 years...Unemployment rate is around 26% with more than 30% of the population living below the poverty line...

MOLDOVA
(3.5 million humans)
Military Forces = 6,100 active duty / $20 million per year

Moldova is west of Ukraine, east of Romania and has been an independent republic since the collapse of the Soviet Empire in 1991 with a long history of being exchanged, bartered, reshaped and occupied by Russia, Germany and the Ottoman Empire for strategic reasons relating to the many wars that have plagued this region of Earth over the past 1,000 years...Linked to Romania for much of it's history, Moldova today is the result of years of political instability with rapidly changing parliaments, presidents, ongoing disputes with Russia over energy supplies and a disputed separatist region with a Slavic majority population and a long period with no president as tensions and accusations between the Communist party (PCRM) and the Alliance of European Integration (AEI) over the future of a nation searching for stability and predictability continued...A parliamentary republic with the president elected to a 4 year term and a prime minister as head of government...Median age is 36 with a life expectancy of 70 years...Almost 5% of children under age 5 are malnourished with a child labor force estimated at 73,000...20% of the population live below the poverty line in a nation very reliant on agriculture and remittances of over a billion dollars from around a million citizens working abroad...One of the poorest nations in Europe with continuing disputes over imports, exports and energy supplies with Russia...

MONACO
(38,000 humans)

This small storied nation on the southern coast of France has been ruled by the same monarchy for hundreds of years, is the second smallest independent state in the world (the Vatican is #1) and is a global tax haven (no income tax) and banking center for foreign corporations and rich individuals...The nation is a constitutional monarchy governed by a Council of Government (5 members) and a minister of state, a 24 seat assembly and the ruling Grimaldi family under the protection of France and Spain at various times in history...A member of the UN, Monaco is well-known for secrecy laws concerning banking and a vibrant tourist economy with lifestyle, a perfect location and a reputation as a gambling mecca bringing in close to a billion dollars every year...Median age is one of the world's highest at 51 years with a life expectancy of almost 90 years for the total population...

MONTENEGRO
(652,000 humans)
Military Forces = 2,000 active duty / $48 million per year

Located on the Adriatic Sea with Bosnia and Serbia to the north, Kosovo to the east and Albania south, Montenegro became independent in 2006 as a result of a vote that ended the union of Serbia and Montenegro initially established by the European Union to replace Yugoslavia...A member of the UN and the euro-zone, Montenegro is a parliamentary republic with a coalition government assembly of 81 seats...The nation's economy is tourism, a large aluminum industry and foreign investment with an unemployment rate close to 20%, a median age of 39 and a life expectancy of 78 years...Geography is stunning with coastlines, mountains and the deepest (4,300 ft) river canyon in Europe...Montenegro is a member of WTO (World Trade Organization) with membership in the European Union dependent on tackling corruption, organized crime, gender rights and media freedom...

Dutch East India Company
(Vereenigde Oost-Indische Compagnie)

Very successful trading company founded in 1602 to maximize trade in the Indian Ocean while the Dutch were fighting with Spain for independence...They were allowed to negotiate trade agreements and treaties with the authority of the Dutch government and became the largest transport and trade business on Earth...Before the end of the 17th century V.O.C. had a network of bases throughout Asia with warehouses, administration facilities, housing and militaries controlling trade and travel...The V.O.C. was dissolved by the Dutch government in 1799...

NETHERLANDS
(16.8 million humans)
Military Forces = 48,000 active duty / $9.8 billion per year

A commercial power in the 17th century after declaring independence from Spain in 1579, the Dutch established colonies worldwide with a renown seafaring reputation that influenced trade policies in Europe and around the world...Located on the North Sea in the far northwest corner of the European continent with some 25% of the country below sea level, the people in this region have a reputation for building dykes and water control systems dating back almost 2,000 years...The famous windmills were originally designed to pump water off the land and are still one of the Netherlands most popular tourist attractions...Median age of 42 years with a life expectancy of 81...A constitutional monarchy with the prime minister as head of government ruling a large multiparty coalition led by the People's Party for Freedom and Democracy (VVD), a liberal party that's enjoying a small majority in the changing Dutch politics dealing with foreign wars, terrorism, immigrant rights and migration challenges...The 6th largest economy in the euro-zone and a large transportation hub with the Port of Rotterdam ranking as the largest port in Europe and the 3rd busiest port in the world...Large industry in food processing with a highly mechanized agricultural industry and exports in machinery, chemicals and fuel...One of the most densely populated nations in the world and a strong participant in global affairs with membership in many organizations including NATO, EU, WTO and the UN...Energy reserves are an estimated 144.7 million barrels of crude oil and one trillion cubic meters of natural gas...The Dutch are a structured and intelligent culture with a reputation for envisioning and responding to future scenarios predictable in a thousand years or more using patience, technology and practice to find solutions...More than anything on 21st century Earth humans need a population which think ahead 10,000 years to find solutions...

NORWAY
(5.2 million humans)
Military Forces = 26,200 active duty / $7 billion per year

Norway is a geographical oddity wrapped around the western border of Sweden stretching over 1,700 kilometers from south to north touching three named waterways, the North Sea, the Norwegian Sea and the Bering Sea...Norway also claims four islands, two in the South Atlantic (Bouvet and Peter I, both uninhabited), one in the Arctic Ocean (Jan Mayen, also uninhabited) and the disputed Svalbard Archipelago, a grouping of islands north of Norway also claimed by Russia...Along the entire western shore of Norway are over 50,000 small islands, an archipelago of natural barriers between the open ocean and the coast called Skjaergaard...After 400 years as part of Denmark and almost 100 years in a forced union with Sweden, Norway finally gained independence in 1905 and in 2014 Norway was the largest producer of oil in Western Europe and the 3rd largest exporter of natural gas in the world (Russia is #1)...A member of NATO with a constitutional monarchy led by King Harald V, a prime minister and a parliament of 169 seats...Norway's economy is oil and gas with more than 6.4 billion barrels of crude oil reserves and 1.8 trillion cubic meters of natural gas reserves, one of the highest per capita GDP's in the world, a low unemployment rate (4.4%) and social programs that are among the strongest on the planet...Median age is 39 with a life expectancy of 81 years...Over 80% of Norway's population live in urban centers with everyone less than 200 km from the coast ...Challenges include a disputed territorial claim in Antarctica and international disagreement on whaling operations that resumed in 1993 (the International Whaling Commission (IWC) issued a ban in 1986) threatening minke whales with quotas rising in recent years and a reported increase in the killing of breeding females...

Norway increased CO2 tax on energy industries to help set up a $1 billion climate change fund for programs in developing nations among other green projects...In 2012, the nation was the world's 13th largest oil producer and third biggest oil exporter, yet has been one of the most active nations in funding climate change projects...

POLAND
(38.5 million humans)
Military Forces = 118,000 active duty / $9.4 billion per year

Poland is located on the southern coast of the Baltic Sea bordered on the west by Germany and the Czech Republic, in the south by Slovakia, bordered on the east by Ukraine, Belarus, Lithuania and an odd small territory of Russia to the north...Established as the Kingdom of Poland in 1025 the nation has endured multiple partitions and occupations over a history of nearly a thousand years and today is a republic with the president as chief of state elected by popular vote and a prime minister appointed by the president with confirmation...A labor force of over 18 million with a large agricultural sector and industries in metals, shipbuilding (three major ports on the Baltic Sea), mining with zinc, silver, sulfur and large deposits of coal...Unemployment is around 10% and an estimated 17% of the population live below the poverty line...Median age is 39 with a life expectancy of 77 years...A member of NATO and the European Union with an active role in both Iraq and Afghanistan...Energy wealth is 92 billion cubic meters of natural gas reserves and 142 million barrels of crude oil reserves...Challenges include illegal immigration along the eastern border and the viability and undeniable political impact of a proposed missile defense system in concert with the United States...

PORTUGAL
(10.8 million humans)
Military Forces = 40,000 active duty / $3.8 billion per year

The westernmost nation on the European continent is located on the Iberian peninsula and was a maritime power for hundreds of years and one of the last colonial powers to grant independence to their African colonies (1975 and 1999)...Portugal's history of colonization is still evident in over fifty nations worldwide, the byproduct of exploration that sought to "civilize" native populations and bring wealth back to the homeland...In Brazil (their largest colony, freed in 1822) where Portugal imported slaves to grow commodities that were returned to Europe and in Africa where they enslaved humans in Angola, Mozambique and Guinea-Bissau to harvest gold and diamonds responsible for the deaths of millions when they fought colonial wars to hold on to the properties...Today Portuguese is the 6th most popular language in the world (over 250 million) and Portugal's culture is evident in the many societies affected by this once powerful European nation without limits...Founding member of NATO and the European Union, today Portugal struggles to recover from the 2008 recession with help from the IMF and the European Union led by a president elected by popular vote and a multiparty coalition from a 230 seat assembly...Median age is 41 with a life expectancy of almost 80 years...Unemployment is around 14% in an economy supported by fishing, textiles, tourism and agriculture (the underground economy is said to be as much as 20% of GDP)...20% of the population live below the poverty line and 35,000 children are in the labor force...Challenges include a dispute with Spain over the territory of Olivenza and the implementation of austerity measures that are a part of the loan package from IMF/EU...

ROMANIA
(21.5 million humans)
Military Forces = 73,000 active duty / $2.1 billion per year

As with many of the Balkan countries the region that is now Romania has been ruled, occupied and partitioned by various empires throughout time including the Romans, the Habsburgs, the Ottoman empire and the Russians with the fate and governance of Romania's three principalities (Transylvania, Wallachia, Moldavia) a continuing contentious point of challenge and control...After a long, bloody history of collusion with Germany in the Second World War and the reported deaths of millions under Communist rule, Romania is now a member of NATO, the European Union and is still recovering from the 2008 recession in a political reality of entrenched corruption that is still deeply embedded in this society...The eastern border is the Black Sea, Moldova and Ukraine, north is Ukraine and Hungary with

Bulgaria to the south and Serbia west in a republic with a elected president who appoints the prime minister...Over 20% of the population live below the poverty line in an economy supported by multiple loans from the EU and IMF to promote reforms and insure stability (many funds offered were not taken when reforms could not be agreed on)...Crude oil reserves are 600 million barrels and natural gas reserves are more than 105 billion cubic meters...Median age is 40 and life expectancy 75 years...Over 25,000 children in the labor force...Economy is agriculture, industry and the service sector...Beautiful mountains (Carpathian) in the interior with 225 kilometers of coastline on the Black Sea...

SAN MARINO
(33,500 humans)

A tiny nation landlocked in a picturesque region of Italy, San Marino is just over 23 sq miles in area and is recognized as the world's oldest republic (301AD)...Influenced by the policies and politics of Italy, San Marino's reputation as a tax haven grew a substantial financial sector coupled with tourism and exports to create a viable economy for a population of over 33,000...Labor force is almost 22,000 and unemployment is at 8%...Government is a 60 member Great and General Council elected every five years who elect a 10 member Congress of State to wield executive power...Life expectancy is over 83 with a median age of almost 44 years...Landscape is dominated by Monte Titano which rises almost 2,300 feet in the south...Military is a small force of volunteers and regulars costing $10 million a year...

SERBIA
(7.3 million humans)
Military Forces = 50,000 active duty / $830 million per year

Famous for Slobodan Milosevic who became president in 1989 leading this country through a turbulent bloody decade of ethnic violence, political and territorial instability leading to the breakup of Yugoslavia and violence in Kosovo (a republic in 2016 recognized by just 108 UN members)...Serbia separated from Montenegro in 2006 to become a sovereign state with a president elected for 5 years and prime minister elected by a National Assembly (250 seats)...Economy is restructuring with reforms and privatization but faces an uphill battle with high unemployment, widespread corruption, inflated government salaries and benefits, a serious need of direct foreign investment and an infrastructure that needs upgrades and repairs...Labor force is three million, unemployment rate is 20% and 10% of the population live below the poverty line...Aging workforce with a median age of 42, child labor force of more than 36,000 and a life expectancy of 75 years...Some energy wealth with an estimated 77 million barrels of crude oil reserves and 48 billion cubic meters of natural gas reserves...The borders of Kosovo and Serbia are still disputed with peacekeepers under UN Administrative control in Kosovo to maintain stability...A quarter million humans displaced internally from conflict over the past two decades and more than 60,000 refugees from Bosnia, Herzegovina and Croatia...

Kosovo is still unrecognized by many nations in various world organizations after it's declaration of independence from Serbia in 2008...It remains under United Nations administration in 2016 and is not a member of the UN...

SLOVAKIA
(5.4 million humans)
Military Forces = 14,000 active duty / $1 billion per year

Bordered by Czech Republic, Poland, Ukraine, Hungary and Austria, this nation originated in 1993 after Czechoslovakia split into the Czech Republic and Slovakia...First direct presidential election in 1999, joined NATO and the EU in 2004 and adopted the euro in 2009...Parliamentary democracy with the president elected by popular vote and the prime minister appointed by the president...Economy has benefited from the privatization of the banking sector and foreign direct investment in industry (automotive, electronics) facilitated by new business-friendly policies...Unemployment rate is around

10% in a labor force of 2.7 million...Median age is 39 with a life expectancy of 76 years...Slovakia's energy wealth is 9 million barrels of crude oil reserves and 14.1 billion cubic meters of natural gas...A beautiful landscape in central Europe with the Tatry Mountains bordering Poland amid valleys, many lakes and the Danube river bordering Hungary...Challenges include a large ethnic Hungarian population (Slovakia was a part of Hungary for over 1,000 years) and over 80,000 Roma (gypsies) who have declared themselves a "distinct minority" and often live isolated away from general population centers in abject poverty...

SLOVENIA
(2 million humans)
Military Forces = 7, 300 active duty / $780 million per year

A nation east of Italy with access to the Adriatic Sea, Slovenia shares a southeastern border with Croatia, a northern border with Austria and Italy in the west...Another republic of former Yugoslavia escaping violence in the 1990's due to a low ratio of ethnic populations, Slovenia is a growing economy recognized by western European nations joining the EU / NATO in 2004 and the euro-zone in 2007...A prosperous nation in Central Europe with a favorable location, a work force of almost a million and an unemployment rate around 13%...Median age of almost 43 and a life expectancy of 78 years with 13% of the population living below the poverty line...A culture of widespread corruption with the president elected by popular vote, the prime minister elected by a National assembly (90 seats) and a population unhappy with planned austerity measures...Challenges include the banking sector with up to 25% of total loans rated "bad" at a cost of over 9 billion euros and an ongoing dispute with Croatia over land and maritime borders in the Adriatic Sea (both have agreed to binding arbitration to define borders)...

SPAIN
(48 million humans)
Military Forces = 123,300 active duty / $11.6 billion per year

After nearly forty years of dictatorship under Franco, Spain democratized with free elections in 1977 joining NATO in 1982 and the European Union in 1986 with the hope of creating a prosperous and intellectual society to share in the wealth of a large growing European community and an emerging global economy...Located in one of the most strategic geographical areas of the world at the entrance to the Mediterranean Sea with a coastline on the Atlantic Ocean, Spain was a world power in the 16th and 17th centuries creating an overseas empire which ended in 1975 with Spain's withdrawal from their last colonial territory in the Sahara...In 2016 Spain's unemployment rate was about 23% primarily affecting the young with a large underground economy and a reliance on tourism for revenue...The two political parties, Socialists and 'Populares', are thought of as equally inept by a population that is more and more demanding change in an economy kept afloat by exports, growing foreign investment and reformation of fiscal policy in education, labor, pensions, healthcare and taxes...Labor force of 23 million with 21% of the people living below the poverty line, a median age of 42 and life expectancy of 81 years...Crude oil reserves of 150 million barrels and 2.5 billion cubic meters of natural gas reserves...Government is a parliamentary monarchy with Head of State King Felipe and seventeen regions who enjoy an autonomy that is distinct for each region...Nearly 5,000 kilometers of coastline and the Pyrenees mountains in the north with a variable climate of wide temperatures and rainfall variations from north to south...The most visible concern today is a struggling economy but also Catalonia, a prosperous group of four provinces (including Barcelona) in the northeast corner of Spain who have declared the right of self-determination and sovereignty (a Spanish court found this claim illegal), a continuing flow of African migrants in the Strait of Gibraltar (water separates Africa and Spain by only 8 miles) and continuing investigations of the King's daughter, her husband and Luis Barcenas for corruption...

Unemployment rate in 2016 for Spanish youth (under 25 years of age) was over 45%...

SWEDEN
(9.7 million humans)
Military Forces = 14,000 active duty / $6.2 billion per year

One of the largest nations in Europe with it's extreme length (almost 1,000 miles) crossing the Arctic Circle in the north, Sweden has coastline on three bodies of water, the North Sea on the west (south of Norway), the Baltic Sea in the southeast and the Gulf of Bothnia on the east between Sweden and Finland...Bordered by Norway almost it's entire length on the west and by Finland in the northeast, Sweden has some 90,000 lakes, large coniferous forests and a fishing fleet skilled in catching herring, cod, salmon, mackerel and shellfish...Sweden emerged as a nation in the 16th century and had alliances with Finland (lost to Russia in 1809) and Norway (a union that lasted almost 100 years until 1905) that solidified the borders that exist today...One of the most enlightened, stable, post industrial nations in the world, Sweden is a constitutional monarchy led by King Carl XVI Gustaf in a ceremonial role with a 349 seat parliament led by a prime minister selected from the majority coalition facing concerns over immigration and slowing growth in all sectors...Economy is industrial engineering with a highly skilled workforce and a focus on foreign trade for nearly 50% of output and exports...Labor force of five million with an unemployment rate around 8%...Median age is 41 with a life expectancy of 80 years for males (one of the highest in the world) and almost 84 for females...Sweden's social welfare system is renown for its effectiveness and their healthcare system is one of the best in the world...The people rejected joining the euro-zone and as a nation who remained neutral in war for almost 200 years they are reluctant to join NATO...Sweden joined the EU in 1995 sending peacekeepers to Kosovo in 2001 and humanitarian aid to Iraq after the conflict...Most energy is from hydroelectric and nuclear with less than 10% from coal and plans for expansion in natural gas production...

SWITZERLAND
(8.1 million humans)
Military Forces = 143,000 active duty / $4.7 billion per year

A mountainous country in central Europe bordered by France, Germany, Austria, Italy and the tiny nation Liechtenstein, Switzerland's borders and neutrality were established in the 19th century with a high priority on independence remaining neutral in both World Wars and refusing to join the United Nations until 2002...Politics this century have been more contentious and challenging than the previous four decades mostly due to immigration policies and in 2014 Switzerland passed a new proposal to strengthen immigration law setting quotas to stop the continuous stream of humans from EU nations into their country...Over 20% of the nation's population are documented foreigners who impact housing, infrastructure and the social services available here...The EU has criticized these new restrictions...A stunning topography with deep valleys and large mountains of 15,000 feet in a snow filled landscape popular with tourists but prone to flash floods and avalanche...An unusual government structure with a seven member Federal Council (elected by a joint session of parliament) as a collective head of state with a rotating president...Economy is a highly specialized financial service sector, a high technology manufacturing sector and tourism...Labor force is almost five million with an unemployment rate of just 3%...Median age is 42 with a life expectancy over 82 years...Estimates of population living below the poverty line are 7% with excellent healthcare and benefit systems...A respected voice on the world stage founding the Red Cross and Red Crescent in 1919 with almost 100 million volunteers worldwide who are critical in providing vital humanitarian aid and health needs in disasters and crisis anywhere on Earth...Challenges are the nation's immigration policies and a growing spotlight on entrenched banking secrecy practices with demands for increased transparency from many voices in many nations...

In 2016, the world's longest and deepest rail tunnel opened after 17 years of construction at a cost of $12.5 billion to connect northern and southern Europe...Gotthard tunnel is 57 km (35 miles) long and straight allowing trains to travel at high speeds (250km/h) carrying both freight (250 trains daily) and passengers (65 trains a day)...

UKRAINE
(44.3 million)
Military Forces = 159,000 active duty / $4.8 billion per year

The politics of Ukraine for eight years has been contentious and violent against the backdrop of the energy "wars" (Russia cut supplies to Ukraine multiple times before a new agreement in 2009) and unpredictability with members of parliament switching sides amid divided loyalties...After Yanukovych was elected president in 2010 parliament abandoned aspirations for NATO membership, the executive branch curbed media freedom, the constitutional court overturned limits on presidential power and a court jailed Yulia Tymoshenko, a former prime minister in what the world agreed was falsified charges for political reasons...In 2012, a protest over a new Russian language law turned violent, parliamentary elections were viewed as "problematic" and "troublesome" and in late 2013 anti-government protests erupted after Yanukovych rejected an agreement with the EU and aligned himself with Russia and Vladimir Putin...In early 2014 Yanukovych fled the capital after more violent protests, the jailed prime minister was released, parliament elected a new president and within a month Russia moved troops into Crimea (home of Russia's Black Sea fleet) and annexed the region into Russia...Separatists in the east sought alignment with Russia through violence and today the future of Ukraine is unknown...

Ukraine is located in a region populated for thousands of years with a long history of relations with Poland and absorption and domination by the Soviet Union...The present day state gained their independence in 1991 with the fall of the Soviet Empire and during the 1990's they introduced a new currency, a new constitution, signed treaties with neighboring countries and shut down the Chernobyl nuclear plant after the world's worst nuclear accident (1986) which was responsible for the deaths of over 10,000 humans...The first decade of this century has been a tumultuous period of state control and widespread corruption...Median age is 40 with a life expectancy of 71 years...Child labor force is more than 350,000 with 25% of the population under the poverty line...Energy wealth with an estimated 395 million barrels of crude oil reserves and over one trillion cubic meters of natural gas reserves...An agricultural and industrial powerhouse when it was part of the Soviet Union, Ukraine's economy is now one of the worst in all of Europe with entrenched state corruption (Yanukovych is believed responsible for over $60 billion missing amid very questionable transactions and outright theft) and severe austerity measures required by the IMF and the banks who promised billions and then put a hold on the money before Yanukovych left...The 'standoff' in eastern Ukraine is now a global stalemate without resolution changing the country in ways that will take decades to recover...Challenges are widespread organized crime, human trafficking and continuing border disputes with Russia, Belarus and Romania...

UNITED KINGDOM
(64 million humans)
Military Forces = 170,000 active duty / $56 billion per year

In 2016, the United Kingdom is Scotland, Wales, Northern Ireland and England established in 1707 with the union of Scotland, England and Wales sharing a single constitutional monarch and a single parliament in Westminster, England (Ireland joined the United Kingdom in 1800)...The UK was a global power by the mid 18th century and 200 years later had colonies worldwide with a combined population of 500 million humans and a legacy of imperialism...White Europeans in the 18th and 19th centuries considered many native populations throughout India and Africa to be 'sub-standard humans' without a capacity for intelligence or civilized manners in European society resulting in a misguided and ignorant paradigm that measured human beings based on skin color and origins, a guiding principle in the British Empire's expansion across the world...In this century the negative impact of colonialism is still at the core of many problems impacting Africa and subcontinent Asia...A misguided attempt to divide and rule Hindu and Muslim populations in India was unarguably a disaster with three million Indians dying of famine in 1943 leading to the creation of Pakistan, Kashmir and modern day India, a region of intense conflict and growing nuclear powers in a radically unstable part of the world...

The United Kingdom is a permanent member of the UN Security Council (with Russia, United States, France and China), a founding member of NATO and a member of the European Union...The UK has been a prominent voice in European and global affairs in modern times and an undeniable cultural and economic impact globally with one of the oldest, stable, informative and widespread media organizations on Earth (BBC)...In a geographically strategic site between the Atlantic Ocean and the North Sea, England was perhaps the most important nation in stopping the conquest of Nazi forces and turning the tide of Hitler's quest for world domination...A constitutional monarchy with a long serving queen (Elizabeth II) and a prime minister with a bicameral parliament consisting of hereditary seats (House of Lords) and elected positions (House of Commons)...Median age is 40 with a life expectancy of 80 years...Labor force is 33 million, unemployment rate of 6% with 16% of the population living below the poverty line...An economy that is Europe's third largest (behind Germany and France) with a large service sector in banking and insurance, a highly mechanized agricultural sector and industry across a large spectrum of products and exports...Energy wealth is over 2.9 billion barrels of crude oil reserves and 241 billion cubic meters of natural gas reserves...

Nuclear power provides about 21% of Britain's electricity from 15 nuclear reactors...

The UK still has multiple territories worldwide, mostly in the Atlantic Ocean and Caribbean Sea including Bermuda, Falkland Islands, Cayman Islands, Anguilla, British Virgin Islands, Montserrat, the Turks and Caicos Islands, Pitcairn Islands and Gibraltar (in dispute with Spain and the residents of the territory)...The Chagos Archipelago in the Indian Ocean is claimed by Mauritius and Seychelles with ongoing disputes over rights of return and the UK's large marine protection zone around the entire archipelago to prevent resource extraction...

In 2016 the United Kingdom voted to leave the European Union...

VATICAN CITY
(800 humans)

The smallest nation state on Earth with an area of 0.17 square miles, the Vatican is home to less than 900 humans who believe themselves to be the servants and spokesmen of an imaginary deity and responsible for dictating policy, rules and the 'truth' to more than a billion people who are incapable of existing without the guidance and discipline of an exclusive enclave of old men who strive daily to convince believers that the way to 'salvation' is rooted in the traditions and practices of an organization whose morality, ceremonies, rituals, dictates and operational policies are outdated and corrupt...The primary function of the Vatican is survival and a continuing effort to keep itself relevant as a growing transparency reveals deceit, mismanagement, greed and widespread sexual abuse that has continued not for decades, but for centuries in an organization that protects their own without concern for victims of abuse or a continuing discrimination of women embedded in church 'law'...In 2012 the Vatican reported an operational deficit of over $19 million despite revenues of more than $150 million from donations and ticket sales to the holiest Catholic shrines...Over half of this religion's one billion members do not adhere to even half of the 'laws', 'policies' and 'scriptures' demanded by the Vatican but in a very human way the members still believe they are practicing the 'laws of God' and the rulers in the Vatican will continue to believe they are relevant in a reality that long ago passed them by...

The Vatican Bank has a long history of deception, collusion, fraud and mismanagement as an "offshore bank" hiding and laundering money for politicians, mobsters, businessmen and criminals...Millions of dollars in secret accounts...

WORLD POPULATION

If human population continues to increase at its current rate, in less than 500 years there will be over 130 billion humans on Earth with less than three square meters of land per person...*This will not happen*...If the world populations were willing to live at a standard of living equal to the standard of living experienced by humans in West Africa, the Earth could support over 40 billion humans...Current projections put Earth's population at the end of the 21st century at 11.5 billion and this will challenge every resource on this planet as fertility of agricultural land grows less and water necessary for growing food is diverted by multiple nations with violence and barriers being normal in a thirsty world making choices affecting the lives of billions...Earth's productive biological capacity in 2016 based on average productivity and current population numbers is around 1.9 hectares per person in a world that consumes an average of 2.2 hectares per person (many of the developing countries consume less than one hectare per person while America's average consumption is 9.5 hectares per person)...With Earth's population at 10 billion by 2070, the planet's bio-capacity will average just 0.9 hectares per person impacting the survival of many species as an exploding human population destroys natural habitat and species that may be essential for human survival...Humans use the equivalent of 1.5 Earth's to support their current level of activity (resource consumption and waste absorption) and they would use the resources of five Earth's if every person consumed 9.5 hectares...All evidence indicates this planet already has more people than it can support and without a consensual intelligence or awareness of the impact of current growth rates *there are no global agreements to do the many things that could be done to slow down an increase of future populations or deal with the impact of current populations*...With an increase in the development of reality intelligence and a growing awareness of limits within a finite ecosystem, future humans will recognize the impact of population and self-regulate reproduction to maintain a balance necessary for survival...Until that time it is not possible to control birth rates in multiple cultures and societies without draconian measures and violence...Sex is one of the strongest primitive instincts in all species on Earth and only a priority on initial brain growth in the earliest stages of life will create future generations who understand this necessity...

Human Actions Necessary to Sustain Planet Earth

1. Everything on the planet *must be recycled* by every human in every society and governments should do what is necessary to make this happen...

2. Worldwide prohibition against cutting down trees, most important in the tropical rainforests that circle the planet in the equatorial zone...

3. The abandonment of all internal combustion engines for electric alternatives requiring a restructuring of electrical grids and supply lines across the entire planet...

4. The designation of large open spaces both on water and on the land where solar power plants and storage facilities can be built and large-scale wind farms can be constructed to take advantage of the planet's circulatory atmospheric patterns...

5. Solar energy systems on the roof of every structure on Earth and severe restrictions on the use of fossil fuels for anything not essential to human survival...

6. Worldwide targeted planning to design regions with the ability to grow and supply food for each designated region to reduce the wasteful transport of resources all over the world...

7. A global initiative to educate and free oppressed female populations everywhere on Earth that will allow women to decide without coercion how many children they wish to bear...

8. Prenatal care, birth control facilities and medical procedures in every nation to support every female in a decision and choice that they alone can make...

Our species needs to adopt a world view with targeted priorities if humans want a world that is habitable, progressive, self-sustaining, efficient and healthy...There is currently no indication that any of these actions are happening on a large scale or happening at all...

EVOLUTION

The Many Definitions of Evolution

They all say the same thing...

Evolution: Changes in populations that are considered evolutionary are those that are inheritable via the genetic material from one generation to the next...

Evolution: Changes in the inherited characteristics of multiple biological populations over successive generations...

Evolution: Changes in the properties of populations of organisms that transcend the lifetime of a single individual...

Evolution: Precisely defined as any change in the frequency of alleles within a gene pool from one generation to the next...

Evolution: The gradual process by which the present diversity of plant and animal life arose from the earliest and most primitive organisms, a continuing process believed to date back more than 3,000 million years...

Evolution: The development of a species, organism, or organ from its original or primitive state to its present or specialized state; phylogeny or ontogeny...

Evolution: Changes in the genetic composition of a population during successive generations as a result of natural selection acting on the genetic variation among individuals and resulting in the development of new species...

Evolution: Changes in heritable traits of a population of organisms as successive generations replace one another..It is the populations of organisms that evolve, not individual organisms...

There is enough evidence supporting the reality of biological evolution on Earth over the last three billion years to fill a large library while evidence supporting any other explanation of the gradual development of life on Earth does not exist...Alternative ideas from *creationism* (God made everything) to *extraterrestrial seeding* (bacteria from some alien visit contaminated the planet) or *extraterrestrial intent* (aliens dispersed life in specific places on Earth and left) do not survive even the most elemental scientific inquiry...There is no evidence of any kind that offers an alternative to evolution...Evolution is a process of biological diversification and species development on Earth that continues today and no human who understands reality doubts the science that verifies this continuing process...Real questions that remain include the origin of life (*where did the first spark of life arise and what was the cause?*) and an uncertain evolutionary tree (*did life develop in multiple locations or just one?*)...And there are mysteries involving missing links, unknown species yet to be discovered and the future of evolution in a new paradigm where life affects the development of life on the planet more than the planet affects the development of life...The recognition of specialization among species prove in many ways that humans are animals whose evolutionary advantage was an opposable or prehensile thumb, bipedal locomotion and a much larger brain but humans are in no way superior to all creatures in all aspects...

Who wouldn't want to have the slightly asymmetrical ears of an owl to locate noises on a horizontal and vertical grid or the eyes of an eagle for distant and focus combined with the eyes of a cat to see at night?

All species who survive existence and who have a long evolutionary history exist because of a stable balance within their environment, an advantageous physical construction and instinctive skills that allow every creature to survive in their habitat...Curious and interesting is how the proliferation of humans will impact and effect the evolution of other species and even us...With a global focus on initial brain development it may even be possible to structurally change the human brain over thousands of years with capabilities only imagined today...Humans are one of the newest species on this planet and it's unknown what might be possible given the enormous timescale of existence...

500 million years ago in the Cambrian Period there lived "Trilobites" with rudimentary 'eyes', possibly the first light sensing devices to develop in biological animals, arguably our most important sensory input...

Time is the primary variable in this entire process...Most evolutionary changes take enormous time and it continues everywhere on this world without notice...There are countless volumes and books that explain and confirm this proven reality and individuals who cannot understand this evidence and accept scientific proof of a process that has filled this world with life will be unable and unwilling to confront any of the scientific realities that define human existence...Fascinating and complex, random mutations and adaptation of species are why humans exist...

An evolutionary theory states that human genes' relentless' drive to self-replicate is at the core of all our moral decisions on cooperation... "Deceit is a necessity of life's brutal struggle to survive current conditions, acclimate in whatever way necessary and flourish into the future"...There is no evidence that this is true...

Every DNA molecule contains about 100 billion atoms in sequence and code...The blueprint for a complete human being is replicated in every cell of the human body...

"More than 99.99% of the species that have ever existed have become extinct but the planetary patina (surface), with its army of cells has continued for more than 3 billion years...And the basis of the patina, past, present and future is the microcosm, trillions of communicating, evolving microbes..." Microcosmos Dr. Lynn Margulis

Human DNA and chimpanzee DNA are 98% identical having diverged from a common ancestor 5 million years ago...

No single living organism ever discovered shows any evidence of having evolved by non-Darwinian means...

NORTH AMERICA

The trade relationship between Canada and America is more than $1 billion a day, the largest between any two nations in the world...Both nations have a large trade relationship with Mexico as well...

Bordered by the Atlantic Ocean, the Pacific Ocean, the Caribbean Sea south and the Arctic Sea to the north, Earth's third largest continent is over 9.5 million square miles in an almost unassailable geographic location keeping both world wars from crossing the ocean, high defensible and home to the world's most powerful military with America's defense expenditures almost 40% of the world's defense spending in 2015 (World Total = $1.6 trillion)...The population in North America is around 600 million humans, mostly in three nations, Canada, Mexico and America in a geographical landmass stretching from just south of the Arctic Circle to just north of the Equator...This continent has been home to diverse cultures and civilizations for thousands of years including Aztecs, Mayas, Olmecs, the Inuit, Nootkans, Makahs and Tshimshians, all ancestors of American Indians and many of the indigenous cultures that still exist in Mexico and Latin America...In a well-known history European soldiers and settlers began to arrive in the 16th century bringing disease and war that killed an estimated 90% of the native population and in the three centuries that followed settlers and soldiers killed most of the rest as territorial instincts demanded conquest with a prevalent, dominating European viewpoint that all these indigenous natives were inferior savages...In addition to destroying native populations the Spanish and the other European 'conquerors' destroyed ancient infrastructure, artifacts and languages while pillaging and eventually decimating entire cities to gain wealth for themselves and their kings...By the late 19th century America was well established and the remaining native populations were forced into legally designated reservations representing a small fraction of the land they roamed in freedom for a thousand years...In 2016 this situational reality of America's native populations remains largely unchanged...

North America is home to one of the largest canyons on Earth (larger canyons exist below the sea and are mostly inaccessible) and the Appalachian Mountains, one of the oldest mountains ranges on Earth (500 million years old)...By contrast the Rocky Mountains are around 150 million years old, the Alps just 60 million years old and the Himalayas, home to the tallest mountains on Earth, only 50 to 70 million years old...The Sierra Madre mountain range stretches from Honduras into the southwestern United States with volcanoes appearing throughout Mexico and regions of Latin America and there are rain forests in the northwest of America extending north into Canada...Three major desert regions occupy Mexico and the American southwest and there is a large region in Canada known as the Canadian Shield, a raised plateau with thousands of lakes covering an area of more than 1.7 million square miles, the oldest part of the North American crustal plate...The thousands of lakes on this plate were formed by glacial movement repeated over millions of years and is now a region rich in valuable minerals...The tallest mountain on the continent is Mt Denali in Alaska at 20,320 feet...

America grew from a British colonial experiment too far away to be managed or successfully defended when a loose confederate of colonies voted to fight for independence and after multiple wars ending in 1815, America won independence and the right to self govern settling into an uneasy truce with a British government that is America's strongest ally in this century...Canada was established in 1867 by the British government as a solution to the by now familiar problem of governing a colony across 3,000 miles of ocean...Mexico was governed by Spain for 300 years declaring independence in 1810 and a border war with America in 1846 finally established it's northern boundary...An occupation of Mexico (France) in 1861 brought intense opposition from Mexico's population with America helping Mexico win back their republic after the United States Civil War ended...

Today the U.S. dominates this continent as an economic and military powerhouse in a global environment that's becoming more competitive and more connected...A nation known for their military footprint in many nations amid territorial disputes, trade disputes and uncertain objectives, America has become increasingly ineffective and unpopular with a sharply divided political body operating under the influence of corporate money that challenges any effective foreign policy and creates an uncertainty worldwide as to the intentions and pursuits of this once powerful world leader...Infighting, corruption, indecision, a rogue financial market and a seriously ridiculous focus on winning elections with the expenditure of billions of dollars and campaigns that start immediately after the last election have lowered America's standing and effectiveness in the free world...

Canada is a strong American ally with cooperation on the borders, shared resources, a strong trade relationship and joint military objectives spending billions to buy fighter jets and U.S. support for those jets...A new liberal government is continuing the extraction of tar sand oil and still lobbying the American government to approve a pipeline (Keystone) to transport this product after withdrawing Canada's participation in the Kyoto Accord clearly stating they will meet no target goals or binding agreements relative to climate change in a large, mostly uninhabited territory (almost 10 million square kilometers) that stretches from the North Atlantic to the Pacific Ocean...

The 3rd largest nation in North America is Mexico, a nation with high income inequality, widespread corruption in the government, the military and the police force, a broken infrastructure and broken borders...A rising economy in this century, Mexico is a strong ally of the U.S. in areas of trade and military cooperation...Seven Latin American nations lie between Mexico and the South American continent: Guatemala, Honduras, El Salvador, Costa Rica, Nicaragua, Belize and Panama...All were impacted by the military and political policies of America in a 20th century Cold War paranoia which decimated economies and destroyed political and social structures with the many negative effects of these policies still visible today...

North America has a large reliance on fossil fuels for energy despite an abundance of renewable energy resources...

America's use of agricultural resources focus on eight commodity crops (cotton, wheat, corn, soybean, rice, oats, barley and sorghum) producing biofuel and sustaining animals to meet a growing demand for meat...

Canada's policies continue to focus on short-term profit ignoring vast tracts of land that could be used for innovative renewable energy development instead of the environmentally destructive recovery of an oil product (bitumen) difficult to refine and a large contributor to the accumulation of greenhouse gases in Earth's one atmosphere...

CANADA
(35 million humans)
Military Forces = 98,000 active duty / $17 billion per year

Canada in this century is still mostly wilderness and the second largest nation by area with a population of nearly 35 million who mostly live on one of the coasts or within close proximity to the United States...Bordered on the west by the Pacific Ocean and on the east by the Atlantic Ocean with the U.S. state of Alaska located in the far northwest corner, Canada shares a border almost 4,000 miles long with America in the south...Canada is a parliamentary democracy (emulating the British system), a constitutional monarchy (the Queen is head of state) and a federation (of former British colonies)...A prime minister as head of the House of Commons (308 members elected to four year terms) forming a parliament with the Senate (105 members appointed to serve without elections until 75 years of age) governing in a nation with two official languages, French and English, a point of contention for years depending on what city or province you reference...

A sensible, intelligent nation in the international arena for many years, Canada is now the source of some of the most carbon intensive oil on Earth, a major exporter of asbestos to developing nations, a non-participant in the Kyoto Protocol (withdrew in 2011), a major exporter of coal worldwide and the only country on Earth that kills polar bears for profit, a species losing their habitat at an alarming rate...Justification for all activities is profit...

Canada did indicate at the COP21 conference in Paris (2015) that they will aggressively fight climate change by pledging to reduce CO2 emissions 30% from 2005 levels by 2030...

An industrial power in mining, manufacturing and energy, Canada's labor force is 19 million, unemployment rate is 7% and an estimated 10% of the population live below the poverty line...Crude oil reserves are third highest on Earth with 171 billion barrels and natural gas reserves are 1.9 trillion cubic meters...A huge industry in mining with what are believed to be the largest deposits of asbestos in the world and large quantities of uranium, potash, gold, zinc, silver, copper and diamonds, minerals worth billions of dollars in a global economy demanding more of everything...Canada is fifth in world coal reserves producing around 60 million tonnes a year and exporting more than 75% of the coal to Asia...When you add the focus the government has placed on extracting bitumen, activity destroying areas of wilderness at a rate that is alarming to anyone who knows the impact of this process, Canadian industry is a world leader in providing multiple products globally to be burned in increasing quantities (with little regulation) changing this planet's climate in negative, irreversible ways...Always looking for new stuff to burn or sell, Canada is in dispute with Denmark and the United States over the boundaries and rights to waters in the north being opened by the melting of ice that promise abundant resources for recovery and extraction...Median age is 42 years with a life expectancy of more than 81 years...A focus on healthcare has created one of the best health networks in the world

Oil sands operations account for more than 6% of Canada's greenhouse gas emissions...Withdrawing from the Kyoto Protocol in 2011 was a strategic move to avoid any limits on the development of the tar sands...Producing a barrel of oil from bitumen results in greenhouse gas emissions 3x's greater than a conventional barrel of oil...

In the last decade (2006-2016) over 25,000 humans have 'disappeared' in Mexico...

MEXICO
(120 million humans)
Military Forces = 260,000 active duty / $6.8 billion per year

Mexico is bordered on the south by Guatemala and Belize, the western border is the Pacific, the eastern border is the Gulf of Mexico and a border of almost 2,000 miles is shared with America in the north...Mexico is a large nation with widespread corruption, violent drug cartels, ineffective police and ineffective legislation that has defined this nation for over half a century but indications are things are slowly improving...The PRI (Institutional Revolutionary Party) has held power for 71 of the last 83 years and is once again in control after defeating PAN (National Action Party) in 2012 in a national referendum after 6 years of violent murders, corrupt politicians, stagnant incomes, increasingly ruthless drug cartels and uncertainty at every level of existence...A nation enveloped in corruption, poverty, crime and confusion with over 164,000 homicides in the past eight years (60,000 from drug violence), apprehension and conviction almost non-existent and security infiltrated with organized crime and drug cartels that bribe the nation's law enforcement and many believe the military...A single party rule and political impasse means change comes very slowly...The president has pledged to support paramilitary troops to fight cartels and corruption in law enforcement who are defenseless against the illegal drug operatives ($13 billion a year industry) who pay compensation and bribes to police that overwhelm small salaries and poverty...The 14th largest country in the world with over 9,300 km of coastline and rich deposits in minerals and oil, Mexico could be one of the most influential states in the western hemisphere if there were not this long history of corruption and income inequality that has over 50% of the population living below the poverty line...With a labor force of more than 50 million humans, the government is implementing economic reforms in all sectors including energy, education, banking and telecommunications, but any system of fairness and predictability is challenged by long entrenched interests and the organized crime that threatens everything and everyone on a daily basis...Median age is 27 with a life expectancy of 75 years...Over a million children are in the labor force with over 5% of children under age five malnourished...Energy wealth is 10 billion barrels of crude oil reserves and 483 billion cubic meters of natural gas reserves with the possibility of more in territorial waters...The challenges in this country are many with a stream of immigrants from the south moving north to reach the U.S. and violence just south of the Rio Grande...Mexico is the world's second largest opium poppy cultivator with a potential yield of 50 metric tons of pure heroin every year and a major transit point from South America for drugs going to the U.S. with tunnels, submarines, mules and money creating a perfect storm with redundancies to keep supply lines open...

Millions of gallons of raw sewage flow into the Rio Grande from Nuevo Laredo every day ...

The end game of capitalism is corporations controlling government, the definition of America in this century...

UNITED STATES
(321 million humans)
Military Forces = 1.4 million active duty / $587 billion per year

America is one of the richest and most influential nations on Earth increasingly controlled by a corporate culture enormously successful in institutionalizing corruption on a scale not comparable to any nation, one of the most corrupt nations in the world but in America it's mostly legal...Largest prison population on Earth with about 25% of the world's inmates...Most civilian guns in the world with over 300 million non-military guns...Over fifty million humans without healthcare...Over 45 million humans living in poverty (15 million children)...An estimated 50 million Americans living in "food insecure" households...At least $250 billion needed to repair infrastructure in schools nationwide as regional politicians give public money to private for profit schools and shutter or ignore a vast public school system that used to be the best in the world...One out of nine bridges with significant structural defects (numbering over 66,000)...Schools, parks and libraries being closed in mass...Billions of gallons of untreated sewage entering the waterways through ancient and broken sewage systems, some over 150 years old and thousands of dams at risk of failure...Continuing political efforts to punish the poor for being poor with attacks on social safety nets, medical care for everyone, free education for the next generation and food security for children and infants existing in a society where financial institutions and politicians mismanage and steal billions of dollars without regard of consequence to the people or the environment...It's about profit...The top 1% own 33% of the nation's wealth while states and local leaders in districts nationwide cut taxes for multiple corporations and enrich themselves...15% of the citizens live below the poverty line...World's largest consumer of cocaine, heroin and marijuana...

A report released by the American Society of Civil Engineers in 2009 estimated it would take more than $2 trillion to repair this nation's crumbling infrastructure including aviation systems, levees, transit systems, highways and bridges...They estimate the drivers in this country spend 6 billion hours a year stuck in traffic which contributes to the warming of the planet and contributes to many deaths from the subsequent pollution...

America continues to take it's sovereignty to extremes in a shrinking world seeing cooperation as "weakness" and "appeasement", blustering about being a superpower refusing to sign the "Law of the Sea Treaty" signed by almost every nation on the planet, refusing to ratify environmental treaties (Kyoto Protocol), refusing to be a participant in the International Criminal Court (ICC), refusing to ratify The Convention on the Elimination of All Forms of Discrimination against Women (186 nations have ratified the CEDAW), refusing to ratify the Convention on the Rights of the Child along with Somalia and South Sudan (193 nations have ratified CRC), refusing to ratify the International Covenant on Economic, Social and Cultural Rights who insure labor rights, the right to an adequate standard of living, the right to health care, etc.(160 nations have ratified ICESCR), refusing to sign the Convention on the Prohibition of the Use, Stockpiling, Production and Transfer of Anti-Personnel Mines and their Destruction (Ottawa Treaty) with Russia and China and refusing to sign the Convention on Cluster Munitions prohibiting storage, transfer and usage of cluster munitions (along with China, Russia, India, Israel, Pakistan and Brazil) while maintaining postures and policies that sound ignorant and isolated in insisting that "no one sets limits on America"...

The U.S. is the largest weapons exporter on the planet selling billions ($) in arms to other nations...This one country with less than 5% of the global population accounts for 40% of the world's military spending...

The largest American corporations paid an average of 12.6% in taxes in 2010...Religious lobby groups who specialize in legalized bribery influencing political candidates and political officeholders spend almost $400 million annually to secure votes and write legislation...The American Israel Public Affairs Committee spent nearly $88 million in 2008, the Family Research Council spent $13 million in the same year, the U.S. Conference of Catholic Bishops spent over $26 million in 2009 and the National Right to Life Committee spent around $11 million...Some lobbyists advocate for poverty intervention and real issues necessary and immediate but most lobby groups have a self-serving agenda focused on influencing and manipulating America's politicians in setting policies which favor corporate subsidies and deregulation...In America it works very well...

The healthcare industry spends 3x's the money lobbying that the industrial military complex does...Healthcare in America is the most expensive in the world with every business associated with this confusing, opaque network making record profits every year and they have bought Congress to keep it that way...It's legal in America...

The United States gets just 13% of it's electricity from renewable energy sources...EIA

In 2012 Americans spent over $460 billion on gasoline...

America is about money and politics...Little else matters...Elected politicians are by definition the individuals who control the spending of the public's money and in America the end of one election is the start of a new campaign with enormous sums of money being spent by corporate entities to elect favored politicians who can be controlled and influenced for access and profit...Again, in America it works very well...The U.S. Supreme Court validated the most effective form of political corruption in a 2010 ruling that allows corporations to spend unlimited sums of money to control the political process (money is so coveted in this nation that giving money to a politician is now considered exercising one's right of "free speech")...Today almost no politician in the House of Representatives and at the state level actually "worry" about re-elections or doing the will of the people because of "gerrymandering", an obviously corrupt system of defining electoral districts in bizarre patterns to favor the politician who represents that district...Corporate money is responsible for much of the silliness defining the United States Congress and it's unlikely to change...There are no "poor" elected officials in Washington D.C.

BP, Chevron, ConocoPhillips, ExxonMobil, and Royal Dutch Shell combined profits in 2011 were more than $137 billion...These five corporations have made over $1 trillion in profits between 2001 – 2011 and they have spent over $60 million lobbying to influence Congress...They all still receive generous annual subsidies...President Obama has proposed cutting subsides that give oil companies yearly payouts from $13 billion to $50 billion but Congress, in a reflection of the power and deep pockets of corporate America, has never held a vote on the proposals...

The Chief Financial Officers Act requiring a complete audit every year of America's Defense Department was signed into law in 1990...As of 2016 a complete audit has never been done...The Government Accountability Office cited "serious financial management problems at the D.O.D. that rendered its financial statements unauditable"...Due to antiquated computers, malfeasance, ineptness and an ineffective governing body that writes and enforces the laws of this nation, over $285 billion went to contractors engaged in fraudulent behavior over a three-year period, many of which were already suspended for misusing taxpayer funds..The Pentagon insists they are "not blind" and Defense Department Comptroller said they are working toward getting the Pentagon audit-ready by 2017...

United States, Iraq, Iran, Saudi Arabia and China represent the top five nations who execute humans...

BELIZE
(340,000 humans)
Military Forces = 1,100 active duty / $27 million per year

Located on the Caribbean Sea with Mexico to the north and Guatemala to the west, Belize is a parliamentary democracy and a Commonwealth realm to the British with Queen Elizabeth II as chief of state, a prime minister and hereditary monarchy...Tourism and exports of agricultural products support an economy which carries a large debt with over 40% of the population below the poverty line and a rising unemployment rate...6% of children under age 5 are malnourished, more than 27,000 children are in the labor force and a high prevalence of HIV/AIDS is challenging the healthcare network...Home to the 2nd longest barrier reef on the planet and the site of Mayan ruins dating back over 3,000 years, tourism has continued to grow over the past decade employing more than 70% of the labor force in this industry...Media is generally free with exceptions for morality and state security...Recent discoveries of energy resources are expected to stimulate the nation's economy with over 6.5 million barrels of crude oil reserves already being recovered...Challenges include an ongoing border dispute with Guatemala and a high level of drug trafficking in shipments destined for the United States...

COSTA RICA
(4.7 million humans)
Military Forces = No standing Army

Costa Rica has the Pacific Ocean on it's west coast and the Caribbean on the east coast with Nicaragua to the north and Panama to the south...Another Central American nation heavily dependent on tourism with pristine rainforests, tropical waters and popular beaches...Exports include livestock, coffee, bananas, and sugar in an economy with over 24% of the population living below the poverty line...Despite this the standard of living has remained high with a strong social net whose goal is to provide universal access to health, education and clean water...Almost 40,000 children are part of Costa Rica's labor force along with thousands of Nicaraguan's who come here seeking work and economic stability...Government is a republic with a president elected by popular vote and a legislative assembly of 57 seats...Challenges include widespread corruption and crumbling infrastructure in an unstable, earthquake prone region, an ongoing dispute with Nicaragua over river borders and drug trafficking in a geographical location ideal for transshipment from South America to anywhere...

EL SALVADOR
(6.1 million humans)
Military Forces = 16,000 active duty / $160 million per year

El Salvador is located on the Pacific Ocean next to Guatemala and Honduras with a long history of civil war, natural disasters, government corruption and income inequality that continue to challenge a population where 38% of the people live below the poverty line...The smallest country in Central America and the most densely populated, El Salvador's economy survives on remittances from citizens working out of country (over 20% of the population live abroad), exports of food, sugar ethanol and a dependence on foreign aid...A civil war lasting 12 years ended in 1992 with more than 70,000 dead and damage to infrastructure estimated at over $2 billion...Life expectancy is around 72 years and almost 200,000 children are in the labor force...More than 6% of children under age 5 are malnourished in a violent country with one of the world's highest murder rates, mostly by gangs fighting for territory and dominance in an active drug culture...

*In 2015 there were 6,657 humans murdered in El Salvador, an increase of 70% from 2014...*The Nation

GUATEMALA
(15 million humans)
Military Forces = 16,100 active duty / $200 million per year

On the Pacific Ocean with Mexico to the north, Honduras and El Salvador to the south, Belize to the east and and a narrow access to the Caribbean Sea, Guatemala has a long legacy of human rights abuses, government corruption, a rotating leadership of military and civilian rulers responsible for a civil (guerrilla war) war lasting 36 years and killing over 200,000 humans...The youngest population in Latin America with a median age of 21 years, Guatemala is a poor country where more than half of the population (60%) exist below the poverty line, over 40% of children under age 5 are malnourished and a million children are in the labor force...A stunning geography with volcanoes, rainforests, mountains and multiple Mayan ruins, Guatemala's economy is supported by an agricultural sector that exports coffee, sugar, bananas and ethanol with foreign aid and remittances from abroad...Energy wealth is 83 million barrels of crude oil reserves and almost 3 billion cubic meters of natural gas reserves...Transit point of illegal narcotics with gang rivalries and organized crime adding to this nation's instability and insecurity...Volcanic eruptions and earthquakes in the recent past highlight Central America's existence on the edge of a small piece of Earth's crust, the Cocos Plate, slowly being pushed under the Caribbean Plate creating a dynamic that is evolving and unpredictable...

HONDURAS
(8.5 million humans)
Military Forces = 12,000 active duty / $210 million per year

Honduras is one of the most violent countries on Earth with just 2% of crimes solved, a corrupt military, a corrupt police force and a corrupt government...No one seems to care who dies and there are few investigations into any activity...They still receive millions of dollars in aid from the U.S. that does not get to the people...The 2nd poorest country in Central America with a growing inequality in income distribution, the economy survives on coffee exports, bananas and foreign aid...Drug gangs (maras) continue to terrorize the population by controlling multiple regions of the country with brutality and extortion...A democratic republic with the president elected by popular vote, the political process has proved to be ineffective and corrupt no matter who holds office...Located south of Guatemala with Nicaragua on it's southern border, Honduras has access to both the Pacific Ocean and the Caribbean Sea making it a popular transshipment point for drugs...The world's highest murder rate, 60% of the population living below the poverty line, an estimated 300,000 children in the labor force and 27% of children under age 5 malnourished...Death and imprisonment are serious threats against news outlets and journalists trying to report on the reality of a governmental structure that is beholden to itself and does little for the people...Hope lies in new trade agreements to boost industry (textile) and maybe a stronger government but corruption is so prevalent in the security forces that control this nation that any real change in the near future seems impossible...

NICARAGUA
(5.8 million humans)
Military Forces = 14,300 active duty / $47 million per year

South of Honduras and north of Costa Rica, Nicaragua is the poorest nation in Central America and ruled by Daniel Ortega who won another term in 2011 in a questionable election process...A long guerrilla war fueled by the United States funding and backing Contra rebels in the 1980's to defeat the Sandinistas (led by Daniel Ortega) devastated the economy and infrastructure as America's fear of the rise of communism led to destructive policies and unwarranted intrusions into any nation they believed to be a Cold War threat...Over 30% of the population live below the poverty line in another nation with a large disparity in wealth distribution...The economy is supported by exports in textiles, agriculture, processed food products and clothing as well as foreign investment and foreign aid...Median age is 25 with a life expectancy of 73 years...Over 25% of children under age five are malnourished with over 230,000 children in the labor force...Free trade agreement with the U.S. is helping the economy recover and Nicaragua has plans to increase the tourism sector and take advantage of the nation's geographical variations that include two coastlines, rainforests and volcanoes over 7,000 feet in elevation...A planned canal linking the Atlantic and Pacific oceans due to begin construction at the end of 2014 is on hold in 2016 with questionable financing amid social and environmental protests...If built (over five years) the canal would compete with the Panama Canal in a global environment of increasing ocean traffic...

PANAMA
(3.6 million humans)
Military Forces = 12,000 active duty / $145 million per year

A strategic location linking South America with North America and the Atlantic Ocean with the Pacific, Panama is primarily a service economy dependent on the transit of thousands of ships every year through the Panama Canal, the only route from the Atlantic Ocean to the Pacific Ocean without traveling around the tip of South America...Another important sector of Panama's economy is the Colon Free Trade Zone, the 2nd largest "free port" in the world, home to 1,750 companies with no municipal taxes, no local taxes, no capital investment taxes, no taxes on any exports or imports from anywhere on the planet estimated to total over $5.3 billion a year...A constitutional democracy with a president and vice-president elected by popular vote and a cabinet appointed by the president...Government plans to revise the tax code to entice new foreign investors and has expanded the Panama Canal to allow larger vessels and faster transport...A sustained growth rate and an aggressive program to upgrade and repair infrastructure has reduced the unemployment rate and had some impact on poverty but the disparity between wealthy and the poor is glaring with 26% of the population living below the poverty line and an estimated 60,000 children in the labor force...Literacy is high and life expectancy is over 75 years of age...Challenges are clean water and sanitation among the indigenous groups (who suffer the greatest income disparity and the highest malnutrition rates), stable and predictable power generation, adequate schooling in the remote areas of the country and an active and growing illegal drug market that uses the Colon Free Trade Zone (home to over a hundred banking facilities) as a money laundering center for narcotic operations thriving with collusion between organized crime groups and corrupt officials...

ANTARCTICA

The 5th largest continent (14 million square kilometers) is covered with an ice sheet averaging over 2 kilometers thick with a mean annual temperature of -72 deg Fahrenheit (-58 C), no indigenous inhabitants and very few native plant or animals...The United States Navy built the first structure on the continent in 1956 designed for scientific purposes and in this century there are multiple observatories that are manned year round to take advantage of the clearest skies on the planet to measure cosmic background radiation (CMB) and to conduct astronomical proposals in unprecedented detail using the SPT (South Pole Telescope), BICEP2 telescope and the High Elevation Antarctic Terahertz Telescope (HEAT)...The continent is divided into three regions with East Antarctica the largest and most elevated region, West Antarctica lying mostly below sea level and separated from the East by the Transantarctica Mountains and the Antarctica Peninsula, a mountainous region in the north...Estimates are the ice sheet in Antarctica contains up to 70% of the Earth's freshwater...

This is a remote and hostile environment with recorded data existing for only the past fifty years and scientists have yet to fully understand the patterns and trends of warming happening in this part of the world but it's clear that conditions are changing rapidly...A research team using the data collected at Byrd Station for over half a century and comparing it with current data being collected today estimated in February 2013 (*Nature Geoscience*) that the temperature in Antarctica has increased at a rate of 0.47 degrees every decade between 1958 and 2012, a total increase of 2.4 degrees...Global average over the same time period is just 0.13 degrees per decade suggesting that Antarctica is one of the most rapidly warming places on Earth...Ice core samples and summer melting also indicate an unnatural warming of the ice directly influenced by temperature increases in the tropics and increasing ocean temperatures influencing the atmospheric activity which carries the warm air south...Measurements show the oceans surrounding Antarctica have warmed at a rate twice the global average over the last few decades, more evidence of the inter-connectivity of all processes on Earth affecting a dynamic ecosystem in ways humans are just now beginning to understand...

Antarctica has no government...It is the coldest, driest and windiest place in the world and is governed by the Antarctica Treaty signed by 12 nations in 1959 with 50 nations party to the agreement in 2016...The treaty maintains Antarctica shall be used for "peaceful purposes only" and shall "permit" the freedom of scientific investigation and cooperation towards that end with observations and results freely exchanged...Several countries have territorial claims not recognized by most nations and the present status quo is preserved by Article IV which states that no "claim to territorial sovereignty in Antarctica shall be asserted while the present Treaty is in force"...

The Protocol on Environmental Protection to the Antarctic Treaty was signed in October 1991 and was entered into force in 1998 designating Antarctica as a *"natural reserve, devoted to peace and science"*...The Environment Protocol sets forth basic principles applicable to human activities in Antarctica and prohibits all activities relating to Antarctic's mineral resources except for scientific research...Until 2048 the Protocol can only be modified by unanimous agreement of all Consultative Parties to the Antarctic Treaty...

Life

Antarctica's only permanent inhabitant aside from the microscopic invertebrates that live in the soil and in the sparse vegetation is the *wingless midge,* a type of fly and the only true insect found on the continent...This largest land animal (.5 inches) survives by absorbing heat with it's dark, natural color having a higher dehydration tolerance than most insects, having larva that can survive up to four weeks without oxygen and having body fluids that resist freezing allowing it to freeze slowly without damage and thaw out again...Over 100 species of plants survive in the 2% of Antarctica not covered with ice, mostly lichens, algae, mosses and fungi...(Some lichens have been found less than 400 meters from the South Pole)...

There are a variety of seabirds and penguins living in the Southern Ocean around Antarctica on surrounding islands or in the sea with more than 30 species of birds visiting Antarctica and surrounding islands each spring, sea gulls, cormorants, terns and several species of albatross who return to the same sites every year to roost, hunt and breed...Seven species of penguin live around Antarctica including the *Emperor penguin* who breed on the continent but live mostly in the water, the *Adelie penguin* who breed on the shores of Antarctica and on all the surrounding islands, the *Gentoo penguin* (the fastest swimmer of all birds) who are spread over the whole of Antarctica and the *King penguin* who number over 2 million and breed in huge colonies on the islands in the Southern Ocean...

The largest settlement of humans in Antarctica is at McMurdo Station with an average summer population of more than 1,000 people...The impact of this has been pollution with the discharge of food waste and sewage allowed under existing agreements along with the seepage and spillage of petroleum products from human activity...Approximately 10,000 tourists visit the continent each year mostly in the Antarctica Peninsula closest to Chile having relatively strict protocols and guidelines to minimize damage to the environment...

August 2013...In research published in the journal Nature scientists found the glaciers on the East Antarctic Ice Sheet (EAIS) advance and retreat in sync with changes in temperature, glaciers containing enough water to raise global sea levels over 50m making it a priority to understand this impact...Most scientists had previously dismissed concerns over East Antarctica, the world's biggest ice sheet where temperatures drop to minus 30C, believing it was impervious to these small, cyclical changes but this new analysis questions that assumption...Researchers at Durham University looking at declassified satellite imagery dating from 1963 to 2012 used the pictures to detect changes in 175 glaciers as they flow into the sea along a 5,400km coastline and what they found was a strong pattern of ebb and flow...In the 1970's and 80's as temperatures were rising they found 63% of glaciers were retreating and during the 1990's as temperatures decreased they measured 72% of the glaciers advancing..."It is the first study to show there is an acute sensitivity in this particular ice sheet to climate variation," said Dr Chris Stokes who lead the research..The. scientists say there is no immediate threat to global sea levels but they are urging further investigation...

The Future of Antarctica

Fossil records prove this land mass was a temperate climate 50 to 60 million years ago with forests, flora, and many animal species who went extinct as a southern ice cap formed over millions of years and it's unknown what fossil records may exist deep within the ground underneath the ice...Today it's an inhospitable landscape surrounded by an ocean rich with marine life, krill, whales, seals and cod who inhabit water so cold it can cause hypothermia and death to any human in just fifteen minutes...In an unpredictable changing global environment, scientists estimate ocean levels around the world would increase almost 60 meters (almost 200 feet) if all the ice in Antarctica melts and in an unknown future it's possible this continent could become a paradise on a rapidly changing planet...Antarctica is believed to be rich in coal, oil and rare minerals including gold and silver and there will almost certainly be attempts to recover these resources as the continent becomes more accessible and nations worldwide look to remote places for energy...The area above Antarctica is where the depletion of atmospheric ozone manifests itself and it's unknown what might happen to this protective layer in a future when the regulation of ozone depleting gases breaks down amid the competing priorities of an overcrowded Earth...Marine life is protected from overfishing by the Commission for the Conservation of Antarctic Marine Living Resources but illegal and unregulated fishing threatens to undermine any conservation efforts...No one can predict the future of what is the last uninhabited region on Earth but this unique ecosystem fits like a puzzle piece in an inter-connected global biosphere that interacts with every other puzzle piece to create a habitable planet where life can exist...As with every process on Earth, ongoing scientific studies and continuous monitoring of environmental balances, reactionary trends and impacts short and long-term must continue if we are to understand the human balance needed to survive...

Marine Protected Areas

In 2013 a Commission for the Conservation of Antarctic Marine Living Resources (CCAMLR), a consortium of nations established in 1982 to protect the marine life and environment of Antarctica, issued a joint call for the creation of marine reserves in Antarctica...The group has been working since 2005 to establish marine protected areas (MPAs) in this region of Earth and have established only one Marine Protected Area to date...They have designated 11 priority areas in the Southern Ocean from which the MPAs would be created setting a goal of increasing protection in 10% of Earth's ocean...

At a meeting in July 2013 the group submitted plans to establish two giant reserves in the Ross Sea and in east Antarctica that were opposed by Russia and the Ukraine who questioned the legality of a plan that would have more than doubled the size of the world's marine reserves...Other countries including Norway, China and Japan challenged the science and the size of the proposed reserves and wanted the inclusion of a "sunset clause" to review the decision in the future...The Ross Sea proposal would have banned commercial fishing in an area of 1.6 million square kilometers and the other reserve would have protected 1.9 million square kilometers on the Pacific side...After compromise the new Ross Sea plan reduces the size of the reserve by 40% and allows discussion on the permanence of the arrangement...The proposal for east Antarctica was also revised but the size of the protected area is the same...Foreign ministers of supporting nations asked for the support of all commission members arguing that the new proposals are based on sound and best available science and will provide a unique laboratory for marine research with profound and lasting benefits impacting ocean conservation...Many environmental groups remain concerned the opposing nations will demand further concessions before adopting the agreement...In October 2016, CCAMLR (represented by 24 nations and the EU) agreed to designate the Ross Sea as Earth's largest MPA (marine protected area) protecting 1.57 million sq km for 35 years, less than 3% of the Southern Ocean...The agreement goes into effect on December 1, 2017...

A new report in 2014 in the journal Geophysical Research Letters based on three years of observations from the Cryosat spacecraft, an instrument designed to measure the shape of sea ice, reported that Antarctica is losing up to 160 billion tonnes of ice every year...This report was released just weeks after a NASA report in the same journal concluded the retreat of the West Antarctic Ice Sheet has passed the point of no return and cannot be stopped (based on forty years of observations)...The Cryosat data complimented the NASA report in reporting that the biggest annual loss of ice is in West Antarctica (134 billion tonnes) followed by the Antarctica Peninsula (23 billion tonnes) and East Antarctica (3 billion tonnes)...The time frame given by the NASA report is 200 to 400 years before the West Ice Sheet is completely gone raising sea levels over 4 ft...

SPECIES ON PLANET EARTH

In 2016, just over 13% of Earth is designated for the protection of species though many of these sanctuaries, reserves, national parks and conservatories are poorly managed with dwindling resources, shrinking borders, lack of data and underpaid caretakers subject to collusion and bribery...Almost 50% of the world's richest biodiversity zones have no protection and less than 2% of ocean habitats are protected, mostly adjacent to coastlines because of the difficulty in enforcing marine areas outside of a nation's boundaries (even as protected areas of the ocean have doubled in this century)...The problem is success is dependent on a global cooperation among countries with competing interests and territorial mindsets that protect traditional human activities that were vital to human existence in the past and almost all nations and industry are seeking to expand their activity to compete in the 21st century...Still, world cooperation is improving as more humans realize these protected areas are as important for their ecological value in the replenishment of water and vital resources as they are for the protection of the species existing in the area...Earth is an active, growing, breathing ecosystem with many species of plants and animals still unknown and human encroachment continues to grow as populations grow...

The list of endangered mammals has more than 1500 species on it and it's obviously impossible to protect every mammal, every insect and every plant in the world...Such control is not possible and it could be disastrous...There is an ebb and flow of variation, adaptation and mutation in all of the species defining life on Earth, variation over long time periods in an environment that continually changes in subtle and distinctive ways...Humans have the intelligence to understand the need to protect large animal species that are all special and rare and have the scientific knowledge to understand how certain insects, bees and certain types of birds play a critical role in the ecological balance that supports human existence...*The human species must begin to listen to the Earth*...Information recorded for many years with new mapping techniques, new measuring dynamics and 21st century analysis verify in a loud way how critical it is to maintain a sustainable food chain in an ocean that supplies nutrients to billions of creatures and stop this continuing destruction of critical habitat on land..The question in this century is who will listen and who will act...It's important that the people who recognize this reality find ways to stop the destruction of Earth from those who are only focused on profit, from the ignorant who believe animals parts to be mystical medicines capable of curing disease or summoning spirits and from societies that defend the killing of whales and other large creatures out of tradition and a deep fear of change...If human ignorance and careless consumption collapse food chains in the ocean the human species will face a serious crisis without solutions and in the fractured world environment that defines the 21st century this is a real possibility...The extinction of some wildlife and plants under protection today is inevitable but there is hope with science, technology, global communication and a growing human awareness of the need to prioritize the actions necessary to balance natural habitat and human activity in a way that will create space on Earth for a diversity of species...

It's estimated that 25% of all Earth's mammals are endangered...They are dying from the careless polluting of habitat, the expansion of the human species everywhere on the planet and an ignorance of causality...If humans do not provide the security and resources to guard the great animals of the world, populations will slowly go extinct, one by one, products of millions of years of evolution and a vital part of the world's ecosystem...Today, with billions of humans unaware of consequence, the most effective way to protect these animals is a global effort to create protected habitat with modern security, new technology and enough resources to render corruption ineffective...It is unknown at the beginning of the 21st century how many species will survive into the next century...

The world's number one consumer of illegally poached animals is China where many animal parts are consumed in the belief they have medicinal or aphrodisiac properties...

*The number of wild animals on Earth has halved in the past 40 years...*2014 World Wildlife Fund

The deepest known life on Earth is a nematode existing 3.6 km below the surface in an African gold mine...You can take a sample of soil, air or water almost anywhere on Earth and find life...A three billion year old legacy...

Republic of Congo Nouabale-Ndoki National Park

Protects Africa's rainforests which are home to the forest elephants, chimpanzees and lowland gorillas...This park was expanded in 2012 by more than 144 square miles to include a dense swamp forest known as the Goualougo Triangle, a remote forest miles from human settlements that is home to chimps and forest elephants, a smaller species of elephant (maybe 8 feet tall) who can range over 800 miles endangered by the threat of encroaching human populations and poachers who kill the elephants for their tusks...Nouabale-Ndoki National Park is now over 1,600 square miles and contains more than 300 bird species and over a 1,000 plant species in a part of Africa so steeped in violence and ignorance that it will be a challenge to protect these native creatures even in this vast reserve...It needs to be a focused, ongoing world effort...

Kahuzi-Biega National Park in the Democratic Republic of Congo

Over 600,000 hectares in size stretching through rainforests and bamboo forest this park is home to one of the last groups of lowland gorillas who number less than 300, a seriously endangered species that is the victim of bush meat hunters and human encroachment that continues to destroy their natural habitat...Described as one of the most diverse and ecologically rich regions on the continent, Kahuzi-Biega has 349 species of birds, over 1,100 inventoried species of plants and rare transitions of forest development from 600 meters to over 2,500 meters up the slope of two extinct volcanoes, Mount Kahuzi and Mount Biega...There are villages and populations in the park as well as oil exploration, some roads and a threat from the displacement of thousands of people fleeing ignorant wars between governments and Africa's strongmen who seek territory and wealth with little concern for humans or wildlife...Added to the list of World Heritage sites in Danger in 1997, the threat to the gorillas who still survive here is real and imminent...

In 2014 it was reported the gorilla populations in Kahuzi-Biega National Park have been reduced by half with refugees living in pockets of protected land and killing the animals for food...

Volcanoes National Park in Rwanda

A mountain range with nine volcanoes, three active and six extinct, this park became famous for the work of Diane Fossey who spent over two decades working with the mountain gorillas who number just 300 and represent half of the world's population...One of Rwanda's most visited tourist destinations with organized hikes, mountain climbing and gorilla watching...The 1st national park in Africa with a history of British colonialism before 1960 and a complete shutdown during the violence of the 1990's, in the 21st century it exists to protect the mountain gorillas and give these gentle creatures a chance to survive and increase their population in a hostile, unstable region of the planet...

Sangha Tri-National Protected Area

A 9,700-square-mile (25,000-square-kilometer) area designated in 2012 as a World Heritage Site, this protected reserve covers parts of three nations, Republic of Congo, Central African Republic and Cameroon in equatorial Africa, home to large populations of elephants, chimpanzees, gorillas, forest buffalo and bongo antelope among others...Central to the area is the Sangha River, a tributary of the Congo river, and a feature called 'bais', swampy clearings in the forest landscape where these large animals gather in great numbers...Created to preserve the 2nd largest rainforest on Earth, indigenous tribes and native populations live in buffer zones surrounding the park with agriculture, subsistence farming and controlled logging being the primary activities supporting commerce and trade...

Tanzania suspended it's anti-poaching project in October 2013 citing human rights offenses that resulted in 13 human deaths and thousands of arrests...In the last two months of 2013 over 60 elephants were slaughtered...

Elephants

In the early 20ᵗʰ century it's believed between 10 and 20 million elephants lived in Africa...Now it's estimated less than 500,000 survive...Human populations crowd their territory and many elephants are victims of war and conflict suffering violent sudden death from land mines and violent humans with weapons...Poachers poison these creatures to take their ivory to supply a growing trade in Southeast Asia and are relentless in their pursuit of profit from death with many countries offering no protection or safeguards for the largest land creatures on the planet, gentle, intelligent animals who only wish to be left alone to live as they have for thousands of years...The ivory trade flourishes as most things do that are illegal because of corruption at many levels of government and a profit motive that feeds a supply chain from the killing fields of Africa to consumers who buy finished products for investment, personal status and ego...The ban on ivory trade implemented in the Convention on International Trade in Endangered Species in 1989 was effective at the start but has grown weaker every year as corrupt governments ignore ivory carving markets and suppliers who sell the ivory without fear to customers in Japan, China, Hong Kong and the Philippines where the ivory is used to make religious artifacts for the Catholic faith...In a transparent self-serving action less than ten years after the ban, Namibia, Tanzania Zimbabwe and Botswana insisted they have a right to sell their ivory with the caveat that they would put money from the sale of ivory into a conservation fund to protect the elephants while making it legal to hunt elephants wherever they infringe on the rights of people...In October 2012 over four tons of ivory was seized in Hong Kong that shipped from Tanzania and Kenya...Tanzania has large sanctuaries containing over 60,000 elephants with haphazard and inconsistent protection from poachers and this country has been a leading exporter of illegal ivory in recent years...There is overwhelming evidence of corruption within the Tanzania Ministry of Natural Resources and Tourism as bribery, corruption, lack of prosecution, ignorance and the absence of any effort or ability to understand the interconnected life on Earth slowly decimates populations of these unassuming creatures...With the current governments backing the sale of ivory and with the growth of human populations who are mostly unaware of the slow, continuing destruction of all the large animals on Earth, elephants are threatened with extinction before the end of this century...

2012...Charlie Mayhew, the chief executive of Tusk Trust: "What we have witnessed over the last 18 months or two years has been a significant escalation in the poaching of both rhino for rhino horn and elephant for ivory fueled by a sort of dramatic increase in demand from consumers in the Far East...In 2011, we believe that as many as 35,000 elephants may have been slaughtered for their ivory,"

In October 2013, Ugandan officials seized over 4,600lbs of ivory...The Uganda Wildlife Association (UWA) estimates the ivory to be worth up to $6.7million...The UWA says that Uganda is increasingly being used as a transit country by poachers who kill elephants in South Sudan or in the DRC to fuel a rising demand in Asia where the African ivory is used in ornaments and expensive trinkets to decorate the lives of ignorant, unaware humans...

In June 2014, the Convention on International Trade in Endangered Species (Cities) reported an increase in the number of large ivory seizures (over 1,100 lbs) in 2013 with demand for ivory growing in Asian nations...Using data collected from 51 sites across the continent of Africa the report estimated as many as 20,000 elephants were killed in 2013 for the singular purpose of profit from the trafficking and sale of ivory...

The Wildlife Conservation Society estimates over 800,000 elephants have been killed over the past three decades while a new study published in the 'Proceedings of the National Academy of Sciences' reported 100,000 elephants were killed for their ivory in just the last three years leaving a population of around 400,000...2015

The Western Black Rhino was declared extinct in November 2011...The Northern White Rhino is possibly extinct...The Javan Rhino in Vietnam is probably extinct as none have been seen since 2010...

South Africa reported 1,215 rhinos were poached in 2014 for their horns as ignorant humans in Southeast Asia (China, Vietnam) continue to believe in the mystical, medicinal properties of a simple physical structure made of Keratin, essentially a ball of hair...Over two-thirds of the slaughtered rhinos were found in Kruger National Park, one of the largest game reserves in Africa...Another record number of deaths, a 21% increase over 2013...

Rhinoceros

In 2011, 434 rhino were poached in South Africa for their horns...In 2012, 668 rhino were poached (killed) for their horns, an average of one every 13 hours...In 2013, the killing of rhinoceros in South Africa shattered previous records with a total of 1,004 rhino killed as government protection weakened, the demand for the rhino horn by people who believe in mysticism and miracles grew and this greed successfully influenced and corrupted individuals in control of the rhino habitat who could stop this senseless slaughter...

1,338 Rhinos were killed for their horns in 2015, an increase for the 6th year in a row...IUCN

There are around 19,000 rhinos left in South Africa, close to 90% of all the rhinos in Africa and they are rapidly becoming endangered as the murder rate exceeds their birth rate...Science has clearly proven that the rhino horn has no medicinal properties but a large number of humans in China and Vietnam believe in the mystical spirits and magic of the rhino horn, a part of their culture for thousands of years...In Vietnam ground up rhino horn is used as an aphrodisiac for sexual performance (there is no evidence this is effective) and to cure hangovers (again, no evidence supporting this)...Poachers and suppliers with profit and greed as their only motive are able to bribe and corrupt whoever stands between them and the rhinos and it shines a huge spotlight on the South African government in charge of Kruger National Park where most of the killing is done and the other sanctuaries designed to protect these large exotic animals from the consequences of a growing human ignorance...In the first two weeks of 2014, 37 rhino were killed for their horn...Despite the outcry and efforts of more than 100 organizations worldwide, there is no effective collective effort to stop this...

In 2007, only 13 rhinos were killed...

In 2013, the last rhino was killed in Mozambique adjacent to South Africa...

In 2012, Vietnam was singled out by the World Wide Fund for Nature (WWF) as the top destination country for the highly-prized rhino horn...

There are just over 3,200 tigers left in the wild when there were 100,000 a century ago...

Tigers

Of the nine subspecies of tigers, 3 are extinct (Bali, Caspian and Java) and the rest are in danger from poaching and habitat destruction...The Convention on International Trade in Endangered Species (CITES), an organization created to protect these creatures, is currently recognized by less than half of the nations where the tiger lives and many governments turn a blind eye to conservation as growing human populations demand more resources and more land...The tiger has been intricate to the practice of Asian medicine for thousands of years and on a still primitive Earth cultures are unaware and slow to change what has been tradition for generations...Many tigers killed are exported to China whose connection with all things in Asia and many other parts of the world is indicative of how difficult it is to control a population of over a billion people...The Chinese government made it illegal in the 1990's to use tiger parts for medicinal purposes but corruption among local officials in China have deep roots and prices paid on the black market for a large cat were higher in 2012 than at anytime in the past...Of the approximately 3,000 tigers remaining on Earth over 70% are in India in a diminishing habitat and a threatening environment...The Siberian tiger (also called the Amur tiger) is isolated in small sections of Russia and China and is threatened by the expanse of industry and civilization in these regions...These animals are the largest species of tiger on Earth and can have territorial boundaries of more than 200 miles...In 2015 their numbers are only a few hundred with less than 50 known to inhabit China...If humans in Asia continue to believe in the medicinal properties of tigers and governments continue to make these animals a low priority the future of the tiger will be only as a page in a book...

On the eve of a new global survey of surviving tigers (2016), India reported a rise in their wild tiger population of nearly 58% in seven years with a count of 2,226 in country...

Amur Leopard

A leopard subspecies critically endangered with just 50 cats left in the wild...Native to Russia it has little chance against human civilization and the exploitation of resources that reduce the animals habitat, ongoing climate change that alters their environment and poaching by ignorant humans seeking only profit...Estimates are less than 50,000 leopards remain on Earth...Approximately 5,000 leopards are killed annually for their coats...(In 2015 the count of Amur leopards rose to 57)

Lions

Current estimates of the remaining lions in Africa are between 25,000 and 30,000 when just 100 years ago there were more than 200,000...With no natural predators many of these animals are killed by humans seeking a "trophy" with over 500 lions killed every year this way and in today's world it is nearly impossible to stop this senseless activity of killing just to kill, something unique to the human species...A study released in 2014 by Panthera, a non-profit focused on saving the large cats of the world, shows the West African lion is in grave danger of disappearing with just 250 adults left in poorly managed habitats in only five West African countries...Land cultivated for livestock, the expansion of human population and poorly funded efforts in "parks" with no staff contribute to the decline of a creature whose natural habitat has been reduced to less than 1% of what it was just a few hundred years ago...Many are killed by native populations seeking "bush-meat' and ranchers protecting livestock...The largest and most iconic carnivore in Africa will go extinct without dedicated funding to provide safety and habitable territory...Many of the habitats originally set aside for the lions are now occupied by humans as global population increases exponentially on a planet with a fixed amount of space...

Northern Atlantic Right Whale

Even with estimates of less than 500 right whales left in the world this most endangered of all whale species is still hunted by whalers with others dying from being tangled in fishing gear...Habit degradation, climate change and ocean industrial activities all threaten the future of this creature and in this century America's Interior Department is allowing corporations to conduct seismic testing in the whale's only habitat to map the seafloor in the search for energy resources...Over 138,000 marine mammals are expected to be killed or injured by tests using seismic airguns proven to cause permanent hearing loss with extreme distress...Humans do not need to drill for more oil in the ocean...

The U.S. Department of the Interior approved the use of seismic airguns to search for oil and gas deposits in the Atlantic Ocean saying it was "safe"...March 3, 2014

The North Atlantic Right Whale was so named because whalers believed they were the 'right' whales to hunt due to their close proximity to the shore and the fact that they float on the surface when killed...

Of all the species put on the endangered species act less than 1% have been removed from the list...

In March 2014, the IJC ruled that Japan must stop hunting whales in Antarctica...Japan claimed the activity was for scientific research but the court found over 3,600 minke whales have been killed with little scientific output...

A report in the journal PloS Biology estimates there are about 8.7 million species on Earth with 6.5 million species on land and 2.2 million in the ocean...The most rigorous mathematical analysis to date suggests 86 percent of terrestrial species and 91 percent of marine species have yet to be discovered, described and cataloged...Researchers estimate there are approximately 7.77 million species of animals, 298,000 species of plants, 611,000 species of fungi, 36,400 species of protozoa and 27,500 species of chromists (which include various algae and water molds) with just a fraction of all these species identified...

Sharks

A report by the International Union for Conservation of Nature (IUCN) in 2009 estimated over 33% of all shark species are threatened by extinction...Almost impossible to verify because of illegal fishing and a lack of data on shark catches, researchers estimate between 63 million and 273 million sharks are killed each year to satisfy a demand for "shark soup" believed by many in Asian countries to having healing powers...In "finning" the shark's fin is cut off and the animal is left to die in the ocean...Predators with no natural enemies except humans, sharks have slow growth and reproductive rates and are a critical link in the marine food chain...Future effects of this activity are unknown...

In May 2014, the British Virgin Islands became the 3rd Caribbean territory to establish a shark sanctuary banning commercial fishing for shark and rays within a 31,000 mile designated area with a prohibition on the sale and trade of shark products...The other two territories with shark sanctuaries are the Bahamas and Honduras...

Leatherback Sea Turtle

One of the oldest species on Earth with a history going back millions of years, the leatherback is the largest sea turtle (up to seven feet and weighing up to a ton) with an average lifespan of 40-50 years and the ability to dive to a depth of 4,000 ft...Wide ranging with the ability to stay under water for 90 minutes, the leatherback used to populate all the major oceans on this planet but are now seriously endangered...Over the past three decades populations have declined over 95% through entanglement in drift nets miles long and poaching of their eggs and hunters from which they have no defense...More and more of these creatures are found deceased with large amounts of the world's floating plastics in their stomach frequently mistaken for jellyfish, the turtle's favorite food...

Giant Panda

Large creatures weighing up to 200 lbs, the panda live in the mountains of China having been chased from the lowlands by civilization, logging and industrial development...Their numbers are less than 2,000 in the wild with about 300 in zoos and research facilities...They have existed in the bamboo forests for more than a million years, highly specialized animals who reproduce slowly over a 20 year lifespan (35 years in captivity) and are now challenged by a world with few boundaries...Scientists predict that climate change will affect the growth density of bamboo (the panda's primary food source) and the wild habitats which are vital to their existence can be maintained only by the adaptation of the pandas and the bamboo to higher and higher elevations...

The panda cannot migrate to feed in new areas because they are surrounded by human encroachment...They survive off multiple species of bamboo and spend up to 70% of their time feeding each day...

Orangutans

Wildlife experts believe there are only 6,000 to 7,000 of these animals left in the wild with over 1,000 killed each year by the unstoppable invasion of the orangutan's habitat by human population growth and development as corporations seeking profit continue to destroy huge regions of forest that have been home to these creatures for thousands of years...In this century the Earth is losing more than 30 million acres of forest every year through human activity...

Kakapo Parrot

Very few humans care about the fate of this parrot species located in New Zealand whose days are surely numbered even as protectors have sealed their habitat to protect them from predators...So few exist they all have names and the hope is they will breed in their new habitat and multiply...Kakapo is a flightless bird who is the heaviest of all parrots, can live as long as 100 years and were prized in the 19th century for their meat and their feathers...This species of parrot survived for over 50 million years until the arrival of humans and human pets devastated the population...(About 62 exist in 2016)

A study published in the journal Science in May 2014 reported that species on Earth are disappearing at a rate 1,000 times greater than they did before the human species began expanding across the planet...Lead author Stuart Pimm at Duke University said it may be the beginning of Earth's sixth mass extinction event...The results were based on calculations of the "death rate" of current species and a re-calculation of the rate of species disappearance before human population growth...The number one reason for the increased rate of extinction is habitat loss as an expanding human population continues to demand more space...Also cited as primary factors contributing to the rate increase were overfishing and continuing climate change affecting the conditions necessary for survival...

In 2011, the International Commission for the Conservation of Atlantic Tuna adopted measures to protect silky sharks and adopted a modest proposal governing the catch of Mediterranean swordfish whose numbers are 50% below what is sustainable but rejected measures aimed at controlling the harvesting of sharks for their fins as they rejected establishing catch limits for blue and short-fin mako sharks and rejected an effort to require that all sharks be landed with their fins attached...The commission is a consortium of representatives from 48 nations...

Unspoken Menace of the Sea

Spring 2012...Off the coast of northern Peru an alarming number of dolphins (800+) died in a short period of time while months earlier an alarming number of pelicans (1000+) died with no apparent causality as well as turtles and sea lions...Possibilities include pollution, climate change, a marine mammal virus, morbillivirus DMV or sonar equipment used by oil companies in tests which were carried out in 2011...(Energy companies deny any connection between the deaths and their undersea activities)...

September 2012...Seventeen pilot whales died after a mass beaching and several isolated whales beached themselves and died during the same period in which survey ships were implementing high tech sonar scans of the topography of the ocean floor in connection with an offshore wind farm...The Whale and Dolphin Conservation Society claims that loud, low-frequency pulses interfere with the sonar whales and dolphins use to navigate and seismic surveys using loud, low frequency noise are believed to be responsible in some way...

It is recognized by marine biologists around the world that hearing is essential to the survival of whales and dolphins who use this sense for diving, feeding and navigation and it's understood that increased amplification of any loud sound underwater can destroy the hair cells in the mammal's ears critical in their ability to hear...It's also known that natural phenomenon like underwater earthquakes can cause pressure changes and noise sufficient to cause damage but rarely spoken of is human generated noise of military sonar, oil industry airguns, underwater explosives and the enormous noise from sea traffic that can cause the same damage and are in many cases responsible for the mass beaching of whales and dolphins around the world...Still, with clear research and multiple experts who agree this is happening, *military and oil companies continue to deny without reservation the causality of their undersea activities insisting their activities have no negative effects on the creatures of the sea*...In a world of terrorism and paranoia the military continues to install listening stations all over the world to track smaller and smaller vessels while developing weapons that use high frequencies in a focused way to destroy targets...Oil companies are expanding their search for resources in oceans all over the world using high-powered guns and explosives to create and record seismic shock waves to determine the location of these resources and there is little doubt of the effect on the animals in the ocean...What is sad is a lack of honesty, an ignorance of reality and continuous denial that this situation exists with no cooperative effort to look for other ways to accomplish objectives and minimize activity wherever possible to reduce the harm to these animals...Instead in a world that prioritizes profit, greed and territorial control over all else there is obvious ignorance and a determination to maintain secrecy without regard for life in the sea, a critical link in the global food chain...

In experiments published in two Royal Society Journals in 2013 scientists conducted sonar reactivity tests on beaked whales and blue whales who were tagged with tracking and sound recording devices while feeding or hunting and found a definitive response to sonar from both types of whales even though the two whales types use very different frequencies for food gathering and communication...Upon activation of the sonar the beaked whales immediately stopped hunting and swam swiftly and silently away from the area...Blue whales, who use a much lower frequency than sonar range, reacted hardly at all when they were on or close to the surface but were obviously affected when diving below the surface.."As soon as the sound started the animal stopped feeding, maintained a directed heading and moved away from the sound source."...Researchers have linked mass strandings and deaths of beaked whales worldwide to military exercises using mid-frequency sonar...Military's globally use sonar in training exercises against real and imagined dangers without regard for consequence...

In this century human activity is impacting survival and the habitat of every large species on Earth, land and sea...On land it's human encroachment on habitat with exploding populations seeking expansion without regard for the life affected...Coupled with weak governments whose priorities are elsewhere and corruption in protected sanctuaries (many animals last hope) it's impossible to know which creatures will still exist in 2100...In the sea there are three areas of concern which surface over and over again: military activity, overfishing and commercial activities by energy corporations...Sound is amplified in water traveling long distances and an ocean that was quiet 200 years ago is now filled with violent, unpredictable noise with territorial concerns and profit always the priority over the variety of aquatic life that have survived in a stable environment for thousands, even millions of years...What's sad is the lack of responsibility, creativity and honesty in recognizing a proven causality with little focused effort at exploring other ways to accomplish human objectives without damaging stable ecosystems or the many species suffering the consequences of human activity...

In 2013, the World Wildlife Fund compiled a report card ranking compliance with an international treaty (The Convention on International Trade in Endangered Species of Wild Fauna and Flora) regulating global trade in wild animals ranking 23 nations focusing on elephants, rhinoceroses and tigers...The report looked at where these animals originate, the countries they must travel through and the countries where they arrive for sale...India and Nepal received green marks for all three species indicating progress complying with the treaty and enforcing policies to prevent illegal trade but most countries received red marks for failing to uphold their commitments under this treaty signed by 175 nations making almost all commercial trade in rhino horns, ivory, tiger parts and other species threatened with extinction illegal...In Vietnam demand for rhino horn has exploded due to rumors of healing and aphrodisiac properties with no recorded seizure of rhino horn in the country since 2008 and although China has strict laws controlling the sale of elephant ivory it does not have a strong record of enforcing these laws...

The Living Planet Report published every two years by the World Wildlife Fund (WWF) and the Zoological Society of London (ZSL) reports global wildlife populations have fallen 58% since 1970...(2016)

SOUTH AMERICA

Earth's 4th largest continent has a surface area of nearly 18 million square kilometers with 12 sovereign nations and the worlds driest desert (Atacama), the largest river on the planet (Amazon), the world's highest waterfall (Angel Falls), the tallest mountain in the southern hemisphere (Aconcaqua), some of the largest telescopes on Earth, the largest rainforest (Amazon Basin), 400 million humans (over half in Brazil) and the world's 4th largest reserves of shale gas...It's believed the earliest humans on the continent came from Asia over the Bering Strait through North America over 10,000 years ago (Monte Verde, a human habitation site in Chile has been dated to 18,500 years ago) to settle along the mountainous Pacific west coast developing differing cultures and civilizations over many thousands of years with little contact outside of their societies until the arrival of the Spanish in the 16[h] century...It's interesting that the earliest complex societies on an isolated continent creating art, intricate architecture, rituals, religion and cultures flourished around the same time (3,000 BC) as the Egyptian civilization doing many of the same things with enormous structures to unknown deities (using massive stones) and sacrifice to appease the gods in differing societies that were very efficient in similar ways to support the survival of large integrated groups of humans...When Europeans arrive (1500) the Inca civilization was at it's peak stretching from Columbia to Argentina (3,400 miles) with almost 50,000 Incas ruling over an estimated 10 million humans with some 40,000 kilometers of roads and a hierarchical structure that was organized to bring great wealth to those at the top...By the mid 16th century Spain had defeated the Aztecs in Mexico, the Incas in South America, controlled all of Peru and the western coast of South America while Portugal had Brazil and controlled a great deal of South America's eastern coast...In a single generation millions of indigenous humans died from warfare and imported disease (smallpox, measles, typhus) and Europeans controlled much of Earth's western hemisphere largely motivated by two primary goals: greed and religious conversion by any means necessary...The 18th and 19th centuries were chaotic with European wars greatly affecting rule overseas, local uprisings and 'libertadores' who led multiple campaigns for independence liberating almost all of South America from European control in the 1800's and establishing boundaries that define the continent in the 21st century...Today all of the countries in South America are democracies with corruption widespread and traditional among political elite resulting in a huge inequality within populations and forcing multiple 'leaders' to resign in disgrace when confronted with clear evidence of collusion, gross mismanagement and corruption...Second to corruption (and intricately intertwined) is the enormous illegal drug trade with cartels running billion dollar empires paying off whomever necessary to facilitate their illegal activity...The history of military coup and control is a recent one and still a potential threat to struggling democracies (Venezuela) ruled by those who ignore desperate populations to practice crony capitalism without any real awareness of causality, a political pattern of self-enrichment practiced worldwide in almost every nation on Earth...

Glaring in the light of Brazil's vote to impeach President Rousseff for "misuse of funds" is a legislative body (594 politicians) where 60% of the elected members face serious charges including bribery, fraud, illegal deforestation, kidnapping and homicide (NYT's)...It's difficult to find a western democracy that is more corrupted and entrenched than Brazil at every level of governance...Old white males who like the 'old ways' of doing business...2016

ARGENTINA
(43 million humans)
Military Forces = 80,000 active duty / $4.6 billion per year

Encompassing the southern 3,650 km of the South American continent with Chile bordering the west for it's entire length, Bolivia and Paraguay to the north, Uruguay and Brazil to the east and over 3,100 miles of coastline along the Southern Atlantic Ocean...A presidential republic with the president both head of government and chief of state elected by majority vote for four years...The government walks a line between increasing the viability of the poor through new social programs and increasing inflation of the peso...In 2012 the government essentially took over the Central Bank in order to make the bank's reserves available for government expenditures...The 2nd largest nation on the continent with a median age of 31 and a life expectancy of 77 years...30% of the population exist below the poverty line and 430,000 children in the labor force...Energy wealth is 2.3 billion barrels of crude oil reserves and 378 billion cubic meters of natural gas reserves...Challenges include a long standing dispute with United Kingdom over the Falkland Islands (480 km east in the Atlantic), South Sandwich Islands and South Georgia Island (a popular tourist destination for visitors to Antarctica), ongoing border talks with Chile who share the entire western border and drug trafficking on northern borders...

BOLIVIA
(10.5 million humans)
Military Forces = 60,000 active duty / $310 million per year

One of only two landlocked nations on the continent (Paraguay is the other) with around 35% of it's population indigenous in a challenging environment with an average elevation of nearly 4,000 feet (1,192 meters)...The Andes Mountains to the west and the forests of the Amazon Basin east, Bolivia is bordered by Argentina, Chile, Brazil, Paraguay and Peru with a mix of 52% forest and 34% agricultural land used as pasture and to grow coffee, coca, sugarcane and nuts...The 2nd largest proven reserves of natural gas in South America with over 281 billion cubic meters and an estimated 209 million barrels of crude oil in the ground mostly controlled by the state owned energy giant YPFB with plans to increase it's generation of electricity to market to bordering nations...A poor country with most of the population living in rural areas, the median age is around 24 with a life expectancy close to 70 years and almost 50% of the population existing below the poverty line...Over 700,000 children in the labor force...One of the least developed countries on the continent with the population repeatedly coming out to protest a plan to build a highway through the rainforest, drug eradication policies that negatively impact coca farmers, a planned increase in fuel and commodity prices and wage stagnation...Bolivia is the world's 3rd largest producer of cocaine and has addressed the UN to regain access to the sea (through Chile)...

When Portuguese first arrived on the shores of what is now Brazil a massive forest waited for them, not the Amazon Forest, but the Atlantic Forest stretching over 1.2 million kilometers...The Tupi people numbered around a million and now all of this is gone...93 percent of the Atlantic Forest has been converted to agriculture, pasture and cities and the Tupi people have vanished (slavery and disease) along with the forest's mega fauna, jaguars and giant anteaters...

BRAZIL
(205 million humans)
Military Forces = 350,000 active duty / $31 billion per year

Over 50% of the population is black or mixed race but the power structure is firmly in the hands of whites with studies consistently showing blacks earn half as much as whites in the same jobs...Brazil signed a defense compact with France in 2008 to build a nuclear powered submarine and buy fifty military helicopters to protect newly discovered deep water drilling locations and monitor ongoing Amazon deforestation...A political system desperately in need of reform as protests grow in the streets over government corruption and a disconnect from the people that enriches the wealthy and elite and leaves little for the poor in a repeating pattern recognized in governments worldwide...After a period of unparalleled growth Brazil is the 7th largest economy on Earth and is hosting the Olympics in 2016 with an unprecedented spending spree (more than $14 billion) on these events that has brought an intense focus on widespread corruption and income inequality as promises for social services and improved infrastructure are ignored and the public's money is spent on stadiums and cost overruns that enrich the government and wealthy doing nothing to improved lives of the majority...Enormous slums (favela) exist in urban areas without public services, with high crime, violence and extreme poverty...Recent estimates are some 12 million Brazilians live in these slums...

Brazil was the last country in the Americas to outlaw slavery...

80% of Brazil's land, forests and farmland are owned by 5% of the people...Wealthy landowners...

In 2012, a controversial bill was approved by the Brazilian Congress after multiple veto's that regulates how much land farmers must preserve as forest...Brazil's farmer lobby had argued that an easing of environmental restrictions would promote food production while environmentalists oppose the law saying it will lead to further destruction of the Amazon rainforest...Activist group Avaaz said it handed the government a petition with nearly two million signatures collected online from people in dozens of countries demanding a total veto asserting that the bill would accelerate deforestation in the Amazon rainforest...In the end the final bill was criticized by both sides...

The majority of Brazil's CO2 emissions are blamed on deforestation...

Median age is 31 with a life expectancy of almost 74 years...Ranked 13th in the world for the number of humans living with HIV/AIDS (725,000) and 17th in deaths from HIV/AIDS...Over 23% of the population is below the poverty line with almost a million children in the labor force...Over fifteen billion barrels of crude oil reserves and 388 billion cubic meters of natural gas reserves...Vast mineral wealth in iron ore (30% of global exports) and bauxite (aluminum), world's largest producer of coffee, second largest producer of beef (U.S. is number one), largest producer of orange juice and the largest producer of sugarcane...A wealthy nation with an enormously corrupt government...

In 2013 it was reported that deforestation increased by 28% in one year as a result of the passage of new reforms...The government of Brazil committed in 2009 to an 80% reduction in deforestation by 2020...

In 2014, Brazil and the World Wildlife Fund (WWF) announced the creation of a $215 million fund to protect 150 million acres of Amazon rainforest, the largest tropical forest conservation program on Earth...With a multitude of partners this transitional fund is designed to help the Brazilian government provide protection for these ecological regions for 25 years, vital for species and necessary for maintaining the viability of the equatorial rainforests...

CHILE
(17.5 million humans)
Military Forces = 62,000 active duty / $5.1 billion per year

Located along the Pacific Ocean on the western edge of the South American continent, Chile is the world's largest producer of copper with a stable economy and a newly elected president (2014) who is both head of state and head of government with a National Congress elected by the people...Supreme Court is appointed by the president and ratified by the Senate...After cancellation of the controversial hydroelectric dam project known as HidroAysen, a series of five dams to be built on rivers impacting diverse, pristine ecosystems in Patagonia, Chile now is planning a unique hydroelectric project that will use solar power to pump water into reservoirs high in the mountains and release it to generate energy powering three provinces...Another conservation project is an organization, Conservacion Patagonia, who are committed to developing and protecting Patagonia National Parks (660,000 acres) to create an eco-friendly tourist destination with 100,000 visitors a year to protect and preserve natural ecosystems and wildlife...A service based economy with a large industry in copper (world's largest with 5.7 million tons in 2015) and a significant fishing industry with 6,435 km of coastline ranking 6th in the world (1.3 million tons in 2012)...15% of the population live below the poverty line and 81,000 children are in the labor force...In 2015 Chile announced the creation of two new marine reserves with one around Easter Island and the other around the Juan Fernandez archipelago covering over a million square kilometers...

Chile's most significant contribution to the world might be it's astronomical footprint hosting the most powerful ground-based observatories on Earth...The Atacama Desert is the driest desert on Earth with elevations over 18,000 feet and will be home to the European Extremely Large Telescope (2024), the largest optical telescope ever built...

COLUMBIA
(47 million humans)
Military Forces = 445,000 active duty / $13 billion per year

Columbia is the northwest corner of this continent connecting South America to the rest of the Western Hemisphere through Latin America (Panama) with a modern day history of violence, cartels, human rights violations, insurgencies and an inequality within the population that is staggering...Peru and Ecuador border to the south, Brazil and Venezuela to the east and 3,200 km of coastline north with a president who is both chief of state and head of government elected by majority vote for a single four year term...Mineral wealth (platinum, gold, silver, coal) and energy wealth with 2.4 billion barrels of crude oil reserves and 198 billion cubic meters of natural gas reserves, the economy is dependent on the myriad of free trade agreements and energy exports...In a wealthy country over 30% of the population exist below the poverty line with a per capita GDP of $14,000...Median age is 29 with a life expectancy of 75 years...Over 5% of children under age five are malnourished and almost a million children are in the labor force...Recent peace talks with the largest and longest standing rebel group in the country has been successful with Farc (Revolutionary Armed Forces of Columbia) and the government signing a final peace agreement in 2016...Columbia remains a dangerous country with a lot of issues including a large population (400,000) of refugees from Latin American countries, disputes with Venezuela over maritime boundaries and drug cartels who continue to do business as usual...

ECUADOR
(15.5 million humans)
Military Forces = 41,000 active duty / $2.2 billion per year

On the Pacific coast with Columbia to the north and Peru to the south and east, Ecuador has a long history of emigration and cross border wars over territory...A republic with the president head of government and chief of state elected by majority vote with the vice president...Dependent on energy exports with about 9 billion barrels of crude oil reserves and 6 billion cubic meters of natural gas reserves...Economy is coffee, bananas, cocoa and fishing with 2,237 km of coastline on the Pacific Ocean...Over 25% of the population live below the poverty line with a per capita GDP of just over $11,000...Median age is 27 with a life expectancy over 76 years...More than 7% of children under age five are malnourished and over 220,00 children are in the labor force...Ecuador has been contracting with budget cuts and reduced subsidies since oil prices fell during the world recession and is still a drug trafficking transit point for cocaine bound for America...Over 120,000 refugees from Columbia...

GUYANA
(747,000 humans)

A nation with a turbulent history being a Dutch territory and a British territory before acquiring it's dependence in 1966 with a multitude of ethnicities in country that reverberate and influence culture and politics...Wedged between Venezuela, Brazil and Suriname in the north of South America on the Atlantic with tropical rainforests and mineral wealth, Guyana is a poor country dependent mostly on commodity exports (gold, bauxite, sugar, timber, shrimp) and remittances (over 50% of the population live out of country) with almost 40% of the population living below the poverty line and a double digit unemployment rate...Government is a parliamentary democracy with president as chief of state and a prime minister as the head of government...Median age of 25 with a life expectancy of 68 years...Over 13% of children under age five are malnourished with an estimated 30,000 children in the labor force...In 2015 Exxon Mobil said they had found oil off the coast which could hold over 700 million barrels of oil with the potential to make huge changes in a small nation with less than a million citizens but Guyana's huge neighbor to the west (Venezuela) claims the oil is theirs after years of claiming that most of Guyana should be theirs...Tensions will continue to grow...Venezuela has a huge military while Guyana has a small defense force that would be unable stop any incursion...

PARAGUAY
(6.6 million humans)
Military Forces = 11,000 active duty / $145 million per year

In the center of the continent with Bolivia north, Brazil east and Argentina west and south with a history of political turmoil ruled by a dictator for 35 years until 1989 followed by decades of coup attempts, assassinations and unrest...Today Paraguay is a presidential republic with the president both chief of state and head of government elected for a single five year term presiding over a contentious multiparty coalition...5% of children under five are malnourished and over 200,000 children are in the labor force...Economy is industry and agriculture (6th largest soy producer in the world) with a growing service sector and per capita GDP around $9,000...About 40% of the population live below the poverty line...Challenges are money laundering, rebel groups, widespread corruption and drug trafficking...

PERU
(30.3 million humans)
Military Forces = 120,000 active duty / $2.5 billion per year

Geographically stunning on the Pacific Ocean with mountains over 22,000 feet, rainforests, the highest navigable lake on Earth (Lake Titicaca) and a varied topography bordered by Chile and Bolivia to the south, Brazil to the east and Columbia and Ecuador to the north...Mineral wealth as the world's 3rd largest producer of copper and the 2nd largest producer of silver..Energy wealth is 741 million barrels of crude oil reserves and 435 billion cubic meters of natural gas reserves...Government is a presidential republic with the president chief of state and the head of government elected by majority for a five year term...Median age is 27 with a life expectancy of almost 74 years...Some 6% of children under age five are malnourished and an estimated 2.5 million children are in the labor force...As is common with many countries on this continent, Peru has a long history of political instability with assassinations and coups along with rebel groups, a reported 3,000 political murders and insurgency by Shining Path for 20 years (1980-2000) that killed an estimated 70,000 humans...A history of widespread corruption that is ingrained in all the continent's political structures, a leading producer of cocaine, ongoing protests over mining proposals and borders disputes in a volatile earthquake prone region of the world...

SURINAME
(550,000 humans)
Military Forces = 3,000 active duty / $60 million per year

Bordered by Brazil to the south, Guyana to the west and French Guinea to the east with 386 km of coastline north on the Atlantic Ocean...Independence from the Dutch in 1975 and the 1st democratic elections in 1990, Suriname today is a presidential republic with president both chief of state and head of government...This nation has strong ties to the Netherlands and multiple ethnicities that frequently cross the border into Brazil to seek employment...More than 6% of children under five malnourished and over 6,000 children are in the labor force...Economy is mining exports (alumina and gold) as well as oil (89 million barrels of crude oil reserves) and a very active fishing industry...Almost 70% of the population live below the poverty line with a labor force under 200,000...Challenges are widespread corruption (multiple political figures have been implicated in drug smuggling), protest over mining policies and maritime boundary disputes with neighboring countries...

URUGUAY
(3.3 million humans)
Military Forces = 25,000 active duty / $430 million per year

South of Brazil and east of Argentina on the Atlantic Ocean, Uruguay is a presidential republic with president both head of government and chief of state elected for a five year term in a stable nation with education free from primary through university levels and a strong social safety net...Median age is around 35 with a life expectancy of 77 years...Over 50,000 children are in the labor force and 5% of children under age five are malnourished...Economy is a large service sector with growing tourism and industry linked to agriculture in processing foods and beverage for export...A progressive government legalizing same-sex marriage, abortion and the cultivation, sale and use of marijuana...

VENEZUELA
(29.3 million humans)
Military Forces = 135,000 active duty / $4.8 billion per year

Venezuela is the shining example of neglect and government mismanagement with an abundant supply of resources (gold, coal, iron ore), some of the largest deposits of oil in the world and a history of political infighting and rife corruption that has forced millions of citizens to lead desperate lives with little support...Some estimates have 50% of the population living below the poverty line with more than 350,000 children in the labor force...A country that depends on oil exports for over half of revenues and a nation with out of control inflation, the president's (Maduro) 'response' is to implement state control over most of the economy blaming private business and foreign governments for his continuing lack of awareness and glaring incompetence while still managing to block nearly all legislation passed by the parliament to improve the economy for the citizens...Energy wealth is over 298 billion barrels of crude oil reserves and 5.5 trillion cubic meters of natural gas reserves...Economy is mostly services with industry in livestock, chemicals, pharmaceuticals and raw materials...Government is a federal republic without term limits for the president who is both chief of state and head of government...Bordered by Columbia on the west, Brazil to the south and Guyana to the east with disputes over maritime borders and jurisdiction (Columbia and Guyana) in a popular transit region for illicit drugs going to the United States and Europe...In this century Venezuela is aligning itself with Iran and Russia while accusing America of spying and seeding unrest to overthrow the government...

María Gabriela Chávez, the daughter of Hugo Chavez (president of Venezuela from 1999 to 2013), is reported to be worth 4.2 billion dollars while 50% of Venezuela's citizens live in poverty due largely to Chavez's policies...

ECONOMICS

There are thousands of books that try to explain, debate, categorized and model world economic policies and the inter-connectivity of variables too numerous to count which comprise Earth's economy and the advantage every nation seeks in trade, debt, production, commodities, financial markets, free market solutions, regulatory constraints and an undeniable connection and collusion with political systems and corporate structures in almost every nation on Earth...But a deep understanding of equitable realities and the distribution of money, goods and services is mostly impossible in the opaque and corrupt practices that define the large financial structures who control and abuse the 'trust' implied by the nature of their influence and control over the financial health of billions of humans...In 2016 the financial structures the control the world's economies are clearly deeply irresponsible controlled by humans with untouchable wealth and deep connections with lawmakers and regulators who protect and absolve them of serious inconsistencies along with 'boards of directors' who are often handpicked for their lack of oversight...The problem is very simply this...No matter how efficient and fair the tax policies, trade policies and monetary policies written to manage an increasingly complex system, it's success or failure is dependent on the connected web of humans who are responsible for implementing the policies in fair ways and with six billion humans (the poor and the powerful) unable to understand reality, primitive instincts control behavior beneath an exterior of sophisticated intelligence...Humans at the highest levels of responsibility cannot understand the impact of deliberate actions that enrich themselves with the support of those who profit as well and the damage their actions inflict (causality) on billions of lives...

This chapter offers only examples and realities...Nothing will change without a majority to change this ignorance...

An intelligent alien race that took a first glance at Earth would understand immediately that a sentient species with more than seven billion individuals would have to develop a system of record and facilitation to manage, move and distribute goods and services across a wide variety of cultures and societies...The purpose of money...They would also understand this system of record and facilitation was mostly broken and purposely inefficient with primitive instincts and ancient motivations using manipulation, deceit and violence on a global scale to acquire wealth without reason and fragmenting a developing global civilization in devastating ways...Earth is a finite sphere (12,742 km diameter) that has been explored for tens of thousands of years, mapped for thousands of years and in the 21st century is continually monitored and analyzed by multiple satellites from space...It's our planet, our only planet, and in this new time of technology the human species should put the highest priority on designing and implementing strategies that rethink human existence and create global patterns and infrastructure in a sustainable way to insure human survival in a changing (predictably) and challenging environment over long extended time periods...An extraterrestrial intelligence would understand this, a clear reality not understood by more than six billion humans...

If you want to understand the economic absurdity of the world look at JP Morgan...If you want to understand the political corruption in America look at JP Morgan...No institution pays fines of $20 billion dollars unless they've committed crime and fraud on enormous scales and this bank just goes down in the basement and wheelbarrows $20 billion up the elevator to pay whoever is investigating and continues to do business as usual...There have been no extra regulations or restrictions put on this organization to limit the damage they can do to ordinary consumers and no officers in this hierarchical structure have ever been charged with a crime or dismissed from their position...

JP Morgan has fired an estimated 24,000 workers in this decade to "keep expenses down"

"Provincial officials overspent their budget by an estimated $250 million, much of it on questionable or blatantly fraudulent government payments and contracts with private businesses enjoying close ties to the politicians leading the province..." a quote from the NYT's in 2012 in a story about African corruption that sounds exactly like daily business in the U.S. political system...The difference is that in the U.S. *it is legal* to bribe politicians through political donations and guaranteed future employment and it is legal for the politicians to give government money and favorable treatment to the corporations and influential individuals in return for those political donations and the promise of future employment...

Everything on this planet is now about "money" and it has become a thing, not a process for sustaining a species in a complex way...Nations should be merging with ideas and global projects to run the Earth as a whole rather than 192 fiefdoms that operate in secrecy, suspicion, fear and an overblown, primitive, imagined sovereignty marked by a line on a map and maintained with violence through military force...

In 2008, as the economy in the U.S. crashed and unemployment rose, lobbyists in Washington working with special interests were paid over $3 billion (more than any other year on record) to influence and bribe congressional lawmakers to pass laws that met the lobbyists criteria and gave advantage to the organizations they worked for...The Center for Responsive Politics calculated the amount spent by lobbyists averaged $32,523 per legislator, per day for every day Congress was in session...Pharmaceutical companies spent $231 million on lobbying, electric utilities spent $156 million and oil and gas companies spent $133 million...This record will be broken...

In 2013, corporate profits set records while wealthy politicians (there are no poor lawmakers in Congress) cut food subsistence for millions of children and then slashed unemployment benefits for millions of adults...These political decisions effecting day to day existence of ordinary humans were made while the Federal Reserve was giving banks $85 billion a month to stimulate the economy, money that the banks put in their vaults so they could collect the interest and enrich themselves (a bill passed in 2008 allows the Federal Reserve to pay the banks interest on the money they hold)...Even the most remote human can understand this global disregard for fairness when JP Morgan, the largest and most connected financial institution in world ($2.7 trillion in assets), pays out over $20 billion in fines and penalties for fraudulent actions which impacted millions of people in a worldwide recession with no criminal consequences...An enormously wealthy institution with deep political ties who will not be challenged or regulated by a weak government beholden to money...Wall Street is happy, JP Morgan's stock prices continue to do well and in 2014 there were large bonuses for all officers of the bank...The CEO of JP Morgan, James Dimon, is continually referred to as one of the "best" executive officers in the banking business, CEO of the Year in 2011, Time Magazine's list of influential people for four years and a salary+benefits package of over $20 million annually...Even a casual look at the performance of the bank he "controls" paint a realistic picture of fraud, collusion, malfeasance, mismanagement, greed, theft and a complete disregard for laws and regulations in financial decisions that depleted millions of dollars in assets from the accounts of common people worldwide while manipulating interest rates on a global level to enrich the bank and the officers of the bank...Wall Street is happy, JP Morgan's stock prices continue to do well and in 2014 there were large bonuses for all officers of the bank...

It is obvious that banks and corporations worldwide wield enormous power and influence when they commit crime and fraud on massive scales involving millions of dollars, actually stealing money from millions of humans through the manipulation and breaking of financial and criminal law and there is little consequence...A fine and a hearing in front of the lawmakers and enforcers who allow this to happen and business goes on as usual...This is possible in only one way...$$$...Money that flows freely between the politicians that make the laws and create the agencies who are responsible for the enforcement of those laws and the corporate power structures that increasingly control the politicians and the hierarchical structure at the highest levels of government...In the U.S. not only have there been no charges or accountability by the largest banks on the planet for stealing billions of dollars with sleight of hand, collusion, derivatives, insider trading, credit default swaps and bribery, the U.S. government gave these corporate structures billions of dollars in bailouts and loaned the banks $1.2 *trillion* on a single day, 12/05/2008...The government has subsequently "loaned" these banks as much a $6 trillion more at unbelievably low rates that enabled the "banks" to earn more than $15 billion dollars in profit from government / taxpayer money...Entities that invest and hide wealth in offshore havens to avoid paying the taxes they depend upon to remain solvent...In the 21st century the world has entered into an era of corporate and political corruption to a degree never before experienced...

The seven largest banks in America in 2011 were expected to pay compensation including bonuses, salary and benefits of more than $156 billion to their employees...

Federal Reserve Chairman Ben Bernanke argued in 2008 when the crisis hit that revealing borrower details would create a stigma that would have led to more banks collapsing....The Fed has fought to keep the details of the loans, which totaled more than $7.77 trillion, secret for years after this collusion between government and finance...

JP Morgan is the target of more than six investigations initiated by the U.S. Justice Department involving bad mortgages, bribery, improper energy trading and the 'blue whale" scandal in Europe..It is a certainty in America that no executive will be charged no matter what the outcome of the investigations and the largest bank in America will pay easily affordable fines with no admission of guilt in any way...The bank has spent more than $8 billion on legal representation in the past two years...2013

Update...Government authorities in September 2013 imposed a $920 million fine on JP Morgan to resolve ongoing investigations with $300 million going to the Office of the Comptroller of the Currency and about $200 million each to the S.E.C., The Federal Reserve and the Financial Conduct Authority in London...Under the deal with the S.E.C. JP Morgan took the unusual step of acknowledging that it violated federal securities laws reversing a decade-long policy at the S.E.C. that allowed banks to "neither admit nor deny" wrongdoing...$920 million is considered a small price to pay for a financial institution of this size to put this matter behind them and conduct business as usual...

When Walmart started in China 18 years ago they were paying a starting wage of $1 per hour...Since that time the starting wage has increased by only .73 cents...A market decision based on necessity could be understood but the Walton family (6 individuals who own Walmart) are collectively worth over $90 billion...In an intelligent world, some structure would exist to require fair wages for workers (it would not be difficult to establish a fair wage) and some regulation of wealth that tempers the inequality of greed...Tax wealth (from all sources) above $100 million at 90% and use it to provide a level of subsistence for every human on Earth...The Walton's would still be wealthy in unimaginable ways and their stores would do even better with the economic redistribution of great wealth...

Corporate taxes were 33% of the American economy in the 50's...This century most corporations pay less than 10%....

All the pundits and analysts that talk of "irregularities" by the biggest and most powerful corporations in the world continually say that banks and corporations need to "be careful of the cultures they create"...The "cultures" they create is crime, very simple, criminal activity that is overlooked by law enforcement at the highest levels in a "good old boy" network worldwide that favor the wealthy in unbelievable ways...Criminal activity that anyone in another sector of society or anyone farther down on economic ladder would be prosecuted for and jailed...

In front of Congress after Wells Fargo opened 2 million false accounts on unknowing customers ($185 million fine and no admission of wrongdoing) CEO John Stumpf said he will work on improving the bank 'culture'...2016

In the era of high speed trading on the world stock markets, the average human loses $5 per trade (Marketplace estimate Dec 4, 2012) changing the entire focus of finance to high powered rich entities that play the investment game with other rich powerful entities (banks, brokers, hedge funds, etc.) that have little connection to the real world that the majority of humans exist in...The 'average' human has very little to do with the markets...

In Congo, terrorists and militant groups are investing in timber, commodities, minerals and everything else that sells...In a world obsessed with money even the "bad guys" are allowed to function in a market economy...ISIS is making about 3 million a month from oil??? Who facilitates and manages the financial empire of a group who behead innocent humans and post videos for global entertainment...

Rating Agencies S&P and Moody's & Finch were compliant in the theft of billions of dollars from citizens and economies by rating, supporting and championing financial instruments issued by the organizations who paid these rating agencies for their services...Collusion, corruption, nepotism and cronyism that's denied by all involved as they revel in record profits...Paying what's necessary to regulators and politicians that create the legislation that permits this white collar crime at the highest levels is done subversively and openly with "fines" easily paid and zero criminal accountability...

Update...Standard & Poors (S&P), one of the largest financial rating agencies in the world (paid by the financial institutions whose stocks and bonds they rate) admitted they "messed up but broke no laws" while paying an $80 million fine guided by some of the best lawyers in the world...Who pays an $80 million fine if they didn't break any laws??? The U.S. government said it was a MAJOR WIN because it's the first time a rating agency has admitted they 'messed up' and paid a fine...No one went to jail for any of this...If any 'common' human did "something" that warranted an $80 million fine they would also be in jail...Rich, connected, corrupt, powerful...Also, there is a $5 billion lawsuit they are trying to 'settle' but of course they "didn't do anything wrong"...In America it's legal to 'not do anything wrong' and pay millions, even billions of dollars in fines...

S & P (Standard & Poors) Settlement
Agreed to pay a penalty of $1.375 billion but said they violated no laws! Why are they paying a fine of $1.375 billion? Stunning collusion among the rich...*Standard & Poor's Financial Services LLC*

2016...Goldman Sachs agreed to pay $5 billion (as did Bank of America and JP Morgan) for their actions in multiple areas of financial sleight of hand and fraud that caused the recession in 2008 but there is no way to give back the lives that were ruined...No criminal charges...

In 2016, JP Morgan agreed to pay nearly $500,000 to settle another stunning example of greed...Along with the Federal Bureau of Prisons and the U.S. Treasury, the bank issued debit cards to released inmates to help them adjust to life on the outside...The bank loaded the cards with fees and charges not applicable to the general public and the prisoners could not review terms or conditions...The depth of greed in taking advantage of the most vulnerable is stunning...As is always the case there are no charges against the bank and a small settlement...A bank spokesman declined comment...

In 2014, 300,000 humans in Haiti still live in ramshackle tents left over from the earthquake three years ago when the world pledged millions of dollars to help the country rebuild...

$100,000,000,000,000 (trillion) in hidden money by leaders and an elite class of rich everywhere on this world as children go hungry, populations starve and millions die unable to afford healthcare...

A practice that shines an unblinking light into the underdeveloped brain is a common agreement between "enforcers" and the largest corporations in the entire world, companies worth billions who are run by humans worth millions that agree to pay a penalty (usually money) for very obvious clear wrongdoing that in many cases steals millions of dollars from innocents who desperately need the money that was taken from them by sleight of hand or improvised legal technicalities or hidden fees and charges, clear immoral acts by these huge entities who agree to a penalty as long as they can "admit to no wrongdoing"...Transparent ignorance and an insight into a primitive culture of greed that is alarming...They really believe that people will 'understand' they didn't do anything wrong or illegal...Sadder still is the collusion and corruption entrenched between corporations and the enforcers who "protect" the public...

In 2012, the FSB (a task force from the world's top 20 economies) called for greater control of shadow banking, a corner of the financial universe made up of entities that escape regulation..."The objective is to ensure that shadow banking is subject to appropriate oversight and regulation to address bank-like risks to financial stability"...The FSB said shadow banking around the world more than doubled to $62 trillion in the five years to 2007 and had grown to $67 trillion in 2011, more than the total economic output of all the countries in the study...America had the largest shadow banking system with assets of $23 trillion in 2011 followed by the Euro area with $22 trillion and the United Kingdom at $9 trillion...Forms of shadow banking can include securitization, a method to transform bank loans into trade instruments that can then be used to refinance credit making it easier to lend (this slight of hand was responsible for a great deal of the fraud that collapsed the U.S. economy in 2008)...Another example is a repurchasing agreement (repo) where a player such as a hedge fund or a blue chip company sells securities to a bank agreeing to repurchase them later...The bank may then lend those bonds to another hedge fund taking a position on the government debt...The United States is implementing new rules for the $2.5 trillion money market industry which pools money from investors to put in low-risk financial assets that resemble deposits in a bank...With the profusion of lobbyists and corporate interests who often write and determine the final wording on regulation it's impossible to know if any regulation will work as originally intended...(See the Federal Register below)

The Federal Register...after laws are passed and put into the Federal Register they are dissected into small pieces (400 rules and regulations for Dodd-Frank with only half completed after 3 years) as lobbyists and lawmakers write the rules that govern the laws (already passed) to be ambiguous and vague to allow the entities affected by the new laws to avoid compliance and continue business unchecked...(1946 Administrative Procedure Act = Agencies dissect the rules before they ask for public comment)...Lobbyists and trade group representatives have enormous influence in dictating this process...Congress passes a "general law" which may look effective but in the Administrative Procedure the law is watered down to mean little or nothing in the real world...

501C4's are a shining example of corruption and the laws that allow it...These are social organizations whose purpose is to 'improve society'...They pay no taxes, yet raise and donate 95% of their money for political ads...

The financial sector in America represent 4% of labor and collect 25% of all corporate profits...

Easiest country in the world to incorporate anonymously is the United States...

It's estimated over $7 billion was spent on America's 2016 elections...

Thousands of books break down and explain the complexity and efficiency of the world's economic system policies amid a reality of example after example of gross mismanagement and theft...The question is whether the human species is stunningly incompetent or stunningly primitive and the answer is clear...We can manage millions of variables in multiple ways to catch a comet streaking around the Sun but doing such a thing is not about empathy or morality...Operating a fair distribution of goods and services is technically possible but the primitive instinct of self preservation and the subconscious drive to have more than others is very old and very powerful...Nothing will change until a majority of Earth's population understand existence in an intuitive and cognitive way where exploitation for personal gain is unnecessary and unthinkable on a single world circling a single star...The only way this happens is a focus on every baby born for the 1st three years of life to create an intuitive empathy...It will take a thousand years...

ISLAND NATIONS

BAHAMAS
(350,000 humans)

A British colony for over 250 years, this nation is an archipelago with some 700 islands located in the Atlantic southeast of Florida and north of Cuba...A huge tourist industry and offshore banking hub with recently discovered offshore oil and gas whose exploration and recovery are a continuing and lively debate over it's economic impact on tourism...Median age is 31 with a life expectancy of over 72 years...High youth unemployment (30%) with about 10% of the population existing below the poverty line...A popular transshipment point for drugs with many islands and close proximity to markets...Has a small defense force (1,500) and is a commonwealth realm with the Queen of England chief of state and a parliamentary democracy with the prime minister appointed from the majority party...The Bahamas is located on the Atlantic hurricane track and has suffered multiple impacts, the latest in 2004...

COMOROS
(764,000 humans)

An archipelago northwest of Madagascar off the coast of Africa that's been in political turmoil since it's Independence from France in 1974...With four major islands and a circulating presidency, this nation of less than a million people has had peacekeeping forces in country after separatist violence and calls from multiple islands for independence...Median age is just 20 with a life expectancy of less than 64 years...Infant mortality is ranked 19th in the world, 17% of children under five are malnourished and almost 40,000 children are in the labor force...Agriculture, fishing and forestry are the economy in a poor country with nearly 50% of the population living below the poverty line...President is both chief of state and head of government...Small military force with a French alliance for defense...

CUBA
(11 million humans)
Military Forces = 90,000 active duty / $700 million per year

An island in the Caribbean whose history is dominated by the rule of Fidel Castro who stayed in power for almost fifty years with support from the Soviet Union for much of that time to promote a communist agenda that became a prison for the island's inhabitants...Since the dissolution of the Soviet empire and the loss of Soviet subsidies totaling billions every year the economy has struggled with an U.S. embargo in place since 1961 that's created a landscape of cars and buildings from the 1950's and unleashed wave after wave of migration by Cubans seeking a better life...U.S. policy allows any Cuban who touches American soil to stay and thousands have...Median age is 40 with a life expectancy of 78 years in a communist state with Raul Castro Ruz (Fidel Castro's brother) both chief of state and head of government...Energy wealth is 124 million barrels of crude oil reserves and 70 billion cubic meters of natural gas reserves...A long contentious and combative relationship with America which should have been addressed years ago is slowly thawing now that the U.S. has begun a slow renewal of services and contact with the island...The infamous prison at the Guantanamo Bay Naval Base is on land leased to the United States with no stipulation that the property is ever returned to the Cuban government except by abandonment or mutual consent...The U.S. says it will never do either of those things...

A recent study published by the journal Nature Climate says 73% of world island groups face severe drought that could leave many islands abandoned before sea level rise becomes an imminent threat...2016

CYPRUS
(1.1 million humans)

A divided nation since 1974 with the south in the hands of Greek 'Cypriots' backed by Athens and the north in the hands of Turkish 'Cypriots' back by Turkey and UN troops on the 'Green' line of demarcation between the two...Located in the eastern Mediterranean Sea with a thriving tourist industry that's been impacted by the wave of migration from Syria and Africa as well as the economic troubles of Greece with banks on Cyprus holding a majority of Greek bonds in Europe that negatively effected the economy leading to a bailout by the EU in 2013...Some energy wealth with 141 billion cubic meters of natural gas reserves although little has been recovered...Challenges include migration through Cyprus from multiple nations, high youth unemployment rate (35%) and a divided island with an uncertain resolution...Militaries from Greece (12,000 troops) and Turkey (17,000) provide defense...

DOMINICAN REPUBLIC
(10 million humans)
Military Forces = 25,000 active duty / $180 million per year

The other nation occupying the island Hispaniola (with Haiti) is much different from it's close neighbor with a thriving tourist industry, free trade zones and a mining industry (gold and silver) that boosts per capita income close to $15,000 (income inequality makes the distribution of this wealth very unequal)...Over 40% of the population exist below the poverty line with a high unemployment rate (15%) and it's estimated that 180,000 children are in the labor force...Over 70,000 humans living with HIV/AIDS and 5% of children under age 5 malnourished...Government is a presidential republic with the president both chief of state and head of government elected by popular vote...Mountains higher than 10,000 ft inland with 1,288 km of coastline...Cross border migration with Haiti is very common...

EAST TIMOR
(1.2 million humans)
Military Forces = 1,300 active duty / $21 million per year

Located north of Australia in the Timor Sea, this nation achieved independence in 2002 just three years after a 25 year occupation by Indonesia ended which UN administrators believed was the cause of 100,000 deaths...A multinational peacekeeping force organized by the United Nations in 1999 (INTERFET) spent years restoring stability and another peacekeeping force (Unmit) was established in 2006 to protect citizens and property from gang violence and pro-Indonesia militias (UN mission ended in 2012)...Coastline is under threat from rising sea levels but the highest point on the island is 2,963 meters...Life expectancy is 67 years with a median age of 18...Over 40% of children under age five are malnourished and 10,000 children are in the labor force...President is elected by majority to a five year term and the prime minister is appointed from the majority party in parliament...Energy wealth is 200 billion cubic meters of natural gas reserves (revenue shared with Australia) but more than 35% of the population still exist below the poverty line...Ongoing disputes with Indonesia over maritime borders...

FIJI
(900,000 humans)
Military Forces = 3.500 active duty / $79 million per year

Fiji gained independence from Britain in 1970 and has been influenced by divisive politics, coups and military rule for most of it's history...Martial law was put in place in 2009 after President Ratu IIoiIo abolished Fiji's constitution and appointed military chief Voreqe Bainimarama as prime minister for another five years after the courts declared his government illegal...In 2012, Bainimarama declared martial law over but with stricter government controls and oversight the opposition says it's still very dangerous to speak out against current leaders or their policies...A myriad of new laws and regulations challenge civilian participation and stability after elections in 2014 returned Bainimarama to power....Mining, sugar and tourism generate much of Fiji's income though the country still relies heavily on international aid...New media requirements and proposed prison terms for journalists has severely restricted the free press with many citizens getting their information from radio broadcasts originating offshore...Median age is 28 with a life expectancy is 73 years...Over 30% of the population live below poverty line...Located in the southern Pacific Ocean with a tropical climate and hundreds of islands (over 1,100 km of coastline) Fiji is one of the favorite vacation destinations in the world...

HAITI
(10 million humans)

The first independent Caribbean state with a long history of instability who share an island with the Dominican Republic southeast of Cuba...The nation is still recovering from an earthquake in 2010 that killed more than 250,000 humans and destroyed the nations infrastructure...A twenty-nine year dictatorship by the father and son Duvalier did little to usher in a modern age and the first freely elected president was ousted by the military just a year later until the U.S. restored democracy in 1994...Today a UN peacekeeping force is in the country amid threats of political violence and gang violence...Over $5 billion was pledged after the earthquake but reconstruction is ongoing six years later with money never delivered and a great deal of it misappropriated (stolen) by government officials, NGO's and the humans who use these disasters to take advantage of the chaos and prey on the population...Median age is 23 and life expectancy is 64 years...15% of children under 5 are malnourished, 60% of the population live below the poverty line and 2.5 million children are in the labor force...Unemployment is 40%...

ICELAND
(330,000 humans)

A large island (103,000 sq km) in the North Atlantic Ocean northwest of Ireland with the oldest functioning legislative assembly on Earth...Located on the mid-Atlantic ridge with volcanoes, geysers and multiple hot springs, Iceland is a parliamentary republic with a president elected by a majority vote as chief of state and a prime minister (head of the majority coalition) as head of government...Iceland had been a stable economy for years before the world recession in 2007 showed the banks to be holding enormous debt which resulted in collapse and restructuring by the government and the IMF...Economy is dependent on fishing and tourism with growth in energy and manufacturing...A median age of 36 and a life expectancy of almost 83 years...Almost all of the population live on the coast using geothermal energy to supply heat in a well educated society with an extensive social safety net...One of only a few nations who still hunt and kill whales (Japan and Norway) part of a long tradition that dates back to the 12th century...Military defense is provided by the United States and NATO...

IRELAND
(4.5 million humans)
Military Forces = 9,100 active duty / $1.1 billion per year

A society with a large Catholic population and a long troubled history with the United Kingdom that split the nation in two with many senseless deaths over religious identity...The economy was on the edge of collapse following the recession in the first decade of this century requiring a bailout by IMF and the European Union (EU) in 2010...Today Ireland's economy is mostly in the service sector with some industry, a high unemployment rate and around 10% of the population below the poverty line...A long, dark history of abuse by the Catholic Church was recently exposed (similar to many, many places around the world where Catholicism was the majority religion) and it will take years to reconcile truth and lies that permeate this institution...Some energy wealth with 10 billion cubic meters of natural gas and the largest output of mined zinc in Europe...A storied history, troubled past and uncertain future...

JAMAICA
(2.9 million humans)
Military Forces = 3,000 active duty / $30 million per year

In the Caribbean Sea south of Cuba gaining independence in 1962, Jamaica is a parliamentary democracy with the Queen of England as chief of state (a commonwealth realm) and a prime minister who is appointed from the majority party...1,022 km of coastline with mountains over 7,000 ft in a hot tropical climate, the economy is remittances, tourism, a large service sector and a large amount of debt currently being restructured by the International Monetary Fund (IMF)...An island 'discovered' in 1494 by Europeans (Columbus) who exterminated the native population and replaced them with slaves who were freed over 300 years later (1838)...Median age is 25 with a life expectancy of 74 years...30,000 people living with HIV/AIDS, almost 40,000 children in the labor force and over 17% of the people living beneath the poverty line...Challenges are high crime in urban areas, drug shipments in a perfect geographical location to access the American market and sex trafficking of both children and adults...

KIRIBATI
(101,000 humans)

A large grouping of islands in the southern Pacific Ocean northeast of Australia (7,981 km), Kiribati gained independence in 1979 and is a presidential republic ruled by a president who is both the chief of state and head of the government...Many of the 33 atolls are at risk from global warming with the highest point just 81 meters above sea level...Economy is fishing licensing, remittances and foreign aid with an unemployment rate of 30%...Home to the largest marine reserve in the South Pacific...

MADAGASCAR
(23 million humans)
Military Forces = 14,000 active duty / $56 million per year

The fourth largest island on Earth, Madagascar is located 1,079 miles (1736km) east of Africa in the Indian Ocean with a maximum elevation of 9,436 feet (Maromokotro)...A former French territory that gained independence in 1960, this nation has a long history of political instability with a continued presence of the military in the political process...Fishing, forestry and agriculture employ over 75% of the population (deforestation is a serious problem) with a strong tourism sector but more than 70% of citizens exist below the poverty line bringing aid from the international Monetary Fund (IMF) after the presidential elections of 2014...Median age is 19 with a life expectancy of 65...Literacy rate is just 65% with 1.8 million children in the labor force...Government is a semi-presidential republic (fragile at best) with a multiparty National Assembly, president is chief of state and prime minister is the head of government (appointed by the president)...Numerous territorial disputes (islands) with France...

MALDIVES
(385,000 humans)
Military Forces = 5,000 active duty / $42 million per year

Over 1,100 islands at risk from rising sea levels with the highest point on any island just 2.4 meters...A republic since 1968 ruled mostly by one man (Maumoon Abd al-Gayoom) for 30 years with political parties illegal until 2005, Maldives is an Islamic Republic with a volatile political landscape declaring a state of emergency in 2015 fearing crowds and violence in support of jailed ex-president Nasheed...Over 18% of children under age 5 are malnourished with a life expectancy of 75 years...The economy is dependent on tourism and fishing...17% of the population below the poverty line...

In 2015 over a million humans crossed the Mediterranean Sea to reaches the shores of Europe with almost 4,000 dying on the trip from the capsizing of sub-standard boats often overcrowded as the smugglers try to maximize profit on each and every trip...The majority of the refugees came from Syria and Afghanistan...

MALTA
(418,000 humans)
Military Forces = 2,100 active duty / $40 million per year

A parliamentary republic (1974) located south of Italy in the Mediterranean Sea consisting of a grouping of three inhabited islands and numerous small uninhabited islands with the largest island (Malta) just 122 square miles in area...A long history of occupation by various European powers and a central role in the wave of migration to Europe due to it's location in the central Mediterranean...Both the most densely populated state in the European Union (EU) and the smallest, the economy is service based (financial) with some tourism and industry...The median age is over 40 with a life expectancy of 80 years and 15% of the population exist below the poverty line...

MARSHALL ISLANDS
(66,000 humans)

Inseparable from the United States since its capture from the Japanese in 1944, these islands have no military being dependent on the U.S. for defense (large military presence on Kwajalein atoll) with over 1,100 islands in the South Pacific 3,200 km from Australia and approximately 800 km north of the equator...Highest elevation on the islands is only 10 meters...Independence in 1979 with the first woman president (Hilda Heine) in the Pacific Island nation states elected in 2016...Median age is 23 and a life expectancy of 73 years...Economy is lease payments from the U.S. for their military bases as well as fishing, tourism and some commercial agriculture...Unemployment over 35%...This group of islands is famous for multiple nuclear test explosions on and around the Bikini atoll

MAURITIUS
(1.3 million humans)

Archipelago east of Madagascar in the Indian Ocean with a stable government and a dependable economy focused on tourism, industry (textiles) and financial services attracting over 30,000 offshore entities who have extensive business ties in this region of the world...Median age is around 35 with life expectancy over 75 years...Independence from Britain in 1968, a geographical location that experiences frequent cyclone and tectonic activity and a dispute with Britain over ownership of the Chagos Islands that's been ongoing since the nation's independence concerning indigenous people and control...

NAURA
(10,000 humans)

World's smallest republic just 21 square kilometers in size located in the South Pacific 53km south of the Equator, this nation has become the processing point for migrants trying to reach Australia (basically a detention camp with very few humans granted asylum)...The economy is heavily dependent on aid and with no military their defense is provided by Australia...New president (2013) has banned foreign journalists and evolved a draconian legal system abusive of human rights leading New Zealand to stop foreign aid in 2015...Largest source of income on the island is the detention camp...

NEW ZEALAND
(4.5 million humans)
Military Forces = 9,500 active duty / $2 billion per year

Three main islands and 30 smaller islands off the southeastern coast of Australia (4,155 km) with mountains over 12,000 feet (South Island) in a temperate climate with 15,134 km of coastline...A median age of 38 in a mix of European descendants and Polynesian descendants (Maori) with a life expectancy over 80 years...Government is a parliamentary democracy in a Commonwealth realm with the Queen of England as chief of state represented by a Governor and prime minister as head of the government...The economy is agriculture, industry and services (tourism) with an energy wealth of 67 million barrels of crude oil reserves and over 29 billion cubic meters of natural gas reserves...Located on the boundary between the Australian Plate and the Pacific Plate, New Zealand records over 15,000 earthquakes a year (almost all are minor) with the last major earthquake in 2014 (6.3 magnitude)...

PAPUA NEW GUINEA
(6.9 million humans)
Military Forces = 2,000 active duty / $110 million per year

One half of the third largest island in the world (behind Australia & Greenland), this nation was controlled by Australia until 1975 and is another Commonwealth realm with the Queen of England as head of state represented by a governor-general and a prime minister as head of government..A large majority of Papua New Guinea is remote, rural and mountainous consisting of many tribes with many languages (over 850) who have little or no connection to modern life or conveniences...Located in a volcanic region north of Australia the island has unexplored natural gas and oil reserves as well as rich mineral deposits difficult to recover due to terrain and the lack of any organized industry to support this activity...Large, rich rainforests support a logging industry that is of concern to environmentalists and conservationists...Over 28% of children under age five malnourished and a life expectancy of 67 years in a desperately poor country still recovering from a nine year conflict with separatists and a high rate of HIV...An interesting report (Science News 2012) that studied war and peace within the Enga society of Papua New Guinea focused on the adjustments these small isolated groups of humans made using compensation to offset defense killings between hunter-gather groups and village societies, but what is alarming is the violence and warfare that has continued unabated for almost 400 years between people who have little chance of enhancing a baby's initial brain growth that would produce a new generation who would understand the endless futility of this behavior and seek not just a way to compensate but a way to end this warfare and merge with cooperation to bring prosperity to everyone...84 Enga wars in 350 years...In 1990 when guns were given to the Enga by mercenaries there was a spike in offensive warfare that killed over 4,800 people between 1991 and 2010...The other half of the island is a province of Indonesia...The economy is non-existent for the majority of the population who live in isolated ways deep in the jungle without monetary support but newly discovered energy wealth of 175 million barrels of crude oil reserves and 155 billion cubic meters of natural gas reserves could change everything...

PHILIPPINES
(99 million humans)
Military Forces = 220,000 active duty / $2.8 billion per year

A large archipelago east of Vietnam (1,467 km) with the Pacific Ocean to the east and the South China Sea west, the Philippines has more than 700 islands located in an active region of volcanoes and earthquakes with the threat of typhoons and tsunami...Independent in 1946 after the 2nd World War with an occupation by Japan (1941-1944) and a large military presence by the U.S., the nation is seeking to increase it's forces as China occupies and claims new territory in the South China Sea...Politics have been fractured, corrupt and contentious for all of this nation's history and a Muslim insurgency in the south (Moro Islamic Liberation Front) has been active for decades amid violence and peace talks...A life expectancy of 70 years with a median age of 23...20% of children under 5 are malnourished and over 25% of the population exist below the poverty line...An estimated 138 million barrels of crude oil reserves and almost 100 billion cubic meters of natural gas reserves...Economy is remittances from a population of almost 10 million who work out of country, some industry, agriculture and services...The biggest challenges going forward are Chinese claims to territory in the South China Sea, insurgents and rebels (Muslim or otherwise) who proliferate when 25 million people are desperately poor...

Misogynist and violent Philippine president Rodrigo Duterte "I don't care about human rights, believe me" won the presidential office pledging to kill 10,000 drug criminals...As of September 2016 the total is around 2,400...Time

SAMOA
(193,000 humans)

Samoa (Western Samoa) is the 1st Pacific Island nation to gain independence (1962) becoming a parliamentary republic in 2007 with a chief of state elected by the Legislative Assembly who appoints the prime minister from the majority party as head of government...A group of islands located in the South Pacific 4,129 km west of Hawaii with 403 km of coastline, inland mountains over 6,000 ft and tropical rainforests that cover a large part of the islands...Economy is remittances, foreign aid, tourism, agriculture and fishing...Military defense is aligned with New Zealand...Median age is 24 years with a life expectancy of over 73...Essentially a cluster of nine volcanic islands in an enormous ocean...

SEYCHELLES
(90,000 humans)
Military Forces = 1,000 active duty / $12 million per year

Gaining independence in 1976 after over 200 years as a British colony, Seychelles is a popular tourist destination with multiple islands and pristine waters...Located north of Madagascar in the Indian Ocean off the east coast of Africa...A growing economy from tourism and fishing (tuna) after an IMF intervention following the world recession in 2008 (declared a successful market economy in 2013) the nation signed an anti-piracy agreement with the EU that allows for troops on the islands as needed with assistance from the United States...Government is a presidential republic with the president elected by majority vote as chief of state and head of government...A labor force of over 40,000 and a relatively low unemployment rate under 5% with a life expectancy of 75 years...

SRI LANKA
(21.5 million humans)
Military Forces = 160,000 active duty / $1.5 billion per year

An island nation off the southern coast of India, Sri Lanka emerged from 25 years of internal conflict in 2009 between the ethnic Tamil minority (Tamil Tigers) and the ethnic majority Sinhalese, a long, violent civil war between humans (who are essentially the same) that killed 70,000 people...In 2011, the UN issued a report concluding that human rights abuses and war crimes had been committed on both sides of the conflict, a report that the Sri Lanka government strongly denies and dismisses as inaccurate, divisive and polarizing...Suspected human rights abuses continue in country with a security force that acts with impunity and unexplained abductions and disappearances reported by Amnesty International...Independent in 1948 with a median age of 32 and a life expectancy over 76 years...26% of children under five are malnourished...Environmentally dramatic with tropical forests, beaches and a rich biodiversity of animals and plant life, mountains as high as 8,000 ft, many rivers, natural waterfalls and a storied and pivotal role in the East that has a history dating back thousands of year...A tourist destination worldwide with a vibrant economy, rebuilt infrastructure and a welcoming population...In a backlash after Mahinda Rajapaksa won election in 2010 and changed the constitution to eliminate term limits, Maithripala Sirisena won the presidency in 2015 with a pledge to reform the office giving more power to the parliament...A mixed coalition makes sweeping reform problematical...

SOLOMON ISLANDS
(605,000 humans)

A parliamentary democracy gaining Independence from the United Kingdom in 1978 located in the South Pacific made up of volcanic islands and coral atolls...One of the poorest nations in the region, the past two decades have been tumultuous with riots, poor governance and widespread crime that has seen multiple peacekeeping forces in country...A median age of just 21 with a life expectancy of more than 75 years...The economy is agriculture, fishing and forestry with mineral wealth (lead, zinc, nickel and gold) still undeveloped amid the chaos and unrest...The Regional Assistance Mission to Solomon

Islands is a mixed force of police, military and advisors from 15 countries in the nation for stability and to keep the peace...Military defense of the nation is provided by Australia and the United States...

TAIWAN
(23.1 million humans)
Military Forces = 290,000 active duty / $11 billion per year

A point of contention to China since gaining independence in 1950, this free nation is still not recognized by mainland China with continuous military posturing and belligerency by both countries that continues in this century...After four decades of martial law, Taiwan became a democracy in 1996 with the president and vice-president elected by a majority vote...A strong ally of the U.S. has been the primary reason China has not tried to take the island by force although China's insistence that this island is a territory and not a nation has kept Taiwan out of the United Nations...Median age of almost 40 with a life expectancy of 80 years...Economy is strong in manufacturing with a per capita GDP close to $50,000, a low unemployment rate and less than 2% of the population below the poverty line...Major challenges are disputes with nations over sovereignty in the South China sea, various islands claimed by multiple nations and this nonsensical continuing dispute with China over independence...

TRINIDAD & TOBAGO
(1.2 million humans)

A two island state northeast of Venezuela in the Caribbean Sea with a thriving economy of oil, gas and tourism...Government is a parliamentary democracy with president as chief of state and prime minister as head of government...Energy wealth is 728 million barrels of crude oil reserves and over 370 billion cubic meters of natural gas reserves, the majority offshore in shallow and deep water over 5,000ft...Small military defense force (3,000 active) who maintain internal security and an increasing police presence with the rise of violent crime in this century...Almost 20% of the population below the poverty line with an estimated 1,200 children in the labor force...

AUSTRALIA
(23 million humans)
Military Forces = 60,000 active duty / $26.5 billion per year

Australia is the planet's largest island, the only country to occupy an entire continent and the 6th largest nation on Earth...Native populations of Aboriginals and Torres Strait Islanders (from a grouping of more than 100 islands north of Australia) existed on this continent for more than 20,000 years before the arrival of Europeans in the late 18th century brought disease (smallpox) estimated to have killed over 300,000 natives...Today the populations of migrants, Europeans and indigenous are mixed with varying percentages which has challenged this nation's legal and social systems for years in identifying and categorizing Australians, a confusing, controversial process that continues today...This is the driest inhabited continent on Earth (almost all humans live along the coast) and home to the Great Barrier Reef, a collection of more than 3,000 individual reef systems and almost 1,000 islands located along the northeast coast of the continent...Due to a remote location, unique climate and insulation from other land masses, Australia has native plant species and various animals found nowhere else in the world creating a unique environment which generates more than $3 billion a year from tourism...

Australia is maybe the oldest land mass on Earth with the recent find of a zircon crystal dated at 4.4 billion years, the oldest piece of crust ever found...Earth formed approximately 4.5 billion years ago suggesting the planet cooled enough in just a few hundred million years to form a crust, the outer shell of the planet where all life exists...

In 1986, Australia became fully independent of the British parliament and independent of the British legal system although they remained a part of the Commonwealth, a voluntary group of over 50 independent sovereign states with most belonging to the British Empire at some time in their past that make the Queen of England head of state...A parliamentary democracy with a conservative coalition who won elections in 2013, Australia's economy has grown for two decades with investment and trade in Asian commodities and exports of food, energy and many natural resources including coal, gold, iron, uranium and a growing industry in natural gas extraction...Energy reserves include 1.2 billion barrels of crude oil and 1.2 trillion cubic meters of natural gas...Median age is 38 years with a life expectancy of 81 years...Just two years after a controversial "carbon tax" went into effect in 2012 the new government voted to repeal the law in July 2014 (carbon emissions dropped 1.4% the first full year the tax was in effect) and is proceeding with plans to extract billions of dollars in oil and gas from the Timor Sea signing a deal with East Timor to share in the profits...The Australian Coal Industry claims to be the world's leading exporter of black coal with the majority going to just five nations: Japan (115 million tonnes), China (42 million tonnes, Korea (41 million tonnes), India (32 million tonnes), Taiwan (26 million tonnes)...India and China are expected to have an increased demand for coal over the next decade as a world without restraint continues a heavy reliance on fossil fuel energy...

Located between the Indian Ocean and the South Pacific Ocean, the smallest continent on Earth and the only continent without glaciers, Australia is mostly arid desert supporting diverse habitats with rainforests, a small mountain (7,312 feet) and the largest number of reptile species in the world...The Great Barrier Reef is more than 133,000 square miles in area (344,400 square kilometers) with a large list of endangered species and endangered reefs from a changing ratio of acidic sea water that continues to increase from the absorption of carbon-based pollution, a byproduct of human energy production...

Challenges in the 21st century include the continuing migration of immigrants seeking asylum which has polarized the citizenry and resulted in an agreement with Papua New Guinea to process these humans offshore in "holding centers", increasingly destructive wildfires many believe are related to climate change, a growing debate of gay marriage laws (declared legal by the legislative assembly in Australia's capital Canberra and overturned by the Australian High Court just five days later) and the fate and status of Australia's indigenous populations, still a violent point of contention...

"A human life cannot be programmed...it simply happens and continues until it doesn't..."

REALITY INTELLIGENCE

REALITY INTELLIGENCE

Who Are We?

How do you define an individual with *'reality intelligence'*?...How can we know if this property exists in others or even in ourselves?...With the enormous variation of human diversity and the mystery of human singularity this is a difficult question...Humans who understand the complexity and reality of human existence recognize in an intuitive and cognitive way *what* we are and *how* we are...No human who exists or has ever existed knows *why* we are...Most humans with reality intelligence are intelligent in recognized ways as a result of dense neural growth early in life but we will have to wait for a future with technology to map the brain in detail coupled with behavioral, observational and historical studies to understand the importance and impact of early developing brain structure in determining morality, intelligence, empathy and a fundamental intuitive awareness of the physical and spatial reality of our existence...Still, there is an understanding of certain realities that define all humans with this property...

An understanding that all Gods and religions on Earth were created by humans...
An understanding that all humans are the same in the most elemental way...
An understanding that our planet is a singular, finite ecosystem...
An understanding that science is the language of human existence...
An understanding of the possibilities of a single species with a shared awareness of reality...

An understanding that all Gods and religions on Earth were created by humans...It's still unknown if there is an intelligence responsible for creating this reality, a reality which may be a tiny part of another reality that's incomprehensible to lifeforms existing on a single planet in a 'ordinary' galaxy...Many humans believe in a kind of spirituality or life-force that exists everywhere in the Universe or unknown multidimensional realities with an eternity of possibilities or even endless 'universes' each with their own laws and properties...In an existence without boundaries it may be that all things are possible but in 2016 there is no evidence, no indication and no verifiable proof of the existence of any God or any divine intelligence beyond the self-aware sentient intelligence we know ourselves to be...Intelligence seeks evidence to explain existence and there exists no evidence of any 'supreme being' or deities...It's becoming increasingly clear to populations everywhere on Earth that all world religions were created, promoted and practiced on this planet by a developing species trying to understand the nature of their own existence and one day all religions on this world will be fascinating histories of an ancient time...

An understanding that all humans are the same in the most elemental way...The inability of a majority of humans to understand this simple statement has been the cause of more death and violence on this planet than is imaginable, millions of humans throughout the entirety of human existence who have died because of a deep-seated primitive fear of any creature believed to be 'different' in some fundamental way...And it continues...All ethnicities are simply variations of the same creature...No matter where any human is born on this world they are the same as every other human born on this world and every human who has ever existed has been born on this world...*We are all the same creature*...Women are equal to men in every way except size yet discrimination all over the planet continues against females because of ignorance and fear by men who live with a primitive instinct that feels threatened and weak if they do not 'control' the female as dictated by tribal, cultural and religious nonsense that subverts a reality they are unable to recognize, unable to understand and a continued teaching of the ignorant by the ignorant...Hutu and Tutsi are the same species...Shiite and Sunni are the same species...Jews and Palestinians are the same species...Hindu and Sikhs and Muslims are the same species...Blacks and whites are the same species...Men and women are the same species...The violence, discrimination, ignorance and killing continues...

An understanding that our planet is a singular, finite ecosystem*...*There is an obvious and undeniable balance of nature which has been essential in the development and evolution of our species, a balance vital to our survival and clear to every human who understands reality...Any human who believes the activities of seven billion creatures who burn their planet for energy is not having a serious effect on the planet's ecosystem does not understand reality...*Why the most intelligent species on Earth would actively change the balance of gases in an atmosphere that has allowed them to evolve and thrive is difficult to understand...*Why do we not notice the choking of rivers until they are choked with waste or the unchecked polluting of the oceans until our food is contaminated and our beaches ruined?...Why do we not notice the ongoing extinction of some of the largest animal species on this planet until there are just hundreds or thousands left?...The balance of one planet filled with life is the biggest challenge in a future that must be addressed in the present and with a lack of reality intelligence and understanding among the humans who control energy resources and set current environmental policies coupled with a continuing denial of science and ignorance of consequence on a world this species continues to burn, little is being done to change a 21st century paradigm of "greed and profit" as multiple corporations and governments promise to burn more fossil fuels to satisfy world populations, create more wealth and let the future deal with consequences both predictable and unpredictable...With the fragmented world of today and an exploding population without constraint, it's difficult to predict outcomes..

An understanding that science is the language of human existence*...*Science is the written record of human reality and the process humans use to determine reality and over thousands of years thousands of people have contributed to the development of scientific techniques and processes humans now use to discover, explore, investigate, verify and confirm this reality...Science is the language of a growing intelligence seeking to know itself...Science organizes our reality in a way that can be tested, verified and confirmed and *every truism taken for fact can be rethought and redefined if new scientific methods or new discoveries show it to be another way...*There is no evidence for a god but if a god did appear on Earth science would test all the parameters in every way possible with multiple measures and multiple means and if the testing confirmed the deity the god would be known to be real and we would alter our construct of reality...Quantum theory is a great mystery operating in a way contrary to any realistic world view of reality and has been tested thousands of times never failing to manifest itself as predicted...We don't understand how it works or why it works but it works...The same rigorous testing of general relativity has proven it to be 'real' in every way tested although humans have a very tenuous connection to the bending of space and the fluctuation of time...One of the most important tools of scientific inquiry, mathematics, continues to take science in directions unanticipated and it's unknown what lies in this direction of inquiry but what is alarming is the unexplained trend now happening in a world controlled by isolated governments and corporate greed: *the censorship of science...*In many institutions and organizations around the globe scientists are not allowed to see or study the work of other scientists with this secrecy being systematically institutionalized out of some primitive territorial instinct...Excepting for the science concerning weaponry, biological and nerve agents in an increasingly dangerous world, science should be transparent to every human on Earth, studied, tested, absorbed, and improved...It is the way we understand ourselves...

An understanding of the possibilities of a single species with a shared awareness of reality*...*This is the promise of humanity...The product of an evolution of intelligence that has taken more than three billion years, this is the next step in our quest to understand ourselves and the amazing reality we exist in...A self-aware sentient species on a blue planet circling a single star in a large galaxy among billions of galaxies, existing in a known reality with definable properties that allow the exploration, habitation and harvesting of an entire star system to sustain human activity for a billion years...*This is possible...*If we can maximize brain growth in every baby born, our species would understand the need to control world population until we can terraform Mars, we would understand the need to sustain our ecosystem and we would have a planet of intelligent humans to make this happen...We'd protect the many species on Earth and know that we cannot save all life in a living, breathing ever-changing world but we can recognize a balance of habitat and urban development to coexist with creatures who may be vital to our existence...We would examine the thousands of unknown plant species that may cure disease creating vaccinations against dangers threatening our children and as master tool builders we would build space ships miles long to travel between the planets, visit the over 170 moons we know exist and build solar collectors in space to power a planet and create a world of limitless, clean energy...It's easy to imagine humans living 300 years or longer in the future and a world where infant mortality is unheard of...No one knows what might be possible...*We are not controlled by deities and superstitions with restrictions and dogma...*We are in control...We are more than seven billion and we are the dominant species in this solar system...Off world exploration has just begun and there is a potential for resources and riches unimagined in the places we know we can go but humans need to cooperate to survive an unforgiving, hostile environment that in no way supports the demands of carbon-based lifeforms...To solve many of Earth's problems and to explore this reality humans must evolve as a species and not as isolated, primitive, territorial nations focused on aggression with our own kind destroying the world with short-term ignorance and apocalyptic visions...It took human beings 200,000 years to explore a planet and establish the 1st World Civilization...How long will it take to explore a solar system?

Imagine trying to explain to a group of zealots from anywhere in the world, Islamic fundamentalists, right wing militia groups in America, rebels who kill in the Philippines, Taliban extremists in Afghanistan, any group with little regard for human life...You are given 3 hours to explain human reality with proof and wit and understanding about the nature of existence and the nature of the world....No heaven, no hell, equal genders, education for all, inclusion of everyone to deliberate the problems of the world, resources for everyone if we work together and a social structure that doesn't abandon anyone in a connected world that understands the possibilities of the future...The zealots serve tea and cookies while they listen, read and watch the proof you provide and after 3 hours they kill you while shouting "God is Great" over and over...Without the dense neural networks developed in the first three years of life necessary to understand that humans exist in a definable reality, they are driven and confused by an inability to communicate and reconcile powerful, primitive and instinctual regions of the brain with an intellect governing reason, logic and understanding...Very dangerous if stimulated in a way causing confusion, disorientation and fear...

For thousands of years there have been questions of how humans with the same knowledge and facts can exist in the same reality and interpret cause and effect so differently without an intuitive or cognitive ability to understand what is proven and obvious, a disparity that continues today in every region of the world despite volumes of scientific conclusions proving physical processes understood by intelligence but questionable to billions of humans unable to recognize the nature of existence...This is not surprising when you realize the scientific community believed our galaxy to be the entire universe less than 100 years ago and it's not surprising when you understand the impact of inequality, abuse and the fragmented isolation of cultures over thousands of years who are now part of an accelerating global civilization...Entire societies without inclusion and entire cultures without the knowledge to understand how essential it is to maximize initial brain growth in babies from the very start of life...

The consequence of this is billions of humans in the 21st century focused simply on survival, without hope, looking for stability and security in a world where inequality grows and a growing global paradigm of interconnected greed and territoriality prevents the effective organization of any kind of world coalition with the strength and resources to remove those who kill their own citizens and those who starve their populations...And no effective world organization with a global awareness or a shared recognition of the extreme need to create programs that can provide safety and nutrition for mothers and babies worldwide...This is also not surprising when you understand that the humans who find themselves in positions of power, wealth and influence in this century were infants just a few decades ago in a world where there was and still is very little understanding of the impact of infant care on the thought processes, intellect and awareness of the adult human...The result is that many of the influential and powerful individuals who control world resources, world military's and the world's wealth are unable to fully understand causality, unable to understand a proven and definable reality and unable to recognize the strengths, needs and possibilities of their own species at this critical time in the history of life on Earth...Threatened by a reality they can't understand they are influenced by primitive instincts of fear, greed and self-preservation (mostly at a subconscious level) without an ability to rationalize the dogma and superstitions that have influenced human activity for thousands of years...

The exclusivity of the frontal lobe can allow individuals to develop a high level of intelligence in multiple areas of life (rocket scientist, entrepreneurs) and still not have this density of neural growth that develops in the first three years of life essential for empathy, morality and a lifetime understanding of the nature of human existence.

There is no way to really know what occurred during the first three years of life in the hundreds and thousands of dictators, politicians, religious leaders and corporate leaders controlling the resources and policies of a global community without a consensual awareness or a unified, cohesive direction...If there are just 1.1 billion humans this century (conjecture without data) who recognize and understand the physical and substantive reality of human existence (but always willing to question if there is evidence) then 84% of Earth's population still believe in dogma and fate (a predetermined or inevitable course of events) without any understanding of what is possible or more importantly, what is necessary for the human species to survive...The thoughts and beliefs of many influential people entrusted with determining a sustainable path for billions of humans are dominated by a belief in God, a belief in destiny, fear of the unknown and ignorance of an existence verified by proven scientific processes...

Subsequently, without a shared intelligence that understands the challenge of sentient life on a single planet in a resource rich solar system, there is little consensus among nations concerning the major challenges that continue to threaten humanity in the 21st century...Instead there is a continuous focus on sovereignty, wealth, power and a ready military force in all the nations of Earth to war for advantage and there will be no intellectual consensus on the nature of human existence until a majority of babies born are nurtured to maximize initial brain growth...In this century billions of humans are unable to recognize what is proven and obvious but there is growing evidence this is slowly changing in a world where global connections and an increasing global knowledge is beginning to influence understanding and outcomes in every society on Earth...A growing awareness that will determine how humans survive an unknown future in an indifferent universe...

Reality Intelligence

A connective neural density developed in the first three years of life that enables a fluidity and volume of communication within the brain merging the rational, logical and most recently evolved regions of the brain with the primitive, instinctive and ancient regions of the brain...This deep, mostly subconscious flow of communication results in a natural morality, a natural empathy and an intuitive, cognitive awareness of human reality and the nature of existence within that reality as each human develops into an adult...Without the technology necessary to observe and count millions (billions) of synaptic connections, individual neurons or pathways of communication within a human brain (the most complex object known) there is only a small body of growing evidence beginning to understand this complex causality relationship implicating initial brain growth in infancy with the ability of every human to recognize and understand the reality of their existence as adults...This will change as new science and technology develop better tools and techniques...(There are multiple projects active in this century to study and map the human brain)...

There is no question more research is necessary to verify and understand this process and this book will be challenged in the emphasis and importance it assigns to this property of the brain but the empirical evidence of case studies, anecdotal evidence of babies from differing situational realities including orphanages and broken families, historical evidence of documented lives from humanitarians to serial killers suggesting the positive effects of early attachment and negative effects of abuse and abandonment and the growth of scientific evidence involving studies on primates, laboratory animals, humans and related brain structure with new technology, new techniques and new tools for observation and measurement all suggest this property is real...In no other way is it possible to explain the disparity of thought and beliefs in individuals and various groups within societies who all exist within the same reality...An unexplained disparity that has continued for all of human history...

Reality Intelligence is a deep permanent connective state of awareness...

The first three years are the most important years of any life...In a limited existence of just one hundred trips around the Sun, the possibilities of discovery, adaptation, change, emotion and creativity are seemingly endless with many lives documented through all of history confirming the intelligence of Earth's most dominant life-form, but the first three years of life make people who they are by creating a framework of internal communication to support a lifetime of intellectual development...A permanently formed structure of neural connections, highways of communication between differing regions of the brain that determine a level of reality awareness and an intuitive, cognitive understanding of the nature of existence...*After age three unused neurons are cleared away, memories are permanent and this framework will determine how each individual absorbs and understands the billions of sensory inputs that create a human life...*This period of development is crucial in creating the organizational structure and neural density that will determine reactions, realizations and behavior...Very simply, what happens from age 0 to 3 determines the possibilities of a human life...This essential connective structure cannot be markedly enhanced after this critical period of brain growth and it explains how people can be very intelligent in different areas of human achievement and still not understand the reality of their own existence due to a lack of early connective density between rational and primitive regions of the brain...

You only have one chance in life to build this network of neurons, a dense communicative construct critical to an understanding of reality and existence intertwined with all aspects of life...However dense and connected this initial construct is will determine how this brain understands existence and reacts to external stimuli...Abuse and indifference restricts growth limiting an individual's realizations and understanding of life...Nutrition, nurturance, sleep and positive stimulation create a dense construct where realizations are almost unlimited...

Evidence and Realizations

The except below from a 2009 publication is an example of a growing awareness among the scientific community of the challenge of understanding the inter-connectivity between different regions of the brain and how critical regional connectivity is in determining cognitive behavior, function and thought...The challenge of imaging and deciphering brain organization with 21st century technology and techniques is referenced multiple times in current scientific literature......

"There has been an emerging focus in neuroscience research on circuit-level interaction between multiple brain regions and behavior. This broad circuit-level approach creates a unique opportunity for convergence and collaboration between studies of humans and animal models of cognition. *Measurement of broad-scale brain networks may be particularly important for understanding changes that occur in brain organization and function during development* (italics mine)... Recent studies in humans have gained much leverage from trying to understand circuit-level interactions among brain regions over the course of development. Such studies use connectivity analysis of functional magnetic resonance imaging both during cognitive activity and during rest and diffusion tensor imaging to measure (respectively) the functional and structural connectivity between discrete brain regions."
Frontiers in Human Neuroscience.2009: 3: 81.Published online 2010 January 20.

In the summer of 2013 various articles began appearing in medical journals and on websites in the United States concerning a condition called *"positional plagiocephaly,"*, flat spots on the heads of babies during infancy caused by resting a growing, flexible skull on one side for too long a time, a condition that is rarely permanent and rarely noticed by anyone except caregivers who are intimately familiar with the baby...It is not implicated in any way with brain development but it illustrates more than anything else the flexibility of the human skull to allow for passage through the birth canal and to allow for the rapid brain growth that occurs during the first years of life...There are six cranial bones that make up the brain cavity and the separations between these bones called *"fontanelles"* fuse together in the infant's 2ⁿᵈ year of life and stay fused throughout the lifetime...

What we do know is studies on animals and humans in controlled environments and a human history of empirical, anecdotal and observational evidence clearly suggest that neuron development and the growth of organizational structure is *maximized* in an environment of secure attachment, positive stimulation, safety, nutrition and sleep and is *inhibited* in an environment of fear, poor diet, insecurity, abandonment and violence...A baby's first days, months and years are a unique period in a human life when communication networks are being grown and developed in ways that will determine brain function for a lifetime...Science has shown that neural growth and change continue in some ways throughout a lifetime in reaction to the experiences and challenges of existence in a ever changing environment, but the initial establishment of the memory centers, organizational structure and neural pathways are mostly complete around age three or four...*Very few humans on the planet can remember events that happened in their life before age three...*Memories earlier than this are absent because the brain is still growing, developing and organizing itself into a memory archive and structure that will remain stable for a lifetime...When this organization is established, memories become permanent retrievable very late into life and almost everyone can remember an event that occurred when they were four or five...Studies have demonstrated many times that memories created in childhood are often more vibrant and retrievable by older adults than are recent memories suggesting that once the brain's structure is finalized around age three it retains an organizational consistency for a lifetime...

My personal history is a clear example of this process that is just now being understood...My father and mother died when I was 976 days old...I was almost three and I cannot remember either one of them...I have pictures, videos, stories and memorabilia from infancy but it makes no difference...My mother held me, fed me, sang to me, played with me and loved me every day for over two and a half years and I can't remember her face, her touch or her voice...Nothing...My brother was eighteen months older (four years old) and he does have memories of our parents...A very human process...

Anyone who has shared life with a two and half year old child knows these small humans have vibrant personalities and are communicating and actively participating in their environment...They light up when anyone they know or love enters the room and they crawl and climb on us seeking attention, love and stimulation...They know your voice, your habits, your touch and your smile...And they have memories...You can leave for several weeks or months and when you come back they know you, call your name and demand a hug...*To realize that if you suddenly disappeared at that moment the child would never remember you even existed is stunning...*It speaks to the complexity of the brain and the development that continues in the first years after birth...Studies continue to show that the brain can grow, adapt and learn throughout an individual's life *but the establishment of the basic structure of your brain that will serve you for a lifetime is largely accomplished during the first three years of life...*

Dense neural growth is the key to intelligence, the key to awareness and the key to an intuitive understanding of the three-dimensional reality of human existence...Science cannot yet count individual neurons and any experimental studies that treat a baby one way to enhance growth and another baby a different way to inhibit growth and prove a hypothesis are unethical and should be...But who will argue that raising every baby with love, nutrition, positive stimulation and secure attachment is harmful or unhealthy?...Babies raised this way laugh a lot and smile, they cry only when trying to communicate and they show endless curiosity in all environmental stimuli...There is a natural morality that develops and you can see in them by age three that they will never kill another human unless it is absolutely necessary...In our violent world it sometimes *is* necessary but babies raised this way grow up aware of reality, the interaction of human society and aware of the world they live in...*How this species grow their babies will determine the future of life on Earth...*A future lasting millions of years if we can network a planet and give all humans a chance to understand the nature of their own existence...

How we grow our babies will determine the fate of the human species...It's that simple...

"After more than a century of sustained progress in biological sciences and medicine, one could argue that mankind has made significant advances in our understanding of how biological systems operate and how different parts of the body function and when damaged, generate disease. At the same time, a comprehensive understanding of the brain remains an elusive, distant frontier." *ACS Nano, 2013, 7 (3) March 20, 2013*

A Measurable Effect

The positive and beneficial results of positive stimulation, diet, sleep and attachment are not limited to the human species and some of the pioneering studies on maternal attachment have shown the importance of security and attachment in animals which determine their ability to survive as infants and form relationships later in life...Experiments by Harlow and Zimmerman in the 1950's with monkeys that were detached from there mothers and given a 'wire mother' and a 'cloth mother' found that monkey's *must* form attachments during the first year of life...Infant monkey's raised in isolation were never able to interact with other monkeys in a normal way and *some died from the stress of being untouched and alone*...They also found that even when the 'wire monkey mother' was the source of food the baby monkeys preferred the 'cloth monkey mother' clinging to it a majority of the time for security and reassurance...When isolated monkeys were allowed to interact with other monkeys for as little as 20 minutes a day it made a recognizable difference in normalizing behavior as they grew to become adults...Studies and behavioral observation over many years have proven beyond any question that social interaction, touch and a sense of safety, security and belonging in infancy are essential for the normal development of intelligence in all mammals, from infancy to adulthood...

A study at McGill University with rats compared the effect of attachment on a group of baby rats who were groomed and licked by their mothers in the first few weeks of life to a group of baby rats whose mothers did not interact with them in any way...Researchers then followed the two groups of offspring into adulthood and found dramatic differences in the the ability of the two groups to cope with their environment...The baby rats who were groomed and comforted by their mother were more confidence, exhibited less stress, were less fearful and did better at mazes than the group of rats whose mothers did not interact with them after birth...The grooming by the parent happened only for a few weeks after birth and after these first weeks of interaction the offspring never saw their parent again but the differences exhibited by the two groups throughout their life was very different...Clear evidence of how crucial it is to nurture and form an attachment with babies immediately after birth...Very important is clear evidence of measurable enhanced growth in the brain of the rats who were groomed suggesting that *just the presence of maternal security and attachment can stimulate brain growth in infants*...This study in 2003 is one of many that document this maternal effect on brain growth...

Maternal care, hippocampal synaptogenesis and cognitive development in rats
Dong Liu, Josie Diorio, Jamie C. Day, Darlene D. Francis and Michael J. Meaney
Developmental Neuroendocrinology Laboratory, Douglas Hospital Research Centre,
Departments of Psychiatry and Neurology & Neurosurgery, McGill University,
We report that variations in maternal care in the rat promote hippocampal synaptogenesis and spatial learning and memory through systems known to mediate experience-dependent neural development...Thus, the offspring of mothers that show high levels of pup licking and grooming and arched-back nursing showed increased expression of NMDA receptor sub-unit and brain-derived neurotrophic factor (BDNF) mRNA, increased cholinergic innervation of the hippo-campus and enhanced spatial learning and memory... A cross-fostering study provided evidence for a direct relationship between maternal behavior and hippocampal development, although not all neonates were equally sensitive to variations in maternal care...

This is just a small window into what is happening to brain growth when babies feel a safe, secure attachment...It shows actual neural growth taking place that is measurable and more than in the babies that do not have security and maternal attachment although the mechanisms and extent are still not known...

In the three excerpts below from a research paper published online in 2009 it's clear human researchers are finally beginning to understand the critical nature of a human baby's initial environment after birth....The authors of this study also note the difficulty in studying brain development in infants and children...Even with new technology that includes structural and functional MRI, MEG, electrophysiological measures of brain activity and advanced techniques for measuring HPA axis function, obtaining these measurements in a reality based environment is often invasive and disruptive to the subject (baby)...

Stress and Early Brain Development

MEGAN R. GUNNAR, PhD
ADRIANA HERRERA, MA
CAMELIA E. HOSTINAR, BS
University of Minnesota, USA
(Published online June 2009)

"In the early years of life when the brain is developing rapidly it is particularly sensitive to environmental influences. Toxic early life stress (ELS) may induce persistent hypersensitivity to stressors and sensitization of neural circuits and other neurotransmitter systems which process threat information. These neurobiological sequelae of ELS may promote the development of short and long-term behavioral and emotional problems that may persist and increase the risk for psychopathology and physical health disorders into adulthood."

"Research in humans increasingly suggests that severe early life stressors (e.g., trauma, maltreatment, neglect) may result in decreased brain volumes, dysregulation of the neuroendocrine stress response system and limbic dysfunction involving regions such as the hippocampus, medial prefrontal cortex and amygdala. Consistent with these findings, animal studies of severe ELS yield evidence of inhibition of neurogenesis, disruption of neuronal plasticity, neurotoxicity, and abnormal synaptic connectivity. Sensitive periods and stages of enhanced brain plasticity are particularly vulnerable to the long-term effects of stress hormones and *may result in altering the typical pathways and organization of the young brain.*(italics mine) Research also suggests that severe ELS may have mental and physical consequences that last into adulthood including increased risk of depression, anxiety, post-traumatic stress disorder, metabolic syndrome, and cardiovascular disease. Notably, research has revealed that the child's access to supportive, attentive, and sensitive adult care plays a salient role in buffering the activity of the HPA system and protecting the developing brain from potentially harmful effects of stressors. Children within secure parent-child relationships learn that when faced with a stressor they can experience distress, communicate their negative emotions, and effectively elicit aid from caregivers. It is likely that this sense of safety prevents activation of the HPA axis and other critical stress-mediating system"

"The early years of life constitute a particularly sensitive period during which chronic stress may lead to dysregulation of the stress system and may compromise brain development. Not all individuals are equally at-risk for developing the neurobiological behavior and health consequences of ELS. It is likely that genetic factors, emotional and behavioral predispositions, stress history, social support, mental health status, age, and sex all play a role in stress reactivity and regulation. Tracing the pathways through which early adversity impacts later development is the key challenge for developmental stress research in the coming decade."

In a 1964 study by Schaffer and Emerson that followed 60 babies for 18 months in the home (a natural environment) they identified the stages of attachment from birth to 9 months...From birth until 3 months of age infants were predisposed to attach to *any* human...The babies responded equally to any caregiver who provided safety, food and touch to meet the baby's needs...*This is very important as it suggests that the human race as a group can rescue and care for any child that is abandoned at birth and provide what is needed to maximize initial brain growth*...The second stage of development from 4 months to 7 months babies were learning to distinguish primary and secondary caregivers but would still accept care from anyone...After seven months babies started recognizing and preferring familiar people, some showing a fear of strangers and distress when separated from a particular caregiver, evidence of secure attachment and comfort with specific humans that can have enormous influence on the development of neuron growth...After 9 months the babies became more independent and were forming multiple attachments with extended family members and others...The researchers noted the babies most often formed attachments with the humans who responded accurately to the baby's signals and more important *the babies formed the strongest attachments with the humans who played and communicated with them* not necessarily with those they spent the most time with...

Awareness and Rationalizations

Scientists and researchers have long known that animals improve in every aspect if they form a 'secure attachment' with a parental unit after birth that remains secure during the earliest growth cycles and many researchers have recognized this need for 'secure attachment' in humans as well but it hasn't been fully understood what a critical difference close continual attachment and stimulation can have on a developing baby's brain in maximizing brain growth and influencing behavior over a lifetime...*Most important is that researchers have not realized how permanently damaging a lack of attachment, love and stimulation can be to a baby's brain during this growth period that can never be recovered...*

What is unique in humans is a complex *frontal cortex* for processing information and skills that can be highly developed throughout an individual's lifetime even if the individual did not develop the dense connective pathways which define *reality intelligence* and it suggests there are many people who exist with intelligence, knowledge and skills without a density of structure critical in the ability of a brain to communicate with all of it's differing regions in a fluid, interactive and integral way...Millions of people who are respected, intelligent, accomplished and informed in multiple ways but instinctively and subconsciously still question existence based on 100,000 years of desperate survival in a hostile world, believing the dogma, myths and legends which have determined human actions for all of written history, unable to recognize reality or understand an unknown 'unknown' that is both frightening and confusing...*This questioning of reality functions primarily at a subconscious level and is reconciled with beliefs of religion and predetermined fate that comfort the individual and allows a 'normal' existence within a framework of societal norms*...In a stable existence billions of humans live their lives in this way with few negative consequences and with the capacity to be productive, generally happy and a part of a community who share in their beliefs...But what happens if the 'stable' functional lives of these individuals are challenged in a way that causes extreme fear, anxiety and confusion in a brain without the dense neural growth necessary to reconcile causality and reality? *What happens when the human brain cannot communicate with itself?*

In 2014 there were 41,149 suicides in the U.S. as reported by the Centers for Disease Control and Prevention...10th leading cause of death...WHO reported in 2014 there are over 800,000 suicides globally every year (1 every 40 seconds) more deaths than conflicts and natural disasters combined...In America the rate of suicide went down during the 1990's with growing prosperity but has steadily risen from 2001 through the recession in 2007-2013 as people were losing jobs and losing their homes (the death rate from suicide is highest in middle-age when people feel they have the most to lose)...Faced with an insurmountable crisis they respond with desperation, hopelessness and anxiety in a reality they can't understand, questioning eternity, the meaning of existence and God...

In America suicide is the the 2nd leading cause of death for ages 10-24...

A survey of California patients found strong relationships between adverse or traumatic childhood experiences and a person's lifetime risk of attempting suicide...S.R. Dube et al/JAMA 2001

In 2014 there were an estimated 14,249 murders in America (FBI)...More than 437,000 humans murdered globally in 2013 (UN)...Why does a human kill another human?...What is the thought process within a brain that rationalizes and justifies ending a human life for whatever reason?...Many humans kill others out of consequence while engaging in illegal activities or when confronted in some way and many of these humans have stated that they 'feel' nothing for the loss of life...No recognition that they ended a human singularity, an individual with thoughts, dreams, memories and maybe insights that no other person can duplicate or replicate...A lack of awareness that becomes a lack of remorse...

In people who kill with malice and intent there is a connection between their first year and their last year of life...

Morality

Existence in this reality is a unique experience that one would think is a *fundamental right* of every individual if that individual does not challenge another human's right to exist...What each individual does with their existence is defined by the laws and properties of this reality in ways still being explored and understood but no person should be damaged, confined or destroyed if that person regards all others in the same way...The right to live life anyway you wish as long as you do not injure or kill others and you recognize their right to live their life anyway they choose would seem obvious but are there fundamental rights in a reality that has no apparent rules or guidelines?

Our species have been trying for 30,000 years to define *morality* spelled out by societal norms or dictated by religion and myth, a process which continues today using the same ancient rituals and logical arguments that have done little to stop killing in multiple societies and diverse civilizations through time...But there is an argument that morality is not a 'code of laws' that can be spelled out by civilizations or by dictates of man, rather *morality is an intrinsic part of the evolutionary process*...Life seeks to know life and within an evolutionary bond that stretches back a billion years, there is a *natural morality* that is a dynamic, underlying property of our existence...Life seeks life...If you and a cat were to go to sleep in a house with 100 rooms filled with all sorts of items to explore, you would wake up in the morning with the cat on your lap because with every other option available to the cat life seeks life with the rhythm of breathing, the sound of a heartbeat and the warmth of another living creature...In babies who are grown with abundant love, positive stimulation, nutrition, sleep and close continual human attachment, there is a *natural morality* that develops to become an instinctive, integral part of that individual's consciousness for a lifetime that understands humans don't kill humans...An intuitive understanding of the nature of human existence that dictates behavior and determines how we interact with this reality based on intellectual reasoning, internal communication and causality...*If a brain is connected with itself through dense neural pathways merging human evolutionary history into a single operating system, an understanding of the uniqueness of life does not need to be taught or proven*...It is understood from a very young age and it is what makes us human...

There is evidence for a natural morality in the history of storytelling, one of the oldest forms of communication in human history...Predominately in every culture and every society on Earth from the very ancient days before religious structure to the present day, in books, movies, plays and music, when good vs evil the good wins...almost always...

Not enough is known about human reality to say death is final...Unknown possibilities in an unknown existence...

Death

The human species is far too young to know if there are any realities outside of the one we know exists...On a physical level death is final and understood, everything alive dies...Many intelligent humans understand this cycle of life and death and it is not mysterious or supernatural but it is sad...If there is some 'dimensional existence' or an 'individual spirit' or some 'essence of being' inside a human which continues after death it is unknown...There is no evidence of this...Many individuals throughout history have thought seriously about this question of spiritual existence after death and the many who have made every effort through experimentation to discover a shred of evidence to support their ideas have failed...Many of these individuals promised associates and loved ones that if an afterlife did exist, in any form, they would communicate with the living after they died but there have been no ghosts, no spirits and no evidence anyone has ever communicated with someone who died...There are parts of the human brain that want desperately to believe there is *something* after death but at this time in human history there is no evidence...Everything living dies and when they do the constituent components of their being (atoms, the essential elements of all matter) are broken down and returned to the cyclic existence of the universe...Death is a reorganization of matter on this planet...Nothing leaves our planet unless we accelerate it to 25,000 miles per hour so everything on this world is eventually broken down into basic elements (atoms) and redistributed within the gravitational field of Earth...Every human alive today is some part of everything else that existed on this world and all of the elements existing on Earth were created in the inflationary beginning of this universe and in the millions of stars exploding since that time seeding space with a 100+ elements that make up essentially everything...

It's difficult to believe we are not 'more' than simple biological beings with a limited life span and a unique consciousness that ceases to exist after death but this is the reality of what is known at this time...And death is sad when you understand humans exist in a stable reality which will continue for billions of years and the human life span is less than 100 trips around the Sun...

Just this awareness makes suicide mostly senseless since we exist for such a brief time but to an individual with *reality intelligence* suicide is an option if one is dying in pain, very old and sick or even as a free personal choice...But humans who understand reality would never kill another human or kill their spouse and children so "they don't have to suffer in this life" or so "they can be with their god"...They would not kill someone they were married to, they would not kill their former employer who fired them, they would not kill the man or woman their husband or wife is sleeping with, they would not kill their daughter or son because they "shamed the family", they would not kill any human unless that human was a direct, immediate threat to them or other innocent humans...People who do this, who kill others out of fear, ignorance, rage, revenge or ideology (thousands around the world) are unable to reconcile a fear and confusion that never ends with a reality they cannot understand...The murderers, suicide bombers, mass murderers, serial killers and self-righteous leaders of genocide have all suffered extreme trauma during the first year of life and have lived a life of pretend trying to act 'normal', trying to fit into an existence they can't define...With a brain dominated by primitive instincts of fear, confusion, vengeance and righteousness without a neural density to reconcile violent emotions and confusion with rationality, logic and intellect, these humans act in what to them is the only solution to their problems, their questions or a lifetime of confusion in a last act of desperation...Many humans who kill others or kill multiple humans 'with purpose' also kill themselves...

There are many levels of reality intelligence, many people with no chance and many on the edge...

Levels of Awareness

In 2016 billions of humans live with questions of God and questions of purpose with a sense of inevitable fate in a questionable reality they rationalize to understand, believing in possibilities of an afterlife and structuring their life in such a way to reflect these beliefs...Most of these people are not violent and many hope for a better world to give to their children, a world with a sense of fairness and equal opportunity, humans who try to adapt to the expectations of the cultural and political framework they exist in but are uncomfortable with the power, greed and ideology of many world communities in an exploding civilization of growing inequality and limited inclusion...Given real options that could improve life for many and given a chance to participate in a future based on science and awareness rather than religion and adversity, it's possible millions or billions would accept a new paradigm that prioritizes an understanding and acceptance of all humans, a single species on a single planet with a chance to create a culture of equality for everyone...Recognizing the necessity of world unity and the possibilities of a world economy that included everyone, there may be enough humans to change the embedded institutions that discriminate and promote violence to control populations and there may be enough to change 21st century economic structures that negatively impact the lives of millions while enriching a minority who control "everything"...At this time the probabilities of a growing "collective voice" with a unity to make a difference in the commerce, poverty and violence of 21st century Earth is impossible to predict...Many nations on Earth profess to hate 'war' while supporting industrial military structures that sell and distribute millions of weapons across the planet and many nations profess to hate 'poverty' while institutionalizing wealthy individuals and corporations who control a majority of the world's wealth while continuing to ignore the desperation of millions who have nothing...

There is a way for a population to exist on this planet with sustainability and progress if the majority understand and accept that all humans have equal standing in every society in every way no matter what gender, ethnicity, race or sexual orientation and if a majority understand Earth's "religions" are products of human invention and accept a logical economic framework which rewards risk but also includes every human on Earth in a market economy to distribute goods and services across a global society...It will take hundreds or maybe thousands of years to change territorial mindsets and ethnic violence *but the intelligence and awareness needed to recognize these realities is a natural, normal developmental process in babies raised in safe, positive environments*...If there were a global focus on providing for every baby from the moment of birth, a consensual awareness of human existence will create a civilization with priorities to feed and shelter every human on Earth and priorities that use the best science to continually measure and interpret the thousands of variables that shape the reality we exist in while creating a world civilization who understand the potential of a global population with a shared interest and a shared intelligence living within the boundaries determined by nature...

The human brain is a compilation of complex processes and ancient instincts that date back millions of years operating interactively with a relatively modern frontal cortex of awareness, thought, logic and reason...

Human Intelligence

Human consciousness is a mystery...No one know how billions of neurons and maybe a trillion connections coalesce to create a sense of self and doctors and researchers have believed for years that the state of human consciousness was a diffuse and global brain process without specificity but this has changed...Recent studies of patients in vegetative states suggest consciousness is networked between three discrete parts of the brain: a section of the prefrontal cortex; a section of the parietal cortex and the thalamus, a primitive structure deep in the brain that mediates signals between the other two parts of this triad...If the connections and communication between any of these three sections is severed or destroyed consciousness is severely effected but it's now believed a human can have significant brain damage and still be conscious and aware if these three sections of the brain continue to communicate through the neural networks remaining intact...

There is evidence the extended developmental period in infants is due to the fact that the human brain grew to large to be birthed fully developed...As a result humans are born with an underdeveloped brain which continues to organize itself and grow for several more years...

A study in 2012 by Olaf Sporns (Indiana University) and Martijn van den Heuvel at Utrecht University Medical Center studied the brain's internal communication network consisting of 12 very interconnected bi-hemispheric hub regions which work together to direct internal communication and process complex behavioral and cognitive tasks...This enormously complex framework or 'backbone' of internal communication processes up to 70% of all neural signals before they reach their destination routing signals through the most efficient pathways to maximize performance and awareness...

The Human Brain = 100 billion neurons with trillions of synapses making trillions of connections...

Studies continue to show the brain's flexibility in rewiring itself to compensate for injury and to utilize unused areas to enhance the brain's ability to understand the reality of the environment outside the body critical for safety and survival...There are noticeable structural changes in the auditory cortex of children who suffer deafness at a very early age suggesting the alteration of auditory structure for other purposes...Brain imaging studies have also shown that blind subjects can locate sounds using both the auditory cortex where sound is processed and the occipital lobe where vision is processed, evidence the brain can grow and rewire itself in a myriad of different ways to adjust to situational realities that challenge the brain's existence...A self-aware structure that recognizes it's own existence...

Many people who read science fiction in the 1950's and 1960's were convinced that 'personal robots' would be plentiful in the 21ˢᵗ century to help with all human tasks but there are very few personal robots today...The almost singular reason for this is human vision...The eye is the only part of the brain that is visible but humans don't see with their eyes, we 'see' with our brain and the volume of information and the relational aspects of the information assimilated in a single glance is stunning...Most robots can't find a ball behind a couch, something a three year old human can do highlighting the complexity of the situational realities that define our existence...Scientists can create an 'eye' that's superior to a human eye but processing visual information is an amazingly complex function...

The average adult brain weighs just under 3 pounds taking up almost 80% of the brain cavity with equal amounts of blood and cerebrospinal fluid occupying the remaining 20%...It consumes over 20% of the body's blood oxygen and more than 25% of your body's blood glucose to feed an incredible landscape created through wrinkling, deep grooves and fissures containing a hundred billion neurons connected in a hundred trillion ways that are impossible to map and understand at this point in time...

Given the enormous computational power of the human brain it's not surprising many people believe humans use just 10% of their brain's capability (a colloquialism expressed by William James and Einstein among others) barely tapping into a potential which seems obvious...But scientists have shown through experimentation, observation and new imaging technology that all areas of the brain are active at one time or another in the living of a life and most believe that humans use all of the brain they have...What may be true instead is that the human species at this early primitive era of social structure and growing awareness only maximize and develop 10% to 50% of every brain's capability at a time when humans are just beginning to understand the intricacies, the possibilities and importance of initial brain growth in every baby born...It might be that we have not gotten close to what a human brain can do if developed to its highest capacity..The promise and potential of combining enhanced neural growth in the first 1,000 days of life followed by a rich environment of cultural and educational studies to enhance intelligence and utilize as much of the brain as possible is intriguing and should be a serious focus of inquiry and understanding...The human brain is the most complex and least understood structure in this reality and it is who we are...All of us...

Extraordinary Possibilities
A glimpse into the extreme possibilities of a human brain...

Consider this capability of a human brain...There exist *savants* who have extraordinary abilities in a single area but are mostly dysfunctional in many other aspects of life that most humans take for granted...Leslie Lemke of Wisconsin can play anything on the piano after hearing it only once, complex and complicated without mistakes...Two identical twin's in the 1960's with an IQ of less than 70 were able to instantly calculate any date and the day of the week that date represented with perfect accuracy past 7,000 AD...Some savants can look at a spilled box of nails on the floor and instantly tell you the number and a savant in the 1850's named Thomas Bethune could play any song after hearing it once *and* correct the mistakes he heard when it was played to him...It's estimated he committed over 5,000 pieces of music to memory...*What is stunning about the* **savant syndrome** *is the realization that if a human brain can do such a thing, no matter what the unique circumstances necessary for this 'ability' to manifest itself, then every human brain has the capability to do these things*...Every human creature has a central control center with the ability and the potential to calculate enormous numbers instantly, to play music flawlessly with no practice at one hearing and a potential for growth and development that will make it possible to interact with reality in ways still not imagined...In a distant future when we understand the brain in ways not yet realized it may be possible to unlock and unleash this ability in every human brain without compromising 'normal' behavior and intelligence...*If these things are possible for one brain to do, no matter what the causality or circumstance, then every human brain has that potential*...We are a young species who have not come close to the operational limits of the brain or the possibilities of human intelligence, the promise of a primitive, sentient, self-aware species just now beginning to discover the true complexity and nature of their existence...

Very exciting are the biological possibilities still to be explored when science achieves the techniques, the technology and a knowledge base that will be integral and interactive with the structure and function of life in the future...If there are animal species on this planet who can grow back limbs and regenerate essential body parts (there are) these things will be possible in humans one day...All life of Earth is built on the same basic blueprint and it may be possible to grow eyes like an eagle or develop hearing with sonar location or even develop gills to breathe under the water...Humanity is so primitive at this time-point in the history of life that it's unknown what will be possible in a million years if our species grows with a consensual intelligence and an awareness of reality...

Imagine telling a small child that he/she is "alive" but we have no idea what that means...We know how something stays alive and we know the mechanisms of life within the foundation of reality but we don't know why you are 'alive' in the first place, we don't know how life starts, why cells divide and create entities that 'beat' and 'pulse', respiration and reproduction...What is the "force" of life that makes a heart beat a billion times? Nobody knows and no one has ever been able to mix chemicals and processes together to create life...Even the simplest life...

Survival in the 21st Century

The first three years of a human life does not create a serial killer, a brutal dictator, a human who kills others without remorse, a jihad terrorist beholden to a violent ideology or any other kind of human monster that disregard life...But these first years of life are critical in creating the organizational structure and neural growth that enhance or inhibit the possibilities of every individual and determine how each human rationalizes existence and their intellectual awareness of that existence...The path any human takes through life is influenced and dependent on the experiences shaping one's thoughts, ideas and beliefs after age three and without question the most influential factors in every child's life are the adults and caretakers responsible for educating, stimulating, motivating and caring for each and every child on Earth...Sadly, there's no standard in this world promising every baby a happy, safe existence with adequate nutrition, abundant sleep, secure attachment and positive stimulation and there's no standard in this world that promises children after age three will be safe, cared for and educated about the realities of existence and the possibilities of a future of their own choosing...But growing research is proving that children whose first years were secure and safe with love, nutrition and positive interaction can adjust to almost any situational circumstance in life as they mature and grow older, even extremely negative situations with intelligence, awareness and a intuitive understanding of reality that maximizes their chances of survival and success...

By contrast some of the same research and the historical, anecdotal and observational analysis of human behavior past and present strongly suggest that inhibiting and retarding neural brain growth during the first 1,000 days through poor nutrition, neglect, brutality and severe survival challenges will create a brain unable to comprehend or understand the reality of it's own existence and a human who on a conscious or subconscious level is confused by the complexities in life, unable to reconcile reality or logic, influenced by primitive emotions and frightened to a point where violence against others is perceived as necessary to guarantee their own survival or extreme violence against all life is acceptable to support an ideology which seems rational and necessary...

This is an extreme example of the possibilities and probabilities of an infant born into poverty, violence, war, neglect and abuse and in between these two extremes are billions of humans with many differing levels of cognitive awareness that shape their lives and beliefs in ways influencing everything they do, but the argument here (and the evidence is growing daily) is the possibility that every human born would understand morality, empathy and human reality if they were raised for the first three years of life in a safe environment of attachment, adequate nutrition, abundant sleep and positive stimulation by multiple caregivers who prioritize the baby's needs above all else...

Taking a child who has survived a desperate infancy and then sequestering them in a mosque, a monastery or any closed ideological system that teaches a singular subject without variation will create a human with no understanding of reality and no amount of explanation, logic or rational discourse will change that...If such a human is indoctrinated, intimidated, cajoled, convinced and bonded to any group with a shared vision or ideology, their inability to understand existence in a cognitive way can create a single reinforced vision of right and wrong where any and every action necessary to accomplish that vision is accepted...Death by suicide is acceptable to kill others and the individual is rewarded with the promise of an eternal afterlife in whatever belief of heaven exists within their mind...The indiscriminate killing of anyone who does not believe in that vision is acceptable and the domination of others to control and dictate activities to promote their vision is acceptable with a lifelong inability to understand the equality and diversity of all people or the reality of the planet that supports their existence...By contrast, taking a child who has survived a desperate infancy and placing them after age three in a secure, stable and positive environment is basically a roll of the dice...Most people who fit this scenario will live a 'life' without incident and merge into an acceptable existence without extreme behavior but there are documented incidents daily where someone has acted "normal" for many years, immersing themselves in their society in acceptable ways and then "snapped" with violence to kill their family, themselves or others out of extreme confusion and fear initiated by a crisis event of some kind...What human mind can "rationalize" killing their own small children in response to internal confusion? How can someone who has survived for decades suddenly kill their family and themselves over financial insolvency, rage over lost love or a misguided belief in gods or demons?

Humans who commit mass murder and often kill themselves have been damaged very early in life, most likely in the first year of life, with abandonment or violence while trying to survive as a baby in an confusing environment of fear and neglect with no understanding of why impeding this crucial period of dense neural growth that cannot be recovered...They have spent their entire life trying to act "normal" in a existence they've accepted but not understood, mostly unaware or unconscious of the conflict in their own mind until a life event triggers a response causing extreme confusion, anxiety, fear and anger they are unable to reconcile...This 'darkness' in people is the brain's inability to communicate with itself and understand existence and without the dense neural nets needed to rationalize logic and reason with the primitive instincts that have determined animal behavior for millions of years, the individual has little understanding of the singularity of life that is unique in every human...

There are billions of people alive today who might respond with violence and many who do respond in violent, unpredictable ways if events happen in their life that cause extreme fear, anger or confusion, people who cannot cope with a life changing event (losing a job, losing a wife, losing a child, losing a house, financial disaster, etc) unable to reconcile intellect, reason, empathy and morality with frightening overwhelming emotion...When these individuals "snap" friends and acquaintances are almost always surprised and again the search is on for meaning, reasons and solutions that will prevent another human from committing such an unpredictable act in the future...What's now being understood is the humans who commit these acts of violence don't suddenly change overnight due to predictable or unpredictable external stimuli...*Every human who kills innocent people has had this possibility, this potential confusion of existence since very early childhood and their path through life has influenced and determined the behavior exhibited at the moment they "snapped"*...

It's interesting that in many cases where a human experiences extreme internal conflict they are unable to reconcile, death is often their only solution...The death of themselves, the death of someone they feel was responsible for whatever they are experiencing or the death of their own family...There are thousands of incidents in the developed world and maybe millions worldwide where a 'normal' human experiences shame, desperation, extreme confusion and reacts with extreme violence and it's almost certain this potential and their reaction is linked to environmental factors in infancy...But when the media, researchers and "experts" look for a cause they almost always begin in early childhood, rarely in the first months or years which are a mystery in most lives, a time of growth known only to family and rarely asked about...It's increasingly clear there is no innate predisposition for extreme good or extreme evil hardwired into the human brain at birth...There is causality and sadly this kind of extreme violence happens daily in cultures everywhere on Earth...

Clearly there are billions of people who do not cognitively recognize existence and causality, people who believe in gods and demons, people who believe in an afterlife and a fated existence and many of these humans live comfortable lives within societal norms without ever reaching a point of extreme internal conflict which might trigger an unpredictable violent response...It's possible and very probable many of these people have strong internal value structures which would make violence as a solution to any problem unthinkable...In a fragmented reality of traditions, ethnicity and many various cultures with multiple value systems it's hard to know, but if this dominant species of Earth can ever learn to care for every baby born from day one, violence against our own species would cease to exist...

There are billions of people who exist between extremes with varying degrees of awareness and beliefs who strive to live a "normal" life with financial security, adequate food, water, safety for their children and acceptance by others within their social order...They try to understand the political and hierarchical structures controlling Earth's populations and resources and most would be 'happy' with adequate housing, education for their children and a fair system of governance but even this is not possible for more than four billion humans on Earth...

Human Monsters

Most frightening is the human who survives horrendous circumstances in the very first months after birth without any kind of human caretaker, with no love, with minimal touch, continuously alone and frightened with no understanding of why...The blueprint of almost every serial killer in the history of humanity...A baby whose normal neural growth was so disrupted and retarded in the first year of life that no internal communication between rational thought and primitive instinct is possible in any substantial way...These humans learn to survive and "pretend" in human society doing what they know will pass for "normal", sometimes desperately wanting to be normal but with no empathy, no morality, no recognition of reality, no feedback loop to understand what life is and no understanding of what being alive means to others...These humans can be extremely dangerous to other living creatures...The monster stories that have been part of human history for thousands of years were not all invented by the storyteller...Many human monsters have existed on this world who kill indiscriminately, murdering women because of one rejection, murdering children because they can, killing and torturing any creature just for the thrill and unable to care, unable to control the impulses that control them...*This will not happen, cannot happen if a healthy baby is cared for from birth with nutrition, nurturance, positive stimulation, safety and sleep*...But the extreme humans described above do exist...This kind of extreme neglect probably represents only a few million humans alive today and it's seems unbelievable that *ignoring a baby can determine a lifetime* but the evidence is becoming undeniable as humans begin to recognize the importance and impact of secure attachment in the earliest stages of life...

Recognizable in billions of humans alive on the planet today is the huge reliance on religious beliefs in all developing nations and many industrialized countries, a continuing skepticism about science and nature, a concern and focus on their own existence without a focused awareness of cause and effect on a global scale, a distrust of many things not traditional or familiar, a growing doubt of the possibilities of human cooperation and a hopelessness about a better future in a world that seems unfair, chaotic and confusing...

The Myth of Evil in Everyone

In many modern nations a mass killing by a lone gunman always brings out the psychologists, psychoanalysts and pundits who ask the same questions "what do we think made the gunman do this act?" and "what happens to normal humans that changes them overnight and drives them to kill without remorse?" Eventually many of these 'experts' conclude that *all humans have this capability in them*, this 'darkness' of the soul waiting to erupt anytime if the circumstances of life provoke this darkness to manifest itself...*This is an ancient common human misconception which is not supported by evidence of any kind...*If a baby is grown with love, security, sleep, nutrition and attachment they will never indiscriminately kill no matter how stressful the situation or unnatural the circumstances...A primary cause of PTSD in the military is the pressure and responsibility on humans with a conscious, a morality and a functional level of reality intelligence who are asked by their country and their fellow citizens to kill other humans without hesitation or consideration...They are told it's the 'right thing to do' and they are often ridiculed, ostracized and punished by a primitive hierarchical power structure if they refuse or question the morality of organized war and systematic killing...Humans are not innately hardwired to kill...Many studies have shown that animal species on this planet do not kill their own for malice or pleasure and their behavior in dealing with members of their own species in acts of aggression are for the most part intentional in avoiding behavior intended to cause injury or death...

The reason so many people are considered 'normal' by friends and acquaintances before they 'snapped' is that humans with an inability to understand empathy or morality are very good at pretending to fit in, pretending to be normal and often believing they are until some event or circumstance creates confusion with violence unimagined...

Many people believe that humans have always waged war and they always will but there is no evidence of an inborn "evil" that's a part of every human with the ability to manifest itself for whatever reason and determine behavior and actions...It's makes no sense that creatures who depend on one another for procreation and survival would instinctively murder their own kind...What is becoming clear is that a damaged or underdeveloped control center in any animal on Earth can manifest itself in unpredictable violence dependent on a history of growth after infancy and the circumstances initiating the confusion that made the animal feel threatened to a point of irrationality...The role of early brain development determining awareness and intelligence is not limited to humans...Dogs and cats and various animals of the world who interact well with humans to learn 'tricks' and perform with some level of understanding have all been raised from birth in an environment of love, attachment and close human interaction which maximized neural growth and the awareness necessary to communicate with caretakers...By contrast, if you isolate any animal after birth and treat it in a inhumane way, it will live a life of fear, confusion and anxiety without the ability to trust humans or other animals...In a primitive species with a written history of less than 10,000 years it's clear, despite the violent past of our species, there exists a natural morality that aligns itself with intelligence in every society and in every culture on Earth...This is the promise of focused early intervention in infancy...A planetwide intelligence coupled with a natural morality that understands human existence, human diversity and human possibilities...

There is much to learn and understand about causality, genetic influences and the environmental impacts that determine brain growth in infants but enough is now understood to make all babies a priority...

In this century, the growth of *reality intelligence* is inevitable in a connected world with the rise of new technology providing stimulating environments for more and more babies with caretakers who are beginning to understand the importance of positive stimulation and continual engagement...Even babies born in third world countries, many in desperate situations, are being exposed to more devices and activities that stimulate brain growth than could be imagined just years ago...Solar powered laptops loaded with software were air-dropped in 2012 into a remote Ethiopian village with no instructions and were being used by children and adults to learn language in less two months, an illustration of the flexibility and power of a child's brain in figuring out devices, puzzles and recognizing patterns that lead to rewards and solutions...This experiment was done by the organization 'One Laptop Per Child' who report 100 million first grade children have no access to schooling, a stunning statistic in a world that considers itself to be modern and inclusive...

Two important determinants in the development of reality intelligence is the absence of negative experiences and adequate nutrition...Baby's are curious and will find their own stimulation but the absence of violence, abuse and environmental distress coupled with essential nutrients are very critical in maximizing initial brain growth...Current conflicts and extreme poverty in many countries make this extremely challenging...

The most exciting aspect of maximizing initial brain growth and the subsequent development of reality intelligence is *it's a permanent physical property of the brain*...If a baby develops in a safe environment with positive stimulation, secure attachment, nutrition and sleep for the first three years of life *there is almost no situation or life changing event after that point in time that will keep them from understanding the nature of human existence for a lifetime*...If their parents die after this critical point of development and the child is raised in a completely different environment with ineffective caretakers or sent to a religious school to be indoctrinated in hate and violence or stranded on a desert island or challenged to survive in a fragmented, unstructured environment, these children will still grow up to understand existence...They will not believe in imaginary beings and they will intuitively understand the nature of existence no matter how fragmented their education or disruptive their life...

Researchers have found that too much stress in young children can lead to permanently low levels of serotonin and high levels of nor-adrenaline...A combination associated with aggression...

The except below, a review of a paper published in 2000, clearly shows that human research is beginning to understand the detrimental and permanent effects of an abusive and neglectful infancy on developing brain structure...What is not clear to the scientific community in 2000 (and 2016) is the severe, irreversible consequences of a desperate infancy that will determine an individual's awareness, intelligence and subsequently their behavior for the remainder of their lifetime...It only takes a few months of severe neglect at this stage of development to determine behavior through a lifetime...

Child Abuse and Neglect and the Brain—A Review
Danya Glaser
Article first published online: 13 OCT 2003
Journal of Child Psychology and Psychiatry
Volume 41, Issue 1, pages 97–116, January 2000

Developmental psychology and the study of behavior and emotion have tended to be considered in parallel to the study of neurobiological processes. This review explores the effects of child abuse and neglect on the brain excluding non-accidental injury that causes gross physical trauma to the brain. It commences with a background summary of the nature, context, and some deleterious effects of omission and commission within child maltreatment. There is no post-maltreatment syndrome, outcomes varying with many factors including nature, duration, and interpersonal context of the maltreatment as well as the nature of later intervention. There then follows a section on environmental influences on brain development demonstrating the dependence of the orderly process of neuron development on the child's environment. Onto-genesis, or the development of the self through self-determination, proceeds in the context of the nature-nurture interaction. As a prelude to reviewing the neurobiology of child abuse and neglect, the next section is concerned with bridging the mind and the brain. Here, neurobiological processes, including cellular, biochemical, and neurophysiological processes are examined alongside their behavioral, cognitive, and emotional equivalents and vice versa. Child maltreatment is a potent source of stress and the stress response is therefore discussed in some detail. Evidence is outlined for the buffering effects of a secure attachment on the stress response. The section dealing with actual effects on the brain of child abuse and neglect discusses manifestations of the stress response including dysregulation of the hypothalamic-pituitary-adrenal axis and parasympathetic and catecholamine responses. *Recent evidence about reduction in brain volume following child abuse and neglect is also outlined.* Some biochemical, functional, and structural changes in the brain that are not reflections of the stress response are observed following child maltreatment. The mechanisms bringing about these changes are less clearly understood and may well be related to early and more chronic abuse and neglect affecting the process of brain development. The behavioral and emotional concomitants of their neurobiological manifestations are discussed. *The importance of early intervention and attention to the chronicity of environmental adversity may indicate the need for permanent alternative caregivers in order to preserve the development of the most vulnerable children...*

Humans are a unique amalgamation of nature and nurture and genetic predispositions combined with the three-dimensional realities and environmental impacts influencing growth and behavior from birth...

BABIES
(An Incomplete Guide to Maximize Neural Growth in Infancy)

There are four essential areas of focus in the first three years of life crucial in maximizing brain growth...*Secure Attachment* = Love, nurturance, touching, holding and a feeling of belonging, security and safety...*Sleep* = As much as they want...*Nutrition* = Adequate nutrients and calories that are critical for a developing brain...*Positive Stimulation* = Play, positive experiences and no physical discipline...

It doesn't matter if you read the Bible or the Qur'an or Dr. Seuss to a baby if it's done in an entertaining, intimate way...No human under age two is going to retain information in an structured and organized way but tell them everything...Sing, explain black holes, constellations, traffic patterns, how to make a pizza and how to plant a tree because every external input has an impact on a baby's developing brain...Memory centers are still growing and neural bridges are still forming and every sensory input is trying to be understood and categorized by the baby's developing brain to understand it's existence and it's relationship to the new external reality a baby experiences every conscious moment...What's important is nutrition, a safe, secure environment, a secure, loving attachment to another human being and positive, interactive stimulation that will promote brain growth and determine a level of conscious awareness and a natural, instinctive morality, empathy and intelligence when the baby grows into an adult...If neural growth and organizational structure is maximized during this time the adult human will understand reality as they mature no matter what teachings or propaganda they are exposed to as children...They will grow up to understand scientific inquiry and scientific proof and they will recognize existence, rejecting the myths, superstitions, ignorance and dogma which have influenced human activity for thousands of years...

It takes dedication and every available adult during a baby's first three years of life to insure that a baby is safe, warm, nutritionally sound and positively stimulated and it's the responsibility of every human being at an age of understanding to interact with babies in positive, beneficial ways to stimulate brain growth and give infants a sense of security and love during this most critical phase of human development...The job of adult human caretakers is to show babies how to exist and thrive in a brand new world through kindness and repetition in all situational environments and this really does take multiple caretakers who understand the balance of sleep, stimulation and nurturance...Every adult who comes into contact with any small human in the first three years of life should respond and interact with the baby in a knowing, caring way to make it a positive growth experience no matter what the situation or the circumstance in our endless variety of real-life scenarios...

Earth's IQ seeks a priority standard of care for all the children of Earth during the first three years of life so they are never hungry without someone to feed them, never lonely without someone to hold them, never hurt without someone to fix them and never awake without someone aware...Every child born on Earth...The first thousand days...

Our vast transparent global civilization is a recent development on a planet that for thousands of years consisted of isolated groupings of humans in isolated environments on a very big world, the reason for many differing races, ethnicities and cultures...In very large cultures (Britain, Japan, Egypt, China, etc) and small isolated cultures a lack of knowledge, adherence to religion, ancient traditional practices, cultural norms, ignorance and unique tribal traditions have determined how babies were nurtured, loved, fed, educated and disciplined during the first years of life...Different cultures with very different beliefs...Only in the past 150 years has real progress been made in understanding the nuances and impact of this early stage of life with an increased focus on the needs of the child as opposed to the needs of the adults caring for the child...In cultures worldwide children were (and still are) viewed as mostly useless until they reach an age of understanding and mobility when they can help the family in some constructive way and many cultures believed children learned very little until school if they went to school at all while strict, inflexible religious practices maintained (and many still do) that corporal punishment was necessary to discipline the child, control behavior, earn respect and fear authority...

There is no 'cookie-cutter' formula...It can be a million little things done in a million little ways as long as the baby has a sense of secure attachment and safety with abundant sleep and nutrition...

In 20th century America psychologists and doctors suggested too much touching and comforting of a baby was not "ideal" and it was good to let babies cry themselves to sleep and be left alone in their own rooms from a very early age to foster independence and self-reliability...These suggestions might work for children above the age of three when these small humans know their boundaries, can easily communicate needs and fears and are aware of the expectations of human caretakers but leaving a infant alone and crying in a world they don't understand is cruel and restricts neuron growth and the development of organizational structure critical during the first 1,000 days of life...*Close continuous contact, a sense of safety and security, no physical discipline and a warm familiar human who always responds to a baby's cry to determine its needs is the formula to maximize early brain growth*...When a baby is engaged in an entertaining activity (no matter how simple and repetitive), when a baby is happy and laughing and when a baby is sleeping, these are all situational realities that enhance brain growth especially in the first and second year establishing a framework of communication and awareness within the brain that will influence behavior and intelligence for a lifetime...

Many religions advocate physical discipline as a way of controlling behavior citing questionable religious text to support a primitive ignorance that's been harmful and detrimental to millions of humans over a thousand year history...It is possible to modify behavior through violence and fear and the number of societies and cultures who still use fear to "control" a small human who is naturally curious and inquisitive about everything in their environment is another clear indicator of the primitive nature of the human species... **Violence should never be used on children for any reason**...*Teach them how to exist in human reality through repetition, diversion, patience, rewards and love...It's very challenging to interact with infants and children this way, especially in a consistent way over three years (and more) in the differing situational realities that represent human existence but it is the formula to maximize initial brain growth and develop a framework of empathy, morality, intelligence, trust and awareness...On a future Earth one day, all infants will have multiple caregivers if this species survives to understand this absolute necessity...*

A study released in 2013 that examined more than 1,300 moms who breast fed their babies found the babies grew up to score higher on intelligence tests than a control group who were fed formula from bottles...The study found that each added month of breast-feeding indicated a 0.3-point increase in intelligence by age 3 and a 0.5-point increase by age 7...Researchers adjusted outside factors in the study that tend to lead to intelligence including parental income, employment, and education and concluded the benefit was greatest when babies were breast-fed "exclusively for the first six months." The irony is this: The measured increase in intelligence had less to do with the quality of milk and much more to do with the interaction between the child and the mother in providing close continual contact and a feeling of secure attachment in the baby stimulating brain growth and creating dense neural bridges and organizational structure critical to the development of intelligence and awareness...

Consistent Care for 1000 Days

Never let a infant cry itself to sleep...There are adults who believe if you let a baby cry itself to sleep it will teach the baby how to be tough, individual and help them survive the rigors of life...*There is no evidence to support this*...A baby should never be left to cry alone without interaction and comfort during the first 1,000 days of life...Crying is the primary way an infant communicates with their world, communication that requires a response by an adult caretaker...It's very different once the child is older than 3 and is beginning to understand the nuance of cause and effect, parental rules, guidelines and has communication skills to express emotion, curiosity and need...During the first one thousand days of existence crying babies should be held, rocked, fed, changed, whatever the communication demands and taken care of until their needs are met and they fall asleep...There is a balance here that is adjusted as the baby grows older...Most caregivers can tell when a baby is 'fussing' and 'vocalizing' as they fall asleep as opposed to a stressful cry that is recognizable and requires a response...

Limit the use of 'baby talk'...Adults should talk to infants distinctively and in a normal way much of the time...Baby is trying to understand and acclimate to your language and your conversations with other children and adults so talk to them like they are tiny adults...It's amazing to watch these small humans pick up words and acquire language in the second year of life unable to enunciate clearly and understanding many more words than they can speak...Talk to them always while pointing out objects and situations...Again, there is a balance in this and it's not too hard to find...

National Sleep Foundation = Newborns should sleep 12 to 18 hours out of every 24 hours...12 to 14 hours for toddlers age 1 to 3...11 to 13 hours for preschool age 3 to 5...

Allow a baby to sleep as much as they want...Abundant sleep is essential in growing neurons in a developing brain and for very young babies this can be from 12 to 15 hours a day...Safe, warm, quiet sleep with an awareness of when they wake up (they will let you know in a familiar vocal way)...

"Baby proof" the environment as much as possible to safeguard this exploratory, curious human from interacting with something they should not...It's impossible to protect a baby from all situations so be vigilant when baby is awake and on the move but *never discipline a baby for exceeding boundaries or exploring anything within reach*...Babies can be taught avoidance behavior through the use of pain but these small humans do not understand the reasoning behind acts of violence on them, even if they change their behavior in response to it...Teach them limits by repetition or diversion, *never violence*...A simple word "no" (not shouted, but spoken in the same manner every time) while stopping the unwanted behavior is very effective over time if the adult is consistent and persistent in this form of behavior modification...It may take many repetitions to modify behavior this way but we are here to 'teach' babies the limits, boundaries and habits of human behavior, not to threaten or injure them...

Even when they grow older and can understand the reasoning of punishment and boundaries, violence and pain are never an acceptable way to modify behavior...Time-outs are very effective on older children if they are done in a consistent and predictable way even if you have to sit with them during the punishment...And always explain to a child why this is happening...As they grow older explain to them the structure of human society (even if you have to do it many times) a paradigm that has grown and changed throughout history in multiple cultures...Explain to them that childhood is less than two decades and in most structured human civilizations the parents are required by consensus to shelter, feed, protect and teach the children in their charge until each child reaches an agreed upon age of independence when they are free to determine their own behavior within the stated laws of that civilization...But until that time the adults are required by the laws of the society to care for their children in all situations and most adults want their children to be interactive as they grow older with many decisions flowing both ways (this works well on teenagers as they realize they are close to freedom)...

There are many things a 3 or 4 year old will understand that babies do not...Never sneak up behind a baby to startle this little human with a popped balloon or a loud noise...A four year old will laugh and understand this surprise that somehow makes the adults laugh but a baby will be frightened, alarmed and not understand...Babies don't understand if you push their face into a cake while you videotape and they don't understand if you force them to do anything that is frightening, strange or uncomfortable...*The first 1,000 days of a baby's life is all about security, trust and a predictability of human behavior* as the baby deals with unknown stimuli in a new, growing environment and interacts with the complexity of the human activity that swirls around them like a storm...Recognizable and predictable patterns of human activity in the continuous, complex interactions taking place every time the baby is awake creates a sense of security, anticipation, understanding and belonging that allows maximum neural growth in an environment of comfort and love...

Positive stimulation (anything that makes a baby smile and interact with humans) can be very simple and is almost always many times repetitive...Make things challenging and fun, never stressful and never forced, and you can return to activities over and over letting the baby explore the world with interest and curiosity...Always talk to your baby when doing activities and tell them everything, even while knowing they can understand very little of what you say...A human voice trying to communicate in a normal way is an enormously positive stimulus for brain growth as the baby tries to assimilate and absorb everything in their environment...Stimulate with toys and games, videos and play, complex workstations and everyday activities that are all equally stimulating and fascinating to a baby...When you bake a cake, pull up the high chair and explain everything you do from measurements to mixing to tasting and cooking...Explain the stars, trees, cars, dirt, water, animals, airplanes, politics, anything and everything...Show them the moon and stars at night (from about 20 months on babies are fascinated with the night sky and we wonder what they must be thinking)...Make it fun, always fun, whatever it is you choose to do because every activity that is positive and makes an infant smile or respond positively is what's needed during this explosion of growth in the brain...Then let them sleep, one of the most important times for neural growth and development and when they awake, do it all over again...

In the journal PloS One, researchers from the University of Rochester published a study of attention patterns in infants age 7 & 8 months and concluded that infants are much more actively engaged in seeking out information from their world than previously thought...Infants do not need fancy or expensive toys and devices to learn, just a positive stimulating environment of whatever is available for them to play with and explore...

I spent a month playing a game with a nine month old baby every time we were together that was simply me putting a cell phone like rubber device that was on a string into a holder attached to a play center he sat in the middle of and him pulling it out of the holder...Over and over again...But every time he flipped it out he looked not at the device but instead at me, with a smile on his face waiting for a human reaction which involved me putting it back in the holder and him flipping it out again...Over and over...This simple activity challenged his spatial coordination, muscle control, his communication skills with another human, his reaction, his concentration and his desire to interact, a simple positive play activity that enhanced and accelerated neuron growth that is continuing nonstop at this age...After a month he was focusing on other activities (walking) and the play center was gone...No activity is wasted, no matter how simplistic if it entertains and challenges a baby in a happy, fun way...

The Realities of a Baby's World
The interactions described above imply that an adult has no other responsibilities in life other than entertaining an infant and that is never the case...It's difficult to make a baby the focal point of your attention in all activities and especially if there are other children...This is why multiple adults and caregivers are essential in the first 1,000 days of human development...*In all cases the formula for intelligence is attachment, nutrition, security, positive stimulation and sleep...*In many developing countries just the work of daily survival demands constant vigilance and unending chores but if daily activities are a part of routine and repetition, farm-work, agriculture, textile, or simply gathering wood and water, a baby can adapt and thrive as long as they feel safe and secure in an environment that is recognizable and familiar...In many families the older children spend a great deal of time helping care for the baby and this can be excellent stimulation for a growing brain if the caretaker child understands the formula and engages the baby in a healthy, happy way...Even babies in very poor situations will do well just interacting with whatever is part of their environment if they receive nutrition, sleep and security...Unfortunately even this is impossible in many world situations and it will require the humans who can help, large religious organizations without conditions of belief and many non-governmental organizations with money and the resources of corporations and governments to provide for what the mother cannot...Perhaps the biggest influence on a baby's brain development and biological health is their environment and the ongoing conflicts and wars in many African nations and other parts of the world are very disruptive to developing brains who do not understand loud noises, fear and stress in adults and children, erratic feeding times, even no food at all and a continuous chaos that can retard the growth of the neural nets necessary for every human to understand the reality of their existence...

It's estimated that *every year* over 230 million children under age five do not reach a level of cognitive and social development needed to understand the complexities of life and the skills necessary to thrive and prosper in adulthood...Many of these babies suffer deficiencies in essential nutrients that are needed in growing brain structure and many babies are susceptible to disease, infection and injury in these forgotten corners of the world...Children suffer as resources are diverted to the military's of the world for fighting at the expense of food, shelter and health facilities that are desperately needed by the populations...In the poorest parts of the world one in five children don't survive until age 5, more than 12 million a year with malaria killing more than 1.5 million children every year...Against such odds it's difficult for a baby to even survive and with almost no chance of growing the neural nets to understand reality these babies will be the adults in this region of the world one day and will do the same thing the adults are doing in 2016, continuously driven by primitive human instincts of fear, survival and greed and trying to survive in a reality they cannot understand with access to almost no resources in a world where nearly four billion humans have access to just 1% of the planet's wealth...Changing this reality for forgotten, abused children and women worldwide will take hundreds of years and a focused effort by all humans on Earth who do understand a reality ignored by billions but there is no way to know in this century if a reorganization of Earth's resources and priorities will happen in time...

History of Childcare

In Europe during the second millennium children were secondary, routinely abused, ignored and the death rate was unimaginable...Crawling was discouraged in some cultures because "man was made by God to walk on two legs" not to crawl like animals along the ground...Only in 1986 were canes and straps outlawed in British schools and it is still legal for British parents to beat their children "as much as necessary if they do not leave marks"...Mothers for most of Europe's history did not feel it necessary to hold or love their children believing babies were essentially useless until they reached an age where they could work and help with labor...A widespread ignorance of reality, cause, effect and consequence that is still prevalent throughout much of the world in the 21st century...

In many societies the importance of continuous positive interaction with babies in safe, secure, and playful ways is simply not known or practiced...Many humans raise their children the same way they were raised and many humans believe babies do not really begin brain growth until they reach school age and can learn from instructors...In ancient times, in the great Roman and Greek cultures, babies were routinely left to die of exposure or killed if the male head of household did not want the child...Surviving children were often nursed by a slave or wet nurse...An ancient Chinese text suggests *"as soon as a baby can recognize facial expressions and understand approval and disapproval, training should begin so that he will do what he is told to do and stop when so ordered...After a few years of this, punishment with the bamboo can be minimized as parental strictness and dignity mingled with parental love will lead the boys and girls to a feeling of respect and caution and give rise to filial piety'*... Dr Spock in his popular book *The Common Book of Baby and Child Care* actually had to tell parents that picking up a baby when they cry will not spoil them because in 1950's America the advice from 'experts' was to feed babies on a schedule and don't touch them too much (you cannot touch a baby too much) and in medieval times throughout European culture babies were frequently *swaddled*, wrapped in cloth with their arms and legs bound to their bodies to help them grow straight and often left in cradles (with ties) to be ignored until they grew older...Even in this century in America, a nation who believe they lead the world, parents do not hold their babies as much as other cultures and many babies are isolated and left to sleep alone in their own rooms in the mistaken belief that this will help them develop independence and an identity as individuals...

Although many cultures interact with their babies more often than Americans there is a global ignorance of how absolutely essential it is from birth to form a bonded relationship with an infant...The first year is more important than the second year and the second year is more important than the third year...After a thousand days these small humans understand some expectations (they are actually very perceptive) understanding humor, nuance, even satire and irony and continue to expand their awareness in a variety of amazing ways but human babies are among the most helpless offspring of all mammals and require constant vigilance...Sweden was the first European country to outlaw corporal punishment (1979) with 31 nations having laws today...Corporal punishment is still legal in America...

Maximizing initial brain growth takes no special actions or special activities...It simply takes love, attachment, nutrition, sleep and the absence of negative experiences...No violence, no abuse and a sense of safety...3 years...

'Include' Your Baby...

What babies and small humans want most of all is to be "included" to do the things that big humans do...When something is going on (big truck fixing a transformer at night with crane buckets, lights, someone coming over, you coming home) small humans are intensely curious about everything in their world so include them in everything possible...At night when I visit a three year old and have 'stuff' to do outside before going in to visit (unloading the car, recyclables, anything), the 1st thing I do is let him come outside with me to hang around while I do stuff...He's included, growing neurons and learning how to be a human in a world full of giants...It's important to small children to be included in the activities humans do every day no matter how routine and that's how they learn the world...It's always easier to leave children inside so they are not 'underfoot' while adults do the things that need to be done but include them whenever you can, challenging but essential to maximize neural growth in the early years of life...When children run around as we do stuff sometimes they can help and most times not but they're included, a part of the activities and it means everything to a developing brain...

The learning objective with all small infants(0-3) is to recognize when they are reaching certain levels of skill and integration into the web of life where they will grow to adults...Clinton at age 2 yrs 10 months is understanding how to pedal a bike...How long would it take for him to know that skill if someone bound his ankles? (unthinkable) At age 3 he has wit, insight and nuance, understanding satire in simple ways and the many aspects of 'wordplay' and their relationship to the reality of the world that swirls around him...Geographical recognition of a 10 mile radius surrounding his existence where he could direct someone to his house, the center of his awareness...He knows Venus, Jupiter, the Moon, Mars, the Sun and that Earth is a planet where everyone lives...Recognizing and reinforcing these thousands of learning points in the first thousand days is crucial in the development of initial brain growth and it takes a village and multiple caretakers, everyone who engages with small humans to make it always positive...

OBEDIANCE IS NOT INTELLIGENCE...Children need freedom to inspect and explore their world and freedom to move in their space as children are meant to do and for their caretakers it takes patience, effort and intelligence to continually balance and adjust this equation...They do not walk in lockstep with the parent afraid of physical discipline and this will be a challenge to many parents...Intelligence is wit, guile, satire, irony, insight, discovery, curiosity and exploration...Continuous interaction with small humans is challenging but children raised with love, nurturance and patience trust their caretakers deeply and communicate with insight, learning through logic, patience, rewards and disappointment, bouncing off the barriers of their existence but exploring everything within those barriers...They go to the edge and that's how they learn and how they know their world...

The energy in a child is infectious and a normal part of development...Children cannot not move...

Spoil your baby (the first three years) with positive reactions to almost everything they do, say 'yes' a lot and redirect activity and complement them for being smart and aware of their world...And continue to explain the world to your baby...Slowly take a leaf apart explaining why it's green, how the sunlight makes it grow, where it attaches to the tree, explain a rock, why the Sun goes down over there and comes up over there, signs on the highway, trucks, cars, helicopters, jets, little planes, food, faucets, people, etc...Never demand that they understand what you tell them...Just talk to them and months or weeks later they will ask you something that makes you realize they've been thinking and assimilating what you explained to them weeks before...The most important thing is secure attachment and *no negative experiences*...Don't hit them, don't yell at them and don't scare them...

The following suggestions from an internet site *WEB MD* are simple, positive, engaging ways to interact and communicate with a small growing human who is trying to assimilate everything in a complex existence that challenges a baby's brain to adapt and grow to understand it's reality...In the first thousand days it is much more about stimulation than assimilation although a great deal of information in the first years is retained and understood as memory centers grow and storage and retrieval areas are developed...Necessary adaptations in order to survive and thrive in this new environment called life...

A Growing Awareness
From the internet site Web MD
In a world that is finally starting to make a connection between babyhood and the rest of life, experts conclude that talking to your baby, playing with your baby, paying attention to what interests your baby and using those interests to foster curiosity lays down the wiring that ultimately stimulates your baby's brain to grow and develop...

Age: Birth to 4 months
Read; make silly faces; tickle the body; slowly move objects in front of your baby's eyes, like a brightly colored rattle; sing simple songs and nursery rhymes with repetitive phrases; narrate everything you and your baby will do, such as "We are going in the car now; we are putting you in the car seat; Mommy is getting into the car."

Age: 4 to 6 months
Help baby hug stuffed animals; stack things (like plastic blocks) and let your baby knock them down (babies love to knock things down); play music with different rhythms; show your baby books with brightly colored pictures; let your baby feel objects with different textures

Age: 6 to 18 months
Talk and interact face-to-face to increase connections between sounds and words; point to familiar people and objects and repeat names; sing songs with repetitive verses and hand motions; play hide and seek.

Age 18 to 24 months
*Play simple recognition games like "spot the yellow car" or " the red flower," or put three objects in front of your child and say "Give me the ..."; **talk directly to your baby as much as possible**; introduce your child to writing tools such as crayons and paper; ask "where and what" when reading to your child; encourage some independent play with favorite toys.*

Age: 24 to 36 months
Lavish your child with praise and encouragement as he or she perfects motor skills; bolster your child's imagination by encouraging new ways to use toys and help your child incorporate 'real life' activities into play...

Humans Caretakers need to use every skill they have to handle these small humans whose brains are going a hundred miles an hour because they're trying to learn everything and they don't know very much about anything...Bedtime, baths, everything is challenging, sometimes very challenging and it takes caretakers with intelligence and flexibility to get these little humans to sleep while building a very close relationship with the child...It's not easy, but it's essential...

A newborn baby can recognize their mothers face just a few hours after birth...

Babies have more brain cells at age 2 than at any other time in life...

There should be pre-school for every child in the world to engage in real-life situations and social interactions...

Finally, in study after study researchers have found an amazing curiosity and flexibility in babies just months old that develops more and more as they grow older...Many suggest challenging these developing brains by not showing them how everything is connected and works but rather introducing a behavior or item to the child showing your interest and curiosity and then letting them explore it on their own...These insightful, curious creatures are amazing when trying to understand their world and there is a balance of interaction that's adjusted as they grow...

Future thinking must prevent wars...The stakes are immense, the task colossal, the time is short, but we may hope, we must hope, that man's own creation, man's own genius, will not destroy him....
Albert Einstein

THE THIRD MILLENNIUM

THE SOLAR SYSTEM

December 1968, the first manned spacecraft escapes the gravitational field of Earth...This is the first time after 3 billion years of evolution that any creature has left this planet, circled our moon and shown our species the reality and fragility of humanity's one world...

In 2006 NASA's Stardust spacecraft returned a sample from comet Wild 2 containing the simplest amino acid, glycine, a compound necessary for life and more evidence the building blocks for life are everywhere in this universe...It is still unknown how life came to be on Earth...

July 1965, the Mariner 4 flies within 6,000 miles of Mars...The first closeup look of what will unquestionably be the most important extra planet in a human future lasting millions of years...

July 1969, the first humans land on the moon and leave a footprint, a used spaceship and a flag...

In a single star system over a light year in diameter, home is a 4.5 billion year old rock covered with life and defined by an incredible variety of plant and animal species, stored energy reserves, large dynamic oceans, violent, unpredictable weather, changing climates, magnetic and atmospheric filtering of deadly radiation from space and a known predictability of motion and behavior that will continue for a billion years...After 200,000 years of survival in a changing environment of exploration, discovery and confusion, humans have finally mapped their world, photographed it from space and launched multiple missions to explore local space and verify the reality of a physical existence unknown just a century ago...We now understand where we are in a universe that is unimaginably big and we now know what we are, the only self-aware intelligence in a solar system that is ours to do with what we wish, a million year focus on exploration, discovery, resource recovery and off-world habitation in a known reality where a predictability of motion and interplay among every object we can observe will make it possible to explore any planet, moon, asteroid or comet for resources needed by Earth...

Mercury is survivable with advanced technology if we land in the perfect spot on a world that circles the Sun in less than three months with a single day lasting almost two months...Venus is nearly identical to Earth in size rotating slowly clockwise (the only planet to spin this way) and sweeping around the Sun (225 Earth days) in less than a Venus day (243 Earth days), a planet whose year is shorter than it's day...A million years from now science will figure out a way to do 'something' with the Venusian atmosphere (primarily CO_2 with almost no oxygen or water vapor) and create another world that will support human existence...Mars is the jewel of the terrestrial planets with an atmosphere of mostly CO_2 that is thinner than Earth's, a history of liquid water and a gravitational force only 38% of the Earth...It will be the first planet our species visits and it's still unknown what might be possible with the explosion of technology in the 21st century...The good news is we've reached this point of awareness just halfway through the life cycle of the Sun and have a billion years of continuous energy output from a G class star to create a civilization that will survive the death of this planet and continue to exist in a vastly changed planetary hierarchy or in other star systems...

Imagine if after Copernicus, Galileo, Kepler, Newton and Einstein had revealed the physical reality of our existence, it was determined the Sun wasn't 4.5 billion years old but 6 or 7 billion years old and the human species had just a few million years to understand their existence, visit the planets, explore the moons and try to find a way to survive the cataclysmic changes of an old dying star...How sad it would be to realize it took the entire main sequence of the Sun for sentient intelligence to arise on Earth and what a desperate race against time to establish a foothold in some other place in the solar system with only the smallest hope of surviving the Sun's unavoidable fate...Our star is middle-aged, only halfway through the main sequence phase and humans still have a billion years to grow and play, to explore and discover, to love and wonder, to question our reality and to expand intellectually while seeking answers to the real mysteries of human existence...

New research modeling the future biosphere of Earth suggest this planet will be uninhabitable for large mammals in about a billion years due to a gradual warming of the Sun with all microbial life disappearing in about three billion years...The Sun will not "die" when it runs out of hydrogen...It will change, eventually becoming a white dwarf that will continue to generate a decreasing amount of heat and energy for billions, maybe trillions of years...

In the 21st century the path ahead is unknowable...The structure of this Universe will be studied in every way in the decades and centuries ahead focusing on identification of energy sources, planetary structure, galactic evolution and detailing with precise maps the location and movement of all things humans can observe and measure...There will be a growing anticipation and priority on finding life some other place than Earth, especially evidence of intelligent life and it's seems almost inconceivable that our species won't detect and possibly interpret an intelligent communication, some kind of signal or recognizable transmission in the next 10 million years (it could be tomorrow)...It's exciting...Science is the language of human existence and the search for a deeper understanding of *everything* is who and what we are, what we will always be...Inquisitive and intensely curious, always seeking, wondering, looking, testing, probing, measuring, questioning, experimenting and finally understanding...

The Solar System makes one complete revolution around the center of the galaxy every 227 million years...

Galaxy = 100 Billion stars = 100 Billion Planets = 6 million liquid water habitable...

Earth

Earth = 32% iron, 30% oxygen, 15% silicon, 14% magnesium, 3% sulfur, 2% nickel and other trace elements...

Everything science defines as alive exists on an isolated sphere of molten rock that is analogous to a cracked eggshell, a single world with millions of plant and animal species, trillions of insects and uncountable microbes that cover this planet like a coat of paint...We are just beginning to search for life elsewhere and there are intriguing possibilities with liquid oceans existing under ice on several moons and technology that makes this kind of exploration possible but on Earth, *life is everywhere*...Humans interact with millions of tiny creatures every day, microscopic life that exists on the highest mountains and in the deepest oceans, billions of creatures that live on us, live inside of us and are essentially the reason we exist (many of these creatures would be the stuff of nightmares if enlarged)...The majority of this microbial life is harmless while many are essential for our continued existence, carbon-based life with RNA or DNA, a connective thread linking all life on this world with patterns of organization that may be common in this universe...At this time in human history we have no comparison...

The outer surface of Earth (crust) is about three miles thick at its thinnest point (ocean floor), about 43 miles at its thickest point (mountains) and is the layer of Earth shaped by weather, oceans, rivers, internal processes and random astronomical events that have continued for millions of years...A cracked eggshell with 17 major tectonic plates moving on top of an 'ocean' of mantle, a layer 2,900 kilometers thick of mostly rock that circulates in giant patterns on geological timescales of millions of years, 80% of Earth's volume...Inside this mantle is an outer core about the size of Mars, a sea of liquid metal that is primarily iron (with trace metals) and similar to a lava lamp with convection currents rising, sinking and circulating 2,000 miles beneath the surface, still active 4.5 billion years after the Earth's creation...The inner core within the outer core is half the size of the Moon, as hot as the Sun's surface (6,000 C) mostly iron and essentially solid because of intense pressure...It's now believed there may be an innermost core less than half the size of the inner core described as a large crystal of nearly pure iron...Movement and dynamics beneath the crust are mostly unnoticed by the creatures who exist on the surface until volcanic activity threaten populations or pressures deep inside the planet move the surface in measurable, destructive ways (earthquakes) but the interplay of liquid metal and rock has been critical in creating an environment for life and essential in creating magnetic fields that surround this planet and make it possible for life to exist in a hostile radioactive environment...In a future with new technology the abundant energy in Earth's interior will be harnessed for power with geothermal power facilities that emit less than 1% of the CO2 released by fossil fuel plants and around 3% of the sulfur compounds (a component of acid rain) released when fossil fuels are burned...Geothermal power plants that use less land than conventional power plants, operate 24/7 with little environmental damage and represent less than 1% of the world's energy generation in the 21st century...

The devastating earthquakes, subsequent tsunamis and volcanic eruptions are mostly the result of crustal movement over time leading to a buildup of pressure resulting in slippage, often measuring just a few feet...There are over a million earthquakes every year on this planet with approximately 5,000 to 6,000 over 4.0 magnitude...

Earth's water is older than the Sun...

Earth is a large rock that's tilted 23.5 degrees spinning at a rate that is slowing down by one or two thousandths of a second every century and circling a common star at an average distance of 92.6 million miles (149,597,887.5 km)...A star formed from the same gas cloud that created everything in this solar system and a stable, continuous energy source essential to life...Less than two million species on Earth have been identified but a recent study estimates there are over 6.5 million species living on land and over 2.2 million species living in the ocean (*PloS Biology*)...Researchers are still discovering more than 15,000 new species every year with incredible diversity, mostly microbial interconnected with all life on Earth, unknown species and research that promise discoveries increasingly important in the development of medicine that could contribute to human existence in still unknown ways...Nearly nine million species on one planet and just one understands the existence of life and the possibilities of a connective intelligence in a single star system...Intended or not, humans will determine the future of life on this world for as long as they exist, an equation with far too many variables to predict...In any of the more than 6,000 languages on this planet, Earth is a unique world...

地球, Země, verden, aarde, maa, terre, Erde, γη, Föld, jörðin, bumi, terra, zemeslode, Žemė, jordkloden, ziemia, pământ, Земля, Zem, zemlja, tierra, Earth...

The Integrated Ocean Drilling Program (IODP), formed in 2003 as an international marine research consortium announced in 2012 it would invest over $1 billion to attempt to drill through the Earth's crust to reach the mantle...Drilling is expected to focus on locations in the Pacific Ocean where the crust is at it's thinnest (about 6 kilometers or 3.7 miles) and it will take up to ten years to reach the mantle...The deepest hole ever drilled in the crust is the Kola Superdeep Borehole on the Kola peninsula of Russia at 12.262 kilometers (7.619 mi)...

A new study with current data from NASA's Kepler Spacecraft using simulations of cloud behavior on alien planets estimates there may be sixty billion planets orbiting in the habitable zone of red dwarf stars....And that is just in the Milky Way...This information doubles the previous estimate of potentially habitable planets orbiting red dwarfs believed to be the most common type of star in the universe...Astrophysical Journal Letters 2013...

The Milky Way contains an estimated 75 billion red dwarf star systems...

2,325 confirmed exoplanets orbiting over 1,000 stars with around 30 existing in the "habitable" zone and over 100 are roughly the size of Earth...Over 4,500 exoplanets waiting confirmation...Scientists estimate the Milky Way galaxy could contain more than 17 billion Earth size exoplanets..Many believe almost every star has a planet...2016

"The science returns of the Kepler mission have been staggering and have changed our view of the universe in that we now think there are planets just about everywhere," Scott Hubbard of Stanford University

Gemini Planet Imager (GPI) is the first fully optimized planet imager designed for exoplanet imaging and deployed on one of the world's biggest telescopes, the 8-meter Gemini South telescope in Chile...The telescope uses the most advanced AO system (adaptive optics) in the world sensing atmospheric turbulence and correcting it with a 2-centimeter-square deformable mirror with 4,000 actuators...The new mirror corrects for atmospheric distortions by adjusting its shape 1,000 times per second with an accuracy better than 1 nanometer...GPI can directly image extra-solar planets 1 million to 10 million times fainter than their host stars...www.llnl.gov

Extraterrestrial Life

The human brain is an enormously complex object with trillions of connections and the ability to recognize itself and understand it's own existence...In a universe less than 14 billion years old the human brain took almost 3.5 billion years (more than 25% of the life of the universe) to develop to a point of sentience with the ability and awareness to define reality...In a universe with the chemicals of life everywhere and trillions of planets, it's almost certain that life exists in many, many places but self-aware intelligence may be very rare...It took extraordinary circumstances (a stable orbit in the "habitable" zone with a Jupiter to protect us from potential impacts) and extraordinary events (the demise of the dinosaurs) for our species to get to this point of awareness...There is no certainty, even in a reality this big, that anything like the human brain exists anywhere else...

Just thirty years ago when Carl Sagan was filming his episodic series *Cosmos* no one had any idea how many planets might exist around other stars...Many speculated from the SETI search and the absence of any detectable alien signals that very few stars had planets or the galaxy would be alive with radio signals...Others believed there were many planets but only in certain conditions that allowed their formation...Few believed there were *billions of planets* just in this galaxy...What's fascinating is what this suggests about the development of intelligent life in a universe where the chemicals for life are everywhere...It seems almost impossible that there is only one intelligence in this galaxy but the development of self-aware intelligence must be relatively rare and require extraordinary circumstances or with billions of planets the sky would be alive with indicators of alien civilizations...

When Sagan worked through the Drake Equation for estimating intelligent civilizations existing in the Milky Way he was amazingly close to the number of planets on which life might develop (100 billion) but he predicted the possible existence of just 10 surviving civilizations due to 'technological adolescence', a doomsday scenario predicated on the proliferation of weapons of mass destruction, the ignorant nonsensical posturing of superpower nations and the lack of a consensual intelligence among differing species concerning what is possible and what is needed for survival far into the future...

We have nothing but the reality of our own existence to reference when trying to understand what might be possible and in a galaxy 13.7 billion years old we have no idea why the sky is not full of intelligent transmissions from other life...When you understand reality, the historical record of the past, the stability and predictability of the physical processes that create the environment we exist in and the ingenuity, flexibility and creativity of the human brain, it seems impossible that our species would not exist for millions of years...How does the progress of science and the rise of technology become the catalyst for a species destruction? And if a billion humans can understand human reality why not seven billion? How can an intelligent species collapse under the weight of their own achievements and cease to exist? To those of us who understand the situational reality of Earth's population in this century, it's clear our species must maximize brain growth from birth without discrimination and with millions of children under age five suffering from negligence and malnutrition, it's not happening...Despite many technological achievements in the last century, humans are not yet a species reaching for the stars...

Just 37 years ago (1977) humans mount a camera (a device perfected only in the 19th century) on two spacecraft, Voyager 1 and Voyager 2, to fly through and photograph their star system for the first time...

Voyager 1 is on course to approach Gliese 445 passing within two light-years of the star after a 40,000 year journey...Gliese 445 is rapidly approaching our Sun and will be just 3.5 light years from Earth when Voyager 1 passes by...The spacecraft is 11.7 billion miles from Earth and moving at 38,000mph...October 2013

Voyager 2 is 9.5 billion miles from Earth and should pass within 25 trillion miles of Sirius (the brightest star in Earth's sky) in about 300,000 years...October 2013

Japan Aerospace Exploration Agency (JAXA) is developing an electrodynamic tether designed to generate electricity that will slow down space-based debris and allow space junk to fall into lower and lower orbits until it burns up harmlessly in Earth's atmosphere...A test of the system in 2014 to extend a 300-meter (900-foot) tether in orbit and observe the transfer of electricity was reported to be successful...Deployment may be possible in 2019...

Space Junk

The Joint Functional Component Command for Space tracks over 23,000 pieces of space junk moving at 17,500 mph in a whirling dance of metal that is a serious threat to any object getting in the way...The organizational structures governing policies dealing with this problem are mostly confusing unilateral policies where nations 'own' their space junk (America has approximately 500 large pieces that are officially junk) with very specific restrictions on affecting or disturbing another nation's space debris but several nations including Switzerland, United States and Japan are initiating programs to build automated ships to change the orbit of multiple targets allowing debris to burn up in the atmosphere...Recently there have been international talks focused on responsibility and technology that would require every item launched into orbit to have an option for disposal when the mission ends...In 2016 this problem is ongoing with little cooperation, little communication and with "fingers crossed"...

NASA officials estimate there are more than 500,000 pieces of debris the size of a marble or larger orbiting Earth with 20,000 pieces of debris larger than a softball along with millions of pieces too small to track...NASA's Orbital Debris Program reports the area of space between 900 and 1000 kilometers from the planet's surface has reached "super critical" debris density levels estimating a piece of space junk larger than 1 cm in diameter will strike a satellite in Low Earth Orbit (LEO) every five or six years...

Real Places

Imagine a human landing on IO in 5,000 years...The closest moon to Jupiter with 160+ active volcanoes maybe the most inhospitable moon in this star system, the closest thing to mythical Hell and a place to be avoided, IO could be a tourist destination in maybe five thousand years...36 years ago humans mounted two cameras on two spaceships and explored the outer planets of the solar system and the many moons that circle these worlds...These spacecraft found diversity, volcanoes, geysers, storms, rings and unknowns never imagined, worlds with predictable physical phenomenon and moons with resources, incredible landscapes, tenuous atmospheres and landing areas easily accessible with current technology...Over 170 moons and millions of asteroids waiting to be explored by a species without any idea of how many resources may be recoverable, resources that could supply important minerals, rare metals and energy to the populations of Earth for millions of years...There are only three other planets in this star system humans can land on and explore: Mercury, small and hot with a very slow rotation that would make it possible to visit the edge and survive: Venus, a destination when humans can build technology to survive on a world with temperatures close to 1,000 degrees in a pressurized atmosphere similar to the bottom of Earth's oceans (One of the biggest challenges in a thousand or a million years will be finding a way engineer a 'different' Venus and make it survivable and habitable): Mars, without question the first planet humans will visit, very survivable with plans already being made for transport, habitat, recovery and return and if this civilization remains relatively stable and progressive, humans will visit Mars within the next fifty years and could have a permanent base and a population on Mars in less than 300 years, a planet that will be home to millions of humans one day...

The first permanent presence off Earth will likely be the Moon which will one day be a personal playground for the populations of Earth, a scientific laboratory with huge telescopes of all frequencies and a stop for any spaceship going to the outer moons...Without a collapse of world governments in the near future it's possible humans will have a growing permanent presence and population on the Moon by 2100 although current political trends and resource distribution on Earth are challenging the future of space exploration at this critical time in human history when the possibilities of expansion into space are endless and unknown...At the start of this century America struggles with a dysfunctional political structure and a growing lack of interest in discovery and challenge but NASA continues to build a heavy-lift rocket for the next generation and private space exploration organizations are making plans that do not need the approval of politicians...China is very interested in a presence on the Moon and with state controlled finances they will be the next nation to plant a flag there...

Japan has a progressive space program and Russia has an active space program that is currently transporting all astronauts to and from the International Space Station (ISS) while sending most of the resupply vessels...Billionaire entrepreneurs are developing space flight for profit with private capital, investment and government subsidies to promote tourist flights in near Earth orbit for the few who can afford it and at this early stage of human development all these activities combine to advance human spaceflight in incremental steps vital to a collective knowledge of what's necessary to explore, survive and return from a dangerous, hostile environment that does not support biological lifeforms...New technologies are working on developing shields to line the inside of spacecrafts (or products painted on or attached to the outside) to prevent radiation exposure during extended spaceflights and it's unknown what might be possible with the growth of technology still to come...Traveling around the solar system with comfort and safety could be a reality in a few thousand years and a round trip to IO in 5,000 years might be a real vacation destination...Climbing mountains (some taller than Mt Everest) in gravity very similar to the Earth's moon while circling Jupiter in less than two Earth days and experiencing the most geologically active place in the solar system with an average temperature of -200 F...

In late 2013, China soft landed a spacecraft on the moon containing a "rover" that has left the lander to explore the northern edge of Mare Imbrium (It ceased to function after 6 weeks...recovery options are still being studied)...A second mission is planned in 2017 to return to Earth with samples of the lunar surface for scientific analysis...

Impacts from Space
In June 2013 NASA unveiled it's Asteroid Grand Challenge, an effort to find and identify all of the asteroids that could pose a threat to Earth and develop ideas to defend the planet against potential impacts...

If the object that hits Earth is large enough it will be a world-changing event and it's important for any global civilization to set up a planet-wide system to search for possible threats from space in real-time with the goal of understanding consequences as far in advance as possible so science can intervene to save the planet...Scientists continue to identify asteroids and large near-Earth bodies that might pose a threat in the future but the worst case scenario involves comets, millions of comets in the far reaches of the solar system which are unpredictable and on orbital paths that may last thousands or millions of years...Comets travel at higher velocities than asteroids and their arrival is unpredictable until they are discovered, sometimes less than a year before they cross the inner solar system and return to deep space...Setting up networks of observational and detection facilities to give early warnings and developing a offensive capability to interact with threatening objects in a relatively short period of time is essential if humans wish to survive into a distant future...

NASA's Wide-field Infrared Survey Explorer has used millions of infrared photographs to identify and classify asteroid groupings in the main asteroid belt between Jupiter and Mars where most Near Earth Objects (NEO) originate analyzing over 120,000 of approximately 600,000 known asteroids and assigning almost 33% of these asteroids to family groupings for future identification and predictability...

The B612 Foundation, an organization focused on protecting Earth from impact events, released a visualization of 26 major asteroids that entered Earth's atmosphere between 2000 and 2013 based on data from the Comprehensive Nuclear-Test-Ban Treaty Organization (CTBTO)...To detect nuclear tests anywhere on the planet CTBTO operates a network of sensors that recorded these 26 major explosions (nearly all the asteroids disintegrated before hitting the surface) with an energy range of 1 kiloton to 600 kilotons...Only one of the twenty-six asteroids was detected before entering the atmosphere...In 2018 there are plans to deploy a B612 Sentinel spacecraft in a Venus type orbit to identify the "hidden" asteroids (hidden by the glare of the Sun) and plot their future trajectories...

AU (Astronomical Unit) is the mean distance from the Earth to the Sun = 150 million Kilometers

In August 2013, NASA released a new map of PHA's (potentially hazardous asteroids at least 140 meters in size) that plots the orbital paths of more than 1,400 asteroids in orbits passing within 4.7 million miles (7.5 million kilometers)...None of the asteroids mapped pose an impact threat to Earth for the next 100 years and this is just the beginning of a deep awareness of our local space environment that is key to the survival of the human species...More dangerous and unpredictable are rogue comets from the Oort Cloud that must be identified and analyzed as early as possible to allow time for deflection or destruction to avoid an impact event...

The Kuiper Belt

Short-period comets that on average take less than 200 years to orbit the Sun (orbiting on the same plane as the planets as opposed to long period comets which can have highly eccentric orbits) are generally believed to originate in the Kuiper Belt, a disk of debris beyond the planet Neptune thought to exist 30 AU (4.5 billion km or 2.8 billion miles) to 53 AU (8 billion km or 5 billion miles) from the Sun...Containing millions of comets and maybe thousands of icy bodies larger than 100 km, the first Kuiper Belt object was discovered in 1992 and now over 1,300 KB objects have been identified and cataloged...Exploring space for just fifty years humans have no idea what might exist beyond what is known and many scientists believe there are objects larger than Earth in the Kuiper Belt or even in the Oort Cloud, objects that could be discovered by space probes or powerful telescopes placed in a distant orbit or on the far side of the moon...In 2004 astronomers released data on an object named Sedna (2003VB12) orbiting the Sun once every 10,500 years in an elliptical orbit that may cross the boundary of the Oort Cloud and in 2005 scientists released data on KBO object Eris (2003UB313), larger than Pluto with an orbit of 560 years (this discovery led to Pluto being reclassified as a dwarf planet)...In 2014 a dwarf planet 280 miles wide that goes beyond the Kuiper Belt was confirmed (2012 VP113) with an orbit ranging 7.4 billion miles (80 AU's) from the Sun at it's closest approach to over 42 billion miles (452 AU's) before returning...It is almost certain new objects will be discovered by the New Horizons space probe as it continues into the Kuiper Belt after visiting Pluto in 2015...

After a successful encounter with Pluto in July 2015 the New Horizons spacecraft will now become the first spacecraft to explore the Kuiper Belt, a region of the solar system beyond Neptune thought to contain million's of comets and asteroids, a large debris field swirling around the Sun leftover from the formation of this star system...

The Oort Cloud

The Oort Cloud is an enormous spherical reservoir of comets surrounding the solar system at a distance estimated from 5,000 AU's to 100,000 AU's representing the outermost boundary of the Sun's influence...This would place the inner edge of the Oort Cloud at about 750 billion kilometers (almost half a trillion miles) from the Sun and the outer edge about 15 trillion kilometers (over 9 trillion miles) from the Sun (A light year = 9.5 trillion kilometers, 5.88 trillion miles)...These enormous distances paint a three-dimensional picture of the 'gravitational well' the Sun creates in space revealing the shape, size, slope and boundaries of the space distorted by the mass of the Sun, the structure of a single star system,...Despite a sense of conquest, invulnerability and destiny, humans are very tiny creatures on a small self-contained world in a very large reality...Many scientists believe the Oort Cloud contains as many as 1 to 2 trillion comets in highly eccentric orbits believed to be responsible for cataclysmic impacts events on all the planets including several documented mass extinction events which changed Earth on a global scale...In this new age of discovery and exploration new technology is beginning to recognize patterns occurring on time scales of millions of years with new theories suggesting these long-period comets may be dislodged from the Oort Cloud by large molecular gas clouds and tidal forces affecting our solar system on a periodic timetable of 30 million years as this entire star system oscillates up and down through the plane of the galaxy...This is a stunning realization...No scientific proof yet but scientists have used iridium deposits to document a record of terrestrial impacts that seem to occur every 28 to 32 million years, impacts corresponding with the transit of the solar system through the densest part of the Orion Spur, Earth's local neighborhood in this galaxy...Although comets account for just 25% of impact events (asteroid impacts account for 75%) comets are believed to have been responsible for some of the most significant collisions in this planet's long history of extinction and recovery and it's important to study and understand these tidal forces, oscillations and relationships between the different variables in this planetary system that determine cause and effect...Measurements and analysis over millions of years to understand what is necessary for survival, to predict and interact with these objects before they become a threat to Earth...

Both Voyager spacecraft carry information about Earth's inhabitants and a map referencing the location of this solar system relative to the position of 14 pulsars...It is debatable as to whether this was a good idea...

The Future

The enormous amount of debris contained within the gravitational field of the Sun represent what is likely a repeated pattern in the formation of many star systems, incredibly violent events over a billion years that stabilize eventually and settle into mathematically predictive patterns until influenced by the energy of other objects in a mostly empty universe...If the universe has a speed limit (the speed of light) that is impossible to exceed in any kind of way (limiting exploration of other star systems), the incredible diversity and size of just this one star system promises new discoveries and exploration for millions of years...There is enough energy to do this and a time scale that would allow humans to find a way to terraform planets, moons and prepare for a far distant future where they can survive the Sun's expansion and survive for another billion years on Mars or the moons of the gas giants while preparing to move to another star system...In this century, only 200 years after discovering electromagnetic force and less than a 100 years after understanding the power of the atom, it's unknown on any scale what might be possible in a future that understands the dynamics of human existence and the balance of life in a reality only recently realized but it may take the rise and fall of multiple world civilizations before a focus on initial brain growth creates a shared intelligence with the awareness and technology capable of managing their star system...There is no knowledge of other sentient species or of how they survived their first technological era to progress to a planetwide intelligence insuring their survivability and it's probable every sentient self-aware species in this universe is incredibly unique with no comparisons to be made between any two diverse histories, but in this 21st century the challenges facing the human species on a single planet circling a single star are real, sobering and in full focus...

This is a pivotal moment in human evolutionary history and the key is the human brain and the development of a planetwide intelligence that recognize the probabilities, possibilities and realities of existence...

Local Space

On an evolving planet with an evolving intelligence, here is more evidence humans exists in an extreme, dynamic environment that will challenge science and technology to understand patterns of reality that could threaten a continued existence or lead to new possibilities...

Scientists examining four decades' worth of data have discovered the interstellar gas "breezing" through this solar system has shifted in direction by 6 degrees suggesting the solar system exists in a surprisingly complex and dynamic part of the galaxy...Charged particles stream off the Sun forming a large invisible shell around the entire solar system called the *heliosphere*...Outside of this shell is the Local Interstellar Cloud (LIC), a haze of hydrogen and helium about 30 light-years across...The Sun is moving through this interstellar cloud at a relative velocity of 52,000 miles per hour (23 kilometers per second) with interstellar winds streaming in from the direction of the constellation Scorpius almost perpendicular to the Sun's path through the galaxy...As these winds interact with the Sun they create a distinctive feature with the "gas being gravitationally focused to create a trail of helium known as the 'focusing cone' behind the Sun as it moves through space," said the study's lead author Priscilla Frisch of the University of Chicago...By studying data gathered by various spacecraft, IBEX, The Ulysses probe and several craft from the 1970's including NASA's Mariner 10 and the Soviet research satellite Prognoz Six, the team found over 40 years the interstellar wind had shifted by 6 degrees...The changing 'wind' could have implications that go beyond understanding this region surrounding the solar system and it will affect studies of the charged particles streaming off the Sun "When we try to understand the past and present heliosphere, we can no longer assume the heliosphere changes only because of solar wind," Frisch said. "Now we have evidence that changes in the interstellar wind may be important in affecting the heliosphere as well"...Published in the journal Science, September 2013

*DNA components and amino acid components have been found in carbonaceous chondrite meteors, carbon-rich meteorites that make up less than 5% of the meteorites found on Earth...*Journal of Chromatography A

Rogue Asteroids

A map of asteroids developed by researchers from MIT and the Paris Observatory charts size, composition, and location of more than 100,000 asteroids throughout the solar system and shows rogue asteroids are more common than previously thought, particularly in the solar system's main asteroid belt between Mars and Jupiter where the researchers found a diverse mix of asteroids...To create the comprehensive map scientists analyzed data from the Sloan Digital Sky Survey that included more than 200,000 asteroids and this new map suggests the early solar system may have undergone dramatic changes before the planets assumed their current alignment...Francesca DeMeo grouped asteroids by size, location, and composition defining composition by the asteroid's surface reflectivity at redder or bluer wavelengths, included nearly every asteroid down to a diameter of five kilometers grouped by size and composition and mapped them into distinct regions of the solar system where these asteroids were observed..."the trickle of asteroids discovered in unexpected locations has turned into a river...We now see all asteroid types exist in every region of the main belt"...*Nature*

Imagine coming back to this Solar System in 13 billion years and finding the human race still existing in a tight, close orbit around a white dwarf that used to be the Sun, existing because of technology so advanced they can create almost anything out of raw materials found in space with the technology to capture comets as they appear to replenish water sources and survive on this one world...The Earth (or Mars?) would be a catacomb of tunnels and underground cities hundreds of kilometers under the surface with survivors still seeking answers to questions asked for billions of years...

"Our species needs and deserves a citizenry with minds wide awake and a basic understanding of how the world works"...Carl Sagan

THE THIRD MILLENNIUM

(2000 – 2999)

How Did We Get Here?

In 1968 the first manned spacecraft left Earth's orbit to circle the moon and humans were able to 'see' their world for the first time...This event and the explosion of technology since that time have verified the reality of scientific progress since Issac Newton's *Principia* and made clear the physical and political reality of planet Earth...For the first time the entire globe could be viewed as a single entity with a finite size and finite resources, a landscape of multiple countries with differing ideologies and diverse populations, a discernible atmosphere with structure and a clear singularity that is stunning against the backdrop of space...Millions of people in the decade that followed this event realized the fragmentation of society without limitations on continuing destructive human activity (war, burning fossil fuels, overpopulation, deforestation, pollution) on a planet of limited size with limited resources was not just unstable and unhealthy in the present but unsustainable and potentially catastrophic in the future...A new kind of human started emerging from the dogma and superstitions of the past, people who recognized the nature and balance of human existence with new ideas, global realizations and a growing awareness in the 1960's and 1970's which had the potential to change human activity globally if leaders in the influential nations across the world could recognize the reality of existence and support policies and actions necessary to modify destructive practices to a sustainability and balance demanded by the limitations of Earth's ecosystem...*This did not happen...*

In America, recognized by many as the global leader in finance and commerce, corporations, governments and military leaders felt threatened by the activities of reality aware humans who were organizing a global voice to curb pollution, find alternatives to industrial waste, end war by any means, find ways to feed hungry children, distribute medicine, combat poverty around the world and control population growth...*The reaction to this growing awareness by corporations focused on profit margins, by military leaders insistent on the continuation of industrialized military growth and by the political lawmakers who recognized the opportunity for personal wealth and power in consort with industry resulted in a rapid explosion of legal corporate special interest organizations created after new laws introduced by politicians effectively made corporate bribery legal*...It worked...In 2016 there are few states, nations or communities where recycling everything consumed by humans is required by law...In America today there are thousands of dumpsters with faded letters reading "recycle only', visible leftovers from an earlier time when humans realized the impact and ignorance of using raw materials for almost everything...These are now trash containers in a global 'throw away' culture with cardboard, plastics and electronics routinely thrown in the trash as a matter of convenience...

America's Corruption

(A Growing World Model)

In 1975, the SUN-PAC decision legalized corporate political action committee's (PAC's) and corporate donations to lawmakers...In 1975, there were less than 90 PAC's...In 1984, there were more than 1,600 PAC's (over 4,600 in 2009) with a high correlation between PAC donations and legislation being passed benefiting these corporations...Corporate structures actively paying politicians to remove regulations, paying politicians to give them the ability to lobby and donate even more political money, paying politicians to give them the tools to undermine organized labor, shift industry oversees to reduce cost and avoid regulation related to proper waste disposal and pollution, paying politicians to create tax shelters to avoid responsibility and accountability, paying politicians to create legislation to undermine consumer protection and paying politicians to lower tax rates on private industry to maximize profit while allowing opaque havens and offshore shell corporations to avoid fiscal responsibility...

It's believed over $7 trillion is in the underground (shadow) economy, unregulated and unreported...

This free flow of unregulated money has been very effective and has quickly spread to almost every nation on Earth with a confluence of private industry, military industry and politicians essentially running governments...In many developed countries the military is heavily invested in private industry and the U.S. is no exception on a planet where only a handful of political systems are not influenced and corrupted by money and power...In America the lobby groups and corporations actually write and create legislation to be introduced by members of Congress (who have received donations from the corporations and are often given a high paying corporate job after leaving Congress) always legislation designed to benefit special interests and corporations and it's somehow legal with limited oversight and almost no enforcement of the laws remaining...Without informed, involved and aware populations who understand the necessity of a true representative government, politicians everywhere on Earth steal elections, use fear and lies to influence populations, pass law to make corporate bribery legal and at this time in history, the majority of Earth's population are defenseless against the individuals in power who manipulate resources and truth for personal wealth and ideology...Ironically, political power might be the best way to achieve serious change without destroying structure in a world without conscience...

With seven billion humans slowly beginning to realize the finite limitations of natural resources and the cyclic balance necessary to sustain life, the greatest challenge in this millennium is uniting all the world governments and populations in ways that recognize the consequence of unchecked industrial processes and a growing waste culture that threatens to destroy sequences, chains and connections of life that make human existence possible...But the challenge of change in the fluid human maelstrom that represent Earth in this century is staggering and maybe impossible...Even with incredible advances in science, communication, technology and medicine, this world has never been more dangerous and unpredictable...There are 195 independent sovereign nations on Earth (some recognize themselves but are not recognized by others?) and over half are ruled by corrupt leaders, corrupt political coalitions or corrupt dictators...Almost every nation has a military and an intense focus on territory and wealth and many world organizations who claim to govern and advise policy and procedure around the globe are bloated and inefficient at anything other than enriching themselves...The largest global nations posture for access to energy resources yet to be recovered and all of them hold "war games" at the slightest provocation just outside of the boundaries of any nation they disagree with...Conferences, nations, summits and international forums continue to talk of "green energy" with promises of a new future but fossil fuel extraction continues in the oceans and on every continent of Earth (Antarctica's protected by international agreement) with money, new innovation and technology promising to burn dirty fuel for decades...With climate change from these unilateral energy policies unstoppable, income inequality growing globally and Earth's population exploding with no serious discussions anywhere of the impact of over 11 billion people by 2100, unrest and protest is growing and violence always follows...

The Gates Foundation and many other organizations work to decrease poverty and feed children and still there are millions malnourished...Doctors Without Borders and the Red Cross work tirelessly to alleviate suffering bringing modern medicine and techniques to the poor and it is only a band-aid on a very large wound...Charities and NGO's rush to disaster areas and years later money has disappeared and thousands still suffer (Haiti)...Corruption grows worldwide and the humans who control wealth and resources with the flexibility and freedom to help are focused on profit and power while using slight of hand, legalities and wordplay to insist they care...In reality there is no "big" solution...Considering the various scenarios, possibilities, trends, histories and realities defining human activity in this century it's difficult to see any way to emerge from Earth's industrial revolution and unchecked population growth to a future of clean energy and a global cooperation that includes every human on the planet in a social and economic revolution that must happen...There are just too many people unaware of human reality and millions who are driven by ancient primitive instincts in dangerous and violent ways...The key is global intelligence, awareness and non-violence but we are years away...Every human who wishes to seek alternatives and possibilities in this existence should be free to do so if they can do so without coercion or violence on others...It would seem obvious but to billions of unaware humans, it is not...

How do you affect population growth when no one is talking about it? And how is it possible to affect violence on a world covered with weapons driven by an industry with an intense focus on profit and a singular purpose of facilitating the killing of humans by humans...Billions ($) in weapons are sold to any nation with the money as corporations, governments and the politicians beholden to these corporations profit from the violence and chaos these weapons make possible...It's clear to all humans with intelligence that Earth needs no more weapons...A violent reality obvious to anyone who is aware but nothing will change in this century...

WORLD HEALTH
Extreme poverty has been reduced in the developing world from 50% in 1981 to 21% in 2010...
Global child mortality rate has been reduced by almost 50% since 1990...
Life expectancy has increased over 10 years in the last two decades reaching 70.5 years in 2014...

At the same time...
Ocean acidity is increasing, coral reefs are dying and dead zones in the sea are increasing every decade...
Water tables in every continent are falling and glaciers are melting with over 50% of the world's topsoil challenged...
Atmospheric pollution from increasing traffic and unregulated power generation affects every major city on Earth...
Over 50% of the world is potentially unstable with population increasing by 4 billion before the end of this century...

One solution that could have immediate and dramatic effects globally is the fair distribution of money...Almost all internal unrest and violence in every nation on Earth is the result of disenfranchised populations who feel they have no voice and no leverage against a growing world paradigm that finds political leaders in every culture working in concert with corporate structures to attain as much wealth as possible with no recognition of the needs of populations, without concern for the sick, poor, babies or the elderly and without any kind of fair distribution of products necessary for survival or essential services to save the next generation while cheating in every way possible to avoid fiscal responsibility...

Africa with money would be a vibrant, growing economy without 50% of their children going hungry...Syria would not be burning today if Al-Assad had shared the nation's wealth with the citizens in a fair way (Al-Assad's wealth is estimated to be over $500 million)...Venezuela has an economy that is rich with energy wealth but the people can't feed their children or even find simple necessities in the markets...An oligarch from Ukraine has one of the most expensive apartments in the world in England, another western country who look the other way and welcome individuals from desperate nations who live in unimaginable luxury while people die on the streets in their nation of origin, a pattern repeated many times in Earth's modern global economy...Ethnic violence is real and will not disappear for years but it would be much less if populations could exist in sustainable predictable ways with an operating infrastructure and some voice in their own societies...It's amazing how reasonable most humans can be when they have safety, food, electricity and a predictability of existence...A clear example of ignorance without vision is Iraq, a nation with vast oil wealth whose former prime minister Maliki loaded his government with Shiites leading a campaign of discrimination and violence against Sunni populations, fixing infrastructure in Shiia neighborhoods while ignoring the needs of Sunni's in all ways bringing violence, suffering and damaging children in ways that will never be recovered...

And it continues...This is the reality of Earth and there is virtually no chance of turning this around in time...The wealthy nations of the world could work together to give every human on the planet $1,000 a year with the obvious exception of known violent individuals, without moral judgment as to what this money is used to buy (no conditions, just include everyone in the world economy) and it would be circulated as it always has...*This will not happen*...The wealthy nations of the world could become very empathic giving humanitarian aid to all the populations in need without conditions and implement security responses that are acceptable and accurate when necessary...If Israel, with the help of the world's best technology, destroyed missile sites (not the humans you think fired the missiles) immediately after an attack with no collateral damage to the native population and gave Palestinians access to the world with intense security (but no restrictions) and allowed the world to deliver needed food supplies, caretakers, clothing, toys and learning devices, these missiles would stop...*This will not happen*...And weapon industries could stop making weapons and use materials and technology to build solar plants, wind technology, power grids, schools and infrastructure across an entire world to begin improvements to benefit our entire species...*This will not happen...*

The United States authorized a $60 billion weapons deal with Saudi Arabia in 2010 of which $48 billion has been finalized as America continues to arm one of the most ignorant, primitive and repressive nations on Earth...

There are many groups of humans on Earth who will not be placated by money or inclusion but they number just a few million and *many inextricable problems across every culture on Earth would be mitigated if there were a focus on including all citizens on this planet in the global economy necessary for the distribution of goods and services around the world*...And if this were a reality people would not need to resort to violence to demand basic necessities and they would be more receptive to any changes needed to reorder and restructure civilization in a way that is more sustainable and healthy for their children...This is possible in this century but it will not happen...There are simply too many humans unable to understand these possibilities...Wealthy nations and individuals will hold on to their money ever tighter building walls of security while blaming the poor for their own plight and hostilities will continue to grow among nations until there is a nuclear event (Pakistan & India?)...Apocalyptic visions will spread worldwide and violence against any population for any reason will be acceptable...Billions will die from hunger and war while the largest, most powerful countries (America, Russia, China) will survive for hundreds of years in geographical, strategic locations that favor survival...It is impossible to predict is how long this violent, fragmented scenario will last but in some future the survivors of this 1st World Civilization (a hundred million or a billion?) will start to build another global civilization on a polluted planet growing hotter every year from a changed atmosphere that will affect life on Earth for thousands of years...This book is written for those survivors...If the next civilization will focus on every baby born and prioritize initial brain growth, there is still a chance for this species to exist far into the future...Earth has recovered from every extinction event in the past and in a few million years there will be another perfect day...Will anyone be here to notice?

Alarming is the rapid advance of cyberwarfare as governments across the world use technology (that has the power and connectivity to facilitate a powerful all inclusive global society) to attack other nations from their own territorial enclaves to gain advantage and steal secrets and wealth...Something made possible by corporations who will sell invasive malware to any nation for profit...Very primitive is the inability of leaders in almost all nations to recognize the reality of our collective existence in a way that encourages communication and cooperation that could move our entire species toward solutions and a peaceful coexistence...

Population

The biggest challenge of human survival...Human population growth will double the current population of Earth to 16 billion by 2150 if it continues to increase without constraint, hard to imagine, but possible despite the science, technology and medical procedures that would give every woman on Earth a choice...On a world with a consensual awareness the majority would recognize overpopulation as a problem with easy solutions and make available medical procedures and multiple choices to men and women everywhere to address this fundamental question before conception preventing unwanted pregnancies and millions of unnecessary abortions...Just this single area of focus would make Earth's survivability challenges more manageable but restrictive laws and policies will not control a growing world population...This can only be solved by increasing the intelligence of our entire species to a point where maintaining a stable population is an understood necessity...Organizations worldwide, mostly religious, protest any control of reproduction as a violation of 'god's laws' while ignoring the fact that intelligent awareness would seriously reduce infant mortality, maternal mortality, abortions and chronic hunger...There are many women on Earth who want no children and many who do...A balance is very possible but there are millions of woman on Earth with no voice controlled by men through violence, intimidation, cultural laws and religion who are not allowed 'choice' in any of the important aspects of life as an exploding population, the greatest single threat to sustainable existence on a single planet, continues without discussion or solutions...*Ignorance, religious dogma and the proselytizations of tired old men should not take precedence over intelligence and science in a millennium that could determine the future of life on Earth*...A millennium that begins with an available global connectivity which allow any member of the human species to communicate with anyone in anyplace at anytime...An explosion of technology that will be an integral part of humanity for the rest of human existence...

The most positive outcome of the industrial revolution and the incredible advances in scientific knowledge over the past 300 years have been in medicine and technology, an understanding for the first time in human history of our place in physical space and time and the ability to recognize the biological threats and necessary responses needed to increase our human lifespan and cure diseases that have plagued the human species for thousands of years...Mysterious 'ailments' that have killed millions without discrimination are no longer mysterious...

In 2016, even in nations beset by continuous conflict and primitive governance, there are small solar power stations on roofs in Afghanistan powering televisions and small satellite receivers...Similar scenarios are seen in South America, the desolate lands of northern and eastern Russia, rural China and India which contain over 2.5 billion humans, islands across the oceans and in Africa, one of the most desperate regions in the world, children playing with laptops in Ethiopia and cell phones available and essential in every part of the world...This growth of technology worldwide will one day connect every human and allow everyone to interact with everything in multiple ways and this exponential increase of information and entertainment in households will stimulate large brains and stimulate small brains as babies get positive reinforcement and knowledge never before available while caretakers network and learn techniques and procedures from a global community to help in growing babies and children with a focus on security, safety and nutrition...Multiple solutions will be explored simultaneously around the planet and intelligence and awareness will grow as transparent information educates a world creating a knowledge base to meet any challenge that threatens our species survival...Technology that amazes us in 2016 has just begun...What might be possible in 1,000 years or in 10,000 years? The human species has only scratched the surface of existence and every indication is there will be no end to progress or possibilities on a connected Earth...Humans control their future and the next thousand years of human activity could determine the fate of life on Earth for the next million years...

Technology
With a new understanding in an invisible world of predictable patterns and energies unknown just 150 years ago, technology has exploded in predictable and unpredictable ways that make clear the survivalist advantage of existing in a reality that can be understood and manipulated on the scale of the impossibly small, creating variable devices, networks and solutions necessary to survive for millions of years...Many view this explosion of knowledge as a double-edged sword giving humans the capability of waging war with increasingly sophisticated killing machines and an ability to locate and recover the fossil fuels that are changing this world in negative ways but technology is here to stay...In a world of ignorance and violence, it's possible new technology will create extreme capabilities which in an ironic twist will destroy the technology that created these capabilities *but the realizations and discoveries of a developing human intelligence will not be forgotten*...Nuclear weapons will be possible for the rest of human existence...Dangerous biological agents will be possible for the rest of human existence...The integration of man and machine, invasive cyber technologies and discoveries yet to made will all be a possibility for the rest of human existence...In a new reality where any human could manipulate science and technology for destructive or 'evil' purposes the only solution is a better human...

In some future an abandoned, crying, ignored, hungry and unhealthy infant will be unheard of and in that future technology will be embedded in the fabric of human existence to a sophistication that cannot be imagined at this primitive stage of development...

To many who see this as a threat to our "humanness" the deeper we explore the reality around us, the deeper we will explore the reality within us...With glimpses and insights into the possibilities of human intellectual development and growing evidence that morality and empathy are intimately linked to early brain growth and highly influenced by environmental realities there is hope...In this century the level of violence on Earth is staggering, the number of people unable to understand causality is very alarming and it's clear the human species isn't close to what is possible...Technology is key in changing this reality...To safely travel through this star system requires complete shielding from all harmful radiation and there is already new research in nanotechnology that promises to develop a light-weight material which will be an integral vital component in every manned spacecraft and spacesuit used for off-world research and exploration...Nanotechnology is also making serious progress in dealing with spent nuclear waste while tons of hazardous waste continue to be stored in many different locations on Earth...And in a distant future electricity will power everything with incredible battery technologies able to store energy quickly with minimal loss and zero pollution...Human technology has just begun, defined by a growing recognition and application of the realities that define this universe and it will be an essential part of human existence for as long as that existence lasts...

By any definition, the purpose of government is to distribute resources in a fair, equitable and realistic manner to create a sustainable society without favor or bribery...

Socialism is not about an equal distribution of wealth but rather an equal distribution of essential protections necessary for survival on this planet...Socialism guarantees the health and safety of all humans through a progressive tax system that does not prevent the ambitious and industrious individuals from accumulating wealth but provides a standard for every human who cannot provide for themselves...A defining feature of an advanced civilization...

Socialism

No system of governance or resource distribution in modern society has been as demonized and criticized as *socialism*...Socialism is an ideology which believes a central government of the people can regulate, innovate and operate a system of social justice and empowerment that functions by the rule of law in a democratic format that makes the protection and welfare of every citizen a priority...For a civilization to survive for millions of years it will have to anchor itself on this principle to maximize resources and eliminate a staggering inequality that is the source of conflict and dissent in almost every nation on Earth...The wealth of a solar system belongs to all humans and as this species becomes more intelligent and aware of their collective reality there will be an understanding of our shared existence and the isolation of a single world reliant on a single star...A world with the resources of a planetary system to supply materials and energy to support a managed population with a sustainable lifestyle...A lifestyle that creates, innovates, plays, works, explores and strives to understand an existence humans have only just discovered...*This will inevitably happen in a distant future if we don't carelessly destroy ourselves in the near future*...There will be no money one day and with an unlimited energy source and a collective consensual intelligence on balance and sustainability, every human alive will have choices and security at all levels of existence...The human species will collectively use available resources to manage this planet and other worlds as isolated territorial posturing becomes ancient history in a world without a need to compete for energy and land...A human population controlled by a global collective awareness of limitations and a necessary balance of survival without superstitions or dogma...

Many people who now believe their activities and choices are negligible would rethink and cooperate if they were included in the world's economy and the world community...

In 2014 there were 1,645 billionaires on Earth with a combined net worth of $6.4 trillion...

Money

A "construct" or "device" used by humankind for thousands of years, money has no real value as a 'thing' but it's an excellent way to track and distribute items essential to all humans and items that are non-essential, various products of human creativity developed by this species for thousands of years to serve other humans as items of comfort, utility or technology, 'things' which are an integral part of existence on this planet...The problem is this advanced method of organizing a world (money) does not work if all the humans on the planet don't have access to a basic number sufficient for a healthy life within a 21st century global society...Today, on a primitive modern world without cohesion, there are billions of people (men, women, children, babies) who suffer every day from a lack of inclusion into a monetary framework that directs all activity on this planet...A lack of inclusion directly responsible for hunger, sickness, desperation and growing violence that could destroy this civilization...In an intelligent civilization with a consensual awareness of reality, every human would be allocated a basic level of existence in a world that recognized a "bottom" to provide essential necessities...In a competitive world with luxury items and new technology people who wanted to use their skills to accumulate money in whatever way they like is acceptable and a necessary stimulus for a regulated free market operating within determined limits, but if there were a world order requiring a basic level of existence for all the people of Earth the wealthy could remain wealthy through work and innovation without creating this unbalanced system of income inequality and desperation that is arguably the cause of almost every conflict on Earth...And a civilization with such a system could allocate necessary funds for necessary things to support the future survival of an entire species...Money and cost should not be a consideration when building nuclear power plants to the standards of safety that guarantee containment and money should never be a consideration when distributing needed medicine and drugs across global populations to eradicate disease and help the sick...Infants and children do not die because there is no food available or no clean water, no medicine or no healthcare...Humans die because they do not have any money to buy food, no money to buy clean water, no money to buy medicine and no money to buy healthcare...

Despite cries of "socialism" "income redistribution" and the danger of "social welfare" that are echoed through the boardrooms of every giant corporation on Earth when such ideas and solutions are opined, if there were an established minimum allocated to every human on Earth the world economy would operate in a more efficient way and industrious humans could still accumulate wealth...

The release of the 'Panama Papers' in 2016 highlights this early stage of human development...

A charity called GiveDirectly gave $1,000 dollars to individuals in Kenya to do with what they wish and the results were stimulating and amazingly effective...In 2012, the GDP of the world was around $71 trillion dollars...If every human on Earth were given $1,000 each year, it would cost about $7 trillion and all of this money would be circulated as money always has been, alleviating great suffering and moving a young species toward a future with possibilities...

Extreme Solutions

A solution to the overabundance of carbon dioxide in the atmosphere being seriously considered in scientific communities is Geo-engineering using human solutions to deflect the heat of the Sun until humans can stabilize their atmosphere and maintain a balance for years...Almost all the ideas involve sunlight deflection or the removal of carbon from the air using techniques which are problematic and controversial...And all solutions are challenging from a technological standpoint...

Among the ideas put forth to reflect the energy of the Sun and give the CO2 in the atmosphere less heat to trap are *space based solar shields* which involve surrounding the Earth with hundreds or thousands of small, reflective spacecraft to "mirror" the Sun's heat back into space (technically more than 100 years away) or *asteroid dust* to shield Earth from the Sun which could cause a host of other problems if all possibilities are not considered...More alarming are ideas like *spraying aerosols or tiny particles* high in the atmosphere with airplanes or tethered balloons...A balloon-aerosol research project called the Stratospheric Particle Injection for Climate Engineering (SPICE) says the biggest question is whether it's possible to get enough particles into the stratosphere via a balloon-lofted hose to make a difference...The atmosphere is fairly calm at an altitude of 12 miles or higher but from the ground up to about 6 miles it's very turbulent and the hose would have to be designed to survive this dynamic (the team is researching Kevlar for it's strength and aerodynamic hoses to overcome the problem)...Another idea uses *cloud seeding ships* that would spray ocean salt-water into the atmosphere to create and enhance white, reflective clouds, ideas inspired by volcanic eruptions which spew sulfur dioxide gas high into the atmosphere and combine with water vapor to form light-reflecting particles and keep the planet cooler...(Clouds can reflect about 50% of sunlight while the ocean surface reflects only about 10%)...This would be done by automated ships about 40 feet long using wind power and guided by satellites to generate the electricity needed to turn the seawater into a spray while using specialized filtration and designer nozzles to keep microscopic plankton from clogging the system...It's estimated approximately 1,500 seawater-spewing vessels spaced nearly 150 miles apart could offset the warming produced by twice the amount of $CO2$ in the atmosphere in 2016...Another possibility is *seeding the oceans with iron fertilizers to prompt phytoplankton blooms* that would absorb the carbon from the atmosphere and sink to the bottom of the sea...

The problem with all of these long-term "solutions" is a possibility (probability) of unintended consequences, especially seeding the oceans with fertilizer (no one can predict the long term effects of this activity) or the cascading effects of thousands of satellites circling the Earth or problems associated with a proliferation of 12 mile long hoses into the sky...There is a history stretching back millions of years governing the cycles of plant life in sync with the predictably of sunlight and radiation essential for the continuation of life and all of these proposals would affect plant growth with unintended and unknown consequences and every scientist and engineer working on these possibilities is hopeful this planet will never reached a point where such drastic measures are needed...But the future of human activity and the viability of Earth's one ecosystem is a unknown variable equation with many possible outcomes impossible to predict in the 21st century and this uncertainty can drive reckless innovation...

Engineering Earth

As master tool builders on a single world, the ultimate goal of this species will be to design this planet with a vision able to recognize the optimum configuration of a single world with many, many contingencies for change...A world that will support a population of billions for millions of years...This is not taken seriously this century but it's what humans will do after generations of maximizing initial brain development creates a single world civilization with a consensual intelligence...Humans will find the way Earth works best with an understanding of the long-term circulatory patterns in the oceans and atmosphere effecting everything while modeling the long-term (really long-term) climate effects this species has been collecting data on for just a few centuries (the measurements increase in accuracy every year)...A collective awareness and technology necessary to engineer Earth to produce maximum efficiency and survivability in this particular reality...Fascinatingly difficult, but necessary...

Transportation conduits will be refined consistent with population distribution and commerce activities...Nuclear waste sites (if necessary) will be few, remote and built with the technology to last a million years...Forests around the world being devastated in the chaos of the industrial age will be reclaimed, multiple regions will be dedicated to agriculture, and power generation facilities will only be located in safe geographical locations with a system of power grids and distribution points unimagined today...And it's very possible that energy production will be localized with very few worldwide power grids necessary in a world of clean, limitless energy with fusion technology or some combination yet to be imagined...Space technology will develop a way to shield space travelers from harmful radiation and make interplanetary travel possible anywhere in the solar system leading to exploration and eventually the colonization of planets and moons...Human's are innovative, industrious, curious, intelligent and at the very birth of understanding the possibilities of new technology and a growing ability to manipulate this reality and survive in a dangerous, indifferent universe...Just a casual look at the exploding growth in battery technology, the new research in the structure of the human brain, energy solutions in solar technology (collection and storage capabilities grow daily) and advances in medicine with possibilities of increasing our lifespan to create a civilization with the ability to prevent all birth defects and repair any human ailment foretell a new future with limitless possibilities...New innovations in agricultural techniques and technology promise a world where real hunger is unheard of and all the waters on Earth (lakes, rivers, oceans, creeks, streams) are safe to humans and safe for the creatures that live there...We are a technological species at the beginning of time and are intimately integrated with all the multiple devices connecting us to a reality with unknown possibilities, a fascinating future if the human species can maximize initial brain growth in every baby on Earth...

Batteries

Envia Systems has already built prototype lithium-ion battery cells that store twice the energy of the best conventional lithium-ion batteries and can be recharged hundred's of times...Researchers have also created the first stable lithium-air batteries which have a potential to store 10 times more energy than the best lithium-ion batteries by replacing the carbon based cathode with a cathode made of gold...Early prototypes are too heavy and expensive to be practical but it demonstrates what mining the solar system could provide for an intelligent species seeking mineral wealth and resources to improve their technology...No one knows how much gold and other minerals might be contained in the asteroids and moons within our reach...

In Nature (01/09/14), Harvard researchers reported the development of a new battery using cheap, organic molecules (quinones) that can be used to store energy from renewable sources using a membrane capable of allowing the passage of ions to store energy and then releasing the energy using a reverse chemical reaction...

Transportation

The world's first gyroscopically stabilized two-wheeled all-electric vehicle (Lit Motors' C-1) made it's debut in 2012 with a range of 200 miles, a top speed of 100mph and untippable...Designed for urban environments with mass production beginning in 2018, the C-1 could ease traffic congestion, decrease fuel use, reduce CO_2 emissions and allow people to get around quickly and efficiently...The extended range of the C-1 is assisted by a KERS regenerative braking system which uses flywheels to store energy kinetically...These flywheels are also part of the gyroscopic stabilization system keeping the vehicle upright in all situations, even in a collision while allowing the control system to dictate the tilt and lean of the vehicle at all times...

There are approximately 1 billion vehicles on the planet...Manufacturing an estimated 100 million vehicles a year by 2020 will push the world total to 2.5 billion by 2050...In 2015 less than 500,000 cars are electric...

India announced is 2012 they have developed technology to build vehicles using compressed air to power tiny "smartcars" at 40mph with zero pollution...The "Airpods" employ pneumatic motors using pressurized air to carry three passengers about 125 miles on a tank of air refueling with an on-board electric motor (or at a specialized fueling station) filling the tank with 175 liters of air...

Environment

MIT researchers have invented a new kind of filtration material for desalination using sheets of graphene, a one-atom-thick form of the element carbon, which is far more efficient and less expensive than current operational desalination systems...Graphene sheets with precise one-nanometer pores have the potential to purify seawater more efficiently than existing methods when water molecules, sodium and the chlorine ions in saltwater encounter a sheet of graphene perforated by holes of the right size allowing the water to pass through while blocking the sodium and chlorine components of the salt...The new graphene system works hundreds of times faster than current techniques with the same pressure or at similar rates to current systems with a lower pressure (reverse osmosis uses membranes to filter the salt from the water but require extremely high pressure to force water through membranes that are thousand times thicker than graphene)...The graphene system operates at much lower pressure and could purify water at far lower cost...The key is very precise control over the size of the holes in the graphene sheet with an ideal size of one nanometer (one billionth of a meter)...Because graphene is such a strong material (pound for pound, it's the strongest material known) the membranes should be more durable than those presently used for reverse osmosis...

Desalination plants operate in more than 120 countries around the world producing over 3.5 billion gallons of potable water a day with approximately 70% targeted for human consumption...

Health

A device called the Second Sight's Argus II is a neuroprosthetic device that uses a camera and microchip implanted on the retina to allow blind individuals to read visual braille...The implanted chip receives a wireless signal from a small computer (worn by the patient) which has processed a signal from a small camera attached to the patient's glasses...The microchip stimulates active cells in the retina to send patterns of impulses to the optic nerve resulting in patterns of light the patient can interpret...

There are about 40 million humans worldwide who are blind...Scientists at the University of Texas Health Science Center reported the discovery of a neural mechanism of conscious perception that could bypass the eyes and use the brain's image generating ability to "see"...Called a *visual prosthetic*, this device would use a web-cam (attached to glasses) to relay information to a computer chip placed in a person's visual cortex activating electrodes that stimulate the brain to create an "illusion" of a flash of light...If a process can be found to create many flashes of light, images could be formed by stimulating the occipital lobe and fooling the brain into perceiving something that is not there...They believe about 27 flashes of light would let patients recognize the outline of a letter...Researchers also found that the electrical stimulation resulted in the illusion of a flash only if there is activity in a region of the brain called the temporoparietal junction...No light flashes were perceived by test subjects if this region of the brain was inactive...The research continues...

The Second Sight Argus II has been implanted in over 50 patients, many who now see color, movement and objects...

In other research related to sight, researchers at UC Berkeley have engineered adeno-associated viruses capable of penetrating multiple cell layers that protect the retina to deliver a corrective gene into cells containing a defective gene...Gene therapy is used to help humans who suffer from inherited defects (retinitis pigmentosa) and degenerative illnesses of old age (macular degeneration)... Previous techniques using needles to inject the needed viruses directly in the eye were invasive and ineffective in reaching all the cells with defective genes and the process often resulted in retinal detachment...The newly engineered viruses (five selected from 100 million variants) are now injected into the liquid vitreous humor inside the eye and deliver the genes to the delicate retinal cells in a uniform way...The technique has been successful in tests on monkeys (eyes are very similar to humans) with clinical trials on humans scheduled in the future..

Researchers at the Center for Regenerative Therapies Dresden found that zebrafish use acute inflammation (as opposed to mammals) to promote central nervous system regeneration (brain repair) in combination with stem cell activity...Continuing research with unknown possibilities...

Human science will one day find a way to end all paralysis...

*Scientists at the University of Texas Medical Center reported in 2014 they had created new nerve cells in the spinal cords and brains of living mammals without the need of stem cell transplants to replenish the lost cells suggesting that it may be possible to regenerate neurons from one's own non-neural cells to repair spinal cord damage...It is still unknown if these created neurons will result in any functional improvements...*Nature Communications

Jonas Frisen at the Karolinska Institute authored a study in the journal *Cell* that found a way to date the birth of neurons in the brain using atmospheric levels of carbon-14 (elevated by nuclear bomb testing more than half a century ago) which have been declining at a known rate for 50 years...The exact atmospheric concentration of the carbon-14 is preserved in the DNA every time a "new" neuron is born and the study carbon-dated the neurons in the hippocampus of deceased humans...They reported over 1,400 new neurons in the dentate gyrus area are added every day during adulthood at a rate that declines only modestly with age suggesting this neurogenesis may contribute to brain function...*2013*

Energy
A study published by the National Renewable Energy Laboratory (NREL) claims 80% of the energy needed by the U.S. in 2050 could be supplied by renewable energy sources with over 50% coming from solar and wind energy using technologies already available today...Maximizing solar power in sun states and wind power in windy states would be the key with a smarter power grid and storage capacities...To increase 2014 wind output from 50 gigawatts to the 439 gigawatts needed in 2050 would require the installation of more than 2,500 wind turbines every year...

LightSail Energy is developing storage systems designed to store energy from solar and wind farms with a compressed air technology and water to double storage capacity and create real options for global regions with crumbling infrastructure, isolated areas and islands...Chief scientist Danielle Fong estimates the potential market for compressed air tanks could exceed $1 trillion in two decades...

Researchers at the University of Tokyo and RIKEN demonstrated a material to eliminate loss in the transmission of electric power using a 'magnetic topological insulator'...The team used an exotic type of semiconductor to exhibit an effect known as the 'quantum anomalous Hall effect' using internal magnetization to create an interaction between ions and the insulator's current carrying particles to effect a transmission with 'zero mass'...Although it requires cryogenic cooling, better material design could allow transmission at higher temperatures and eliminate loss in energy transfer...

In sub-Saharan Africa a solar service company (Azuri Technology) has started distributing solar kits on a pay-as-you-go plan to enable rural families to purchase these devices to produce electricity needed to charge batteries and provide lighting...The buyers pay when they can and after the kit is paid for the electricity is free...The company has distributed over 20,000 kits this way eliminating the use of kerosene, an unhealthy alternative that pollutes the atmosphere, is more costly and is a poison hazard for children...On a continent that desperately needs energy this is a real alternative...*2013*

In 2014 researchers at the University of Maryland found an inventive way to store hydrogen using a "graphene origami" to create the world's highest density hydrogen storage system...By folding graphene into tiny "boxes" that can open and close in response to an electric charge, Teng Li and Shuze Zhu have exceeded established U.S. Department of Energy 2020 storage goals with a hydrogen storage density of 9.5% hydrogen by weight...Very unique...*Published in the Journal ACS Nano*

In development are low-cost solar cells painted on buildings, windows and personal devices to provide electricity...

Stanford University scientists have built the first solar cell made entirely of carbon, a promising alternative to the expensive materials used in photo-voltaic devices today...

Interstellar Space

NASA's Voyager One spacecraft left the solar system in August 2012 emerging into interstellar space 11.66 billion miles (18.77 billion kilometers) from Earth beginning the human exploration of an entirely new region of space with a very used spacecraft, limited power (transmissions available until around 2025) and several working instruments...The key indicators measured suggesting Voyager 1 has left the solar system are a drop in solar particles with a corresponding increase in galactic cosmic rays, an increase in plasma density and a shift in magnetic field orientation...Voyager 1 has observed the first two phenomena and has measured an increase in the strength of the magnetic field but not a shift in direction...The density of the plasma at the edge of the heliosphere is a complex interaction of pressure from the solar wind moving out at nearly one million mph and the interstellar gas which is much colder and denser than the plasma in the outer limits of this boundary...The increase in plasma density was verified by a solar eruption that caused the plasma surrounding the spacecraft to oscillate measuring a plasma density around 40 times denser than the plasma inside the heliosphere...In 2012 a definitive increase in interstellar cosmic rays were detected with a reduction in the solar wind particles emitted by the Sun at a distance of almost 12 billion miles indicating an entry into interstellar space...Scientific theory suggests the interstellar magnetic field is 'draped' around the outside of the heliosphere and the details of the flow of interstellar wind determine how the field lines orient themselves...Scientists will continue to model and study this interaction but are confident from the first two indicators that Voyager 1 is now in interstellar space...

Asteroid Identification

"Mapping the great unknown of the inner solar system is the first step to opening up this next frontier...The B612 Foundation believes humanity can harness the power of science and technology to protect the future of civilization on this planet while extending our reach into the solar system."

"For millions of years we have been a cosmic target with occasionally devastating consequences," said B612 co-founder Rusty Schweickart. "To say that we, as human beings, are going to stop that is a very powerful statement."

Scientists unveiled a plan for the first privately funded deep space telescope to map the inner solar system for potentially dangerous asteroids...The B612 Foundation says the telescope will circle the sun to identify asteroids with orbits crossing Earth's orbit to measure the trajectories of the asteroids and help protect Earth from cataclysmic impacts with an added benefit of aiding mission planners to chart future expeditions deep into the Solar System...The Sentinel Space Telescope will map half a million large asteroids that populate the inner solar system to identify threats decades in advance of an impending collision..."We should be able to establish orbits well enough that we can predict where the asteroids will be in fifty to hundred years," said Stanford Professor Scott Hubbard who is overseeing development of the telescope...A Space X Falcon 9 rocket will send the infrared telescope into orbit around the Sun where it will map the large asteroids that populate the inner solar system while orbiting 30 million to 170 million miles from Earth in a prime position to detect large objects before they come close to Earth...The telescope will scan the entire sky every 26 days with a 24 million-pixel array sending information about asteroid locations and trajectories back to Earth via NASA's Deep Space Network of antennas...With enough warning a well-placed projectile, nuclear explosion, or gravity tractor (a massive spacecraft that could pull asteroids with its own gravitational field) could redirect a potentially devastating impact...Launch is expected within four years (2018-2019)...

In 2012, NASA scientists announced they had successfully tracked about 90 percent of the largest asteroids in orbits approaching near Earth...Data from the infrared WISE space telescope suggests there are about 981 asteroids the size of a mountain or larger on paths that come near Earth and currently 911 of those asteroids have been tracked...

The Galactic Black Hole

Astronomers have found a new star orbiting very near to the super massive black hole at the center of the Milky Way, only the second star observed completing an entire orbit around our black hole which has been calculated to weigh around four million times the mass of the Sun...The new star (S0-102) has an orbital period of 11.5 years, the shortest of any star orbiting the black hole...S0-2, a star with a 16.5 year orbit will make it's closest approach to the black hole in 2018 *(periapsis)* with a stronger gravitational force increasing the red-shift in it's light signature...This will allow researchers to verify Einstein's general theory of relativity by measuring the precise amount of red-shift predicted by the theory...The experiment can be repeated when S0-102 reaches its own *periapsis* in 2021...

Human Intelligence

One of the great challenges in neuroscience is mapping synaptic connections between neurons with the goal of understanding the flow of information in the brain...Incredibly difficult...Do neurons grow independently in abstract ways to form connections or is branch growth controlled by chemical signals to connect with other neurons in specified places? In a landmark paper published in 2012, the EPFL's Blue Brain Project (BBP) identified key principles that determine synapse-scale connectivity using a supercomputer to virtually reconstruct a cortical microcircuit and compare it to a mammalian sample based on unparalleled data about the geometrical and electrical properties of neurons acquired from twenty years of slicing living brain tissue...Each neuron in the circuit was reconstructed into a 3D model on a supercomputer...About 10,000 virtual neurons were put into a 3D space in random positions determined by the density and ratio of morphological types found in corresponding living tissue and researchers compared the model back to an equivalent brain circuit from a real mammalian brain...They found that the locations on the model matched the synapses found in the equivalent real-brain circuit with an accuracy ranging from 75% to 95%...This indicates the neurons grow as independently of each other as physically possible and form synapses at the locations where they randomly bump into each other...A few exceptions were discovered pointing out special cases where signals are used by neurons to change statistical connectivity explaining why a brain can withstand damage and an indication that the positions of synapses in all brains of the same species are more similar than different..."we could vary density, position, orientation, and none of that changed the distribution in the positions of the synapses...it's the diversity in the morphology of neurons that makes brain circuits of a particular species basically the same and highly robust"...The current paper provides another proof-of-concept for this approach by demonstrating for the first time that a distribution of synapses or neuronal connections in the mammalian cortex can, to a some extent, be predicted...

Neuroscientists from the Max Planck Institutes of Psychiatry in Munich, Human Cognitive and Brain Sciences in Leipzig, and Charité in Berlin have identified a specific cortical network associated with *self-awareness*...Using EEG and MRI brain imaging to study "lucid dreamers", individuals who have access to memories during dreaming and are aware of themselves while remaining in their dream state (without awakening), researchers found neural activation in a specific network that is normally deactivated during REM sleep in the right hemisphere...This network consists of the Dorsolateral prefrontal cortex (associated with self-focused meta-cognitive evaluation) in combination with parietal lobules (reflect working memory demands), the Bilateral frontopolar areas (the processing of internal states, one's own thoughts and feelings), the Precuneus (implicated in self-referential processing such as first-person perspective) and the Bilateral cuneus in concert with occipitotemporal cortices (active in conscious awareness and visual perception)...This study was limited to only four subjects who were highly trained lucid dreamers with researchers advising that part of the observed activation may have originated from the eye-signaling and hand-clenching tasks performed during the lucid-dreaming process...In the 21st century the human brain is studying the human brain to understand itself...

The Collapse of the 1ˢᵗ World Civilization

It's difficult to imagine a single solution or any number of multiple solutions with the impact to redirect the political policies, corporate behavior and military structure on a planet where billions of humans compete for resources and territory without the ability to understand the increasingly negative impact of unyielding global policies that are destroying Earth's 1ˢᵗ world civilization...Energy extraction continues without serious commitments to address the environmental impact despite talk of change and despite data confirming the negative effects of this activity...There is no talk of the impact of population growth as the number of desperate humans living with little or no means grows...Millions of women still have no voice and religious fanatics globally are getting more vocal and more desperate for violent action to avenge their God...Russia is building a fleet of strategic bombers with nuclear capabilities and nations are increasingly determined to acquire modern weapons with global capabilities...Almost every technological advance is rapidly incorporated into warfare with sophisticated drones and subversive internet programs designed by governments with territorial concerns and a primitive survival instinct that threaten pre-emptive activity without alternatives...Drought, war and famine are responsible for the premature deaths of millions and responsible for the growing population of malnourished children and babies while tons of food are wasted by developed countries isolated by wealth and prosperity...War is threatened on Iran by nations with hundreds of nuclear weapons to prevent Iran from getting one while Pakistan and India posture with nuclear missiles aimed at each other as both societies continue to be threatened by internal unrest, entrenched corruption and little progress in either area...Multiple conflicts in the Middle-East and Africa continue with millions of humans displaced or murdered to appease the primitive ignorance of whichever human has the most guns, the most money and the most territory...

In Mexico over 60,000 humans have died in drug wars supported by increasing corruption and profit on both sides of the border while America grows more ineffectual with a corrupt political system and their misguided belief of 'exceptionalism' when they are clearly not exceptional...America has an unacceptable infant mortality rate, millions of poor, hungry children, the most expensive health-care system in the civilized world, a systemically corrupt government, thousands of industries shut down and moved overseas to maximize profit for the rich, a public school system that has failed as resources are given to private schools, a corrupt financial system with banks given billions by the government after stealing from investors and taxpayers, a broken unregulated stock market that favors insiders and lawmakers with a global reputation of enriching themselves and their donors while gutting programs to help the poor, the sick and the elderly...A nation who still believe they lead the world...

The impact of self-serving corrupt structures defined by multinational corporations and political ideology will continue to fragment a world already fragmented as conflicts will continue to grow until a nuclear weapon is used with predictable military reactions...In the aftermath of a nuclear event nations will implement increasingly draconian measures to control citizens, individual freedom will be a luxury of the past and the small percentage of humans with the majority of wealth will control economies with military aggression and alliances that ignore the majority to protect the few...Environmental progress will cease with growing pollution and species destruction having no priority and energy consumption having the highest priority...Political leaders will talk of change but without the awareness to recognize reality or understand causality, nothing will happen that doesn't solidify power and crush dissent...In the 21ˢᵗ century the predictable chaos of this scenario is obvious and the future is unknowable...In the next 200 years this civilization could collapse with billions dying as society erupts into hundreds of isolated wars and survivalist enclaves, technology shattered but not gone and food disruption everywhere with drought, famine and disease destroying populations on a planet that still rotates and circles a Sun that will shine for a billion years...With no impacts from space humans will have a chance to build again and when communication is re-established globally, the 2ⁿᵈ World Civilization will begin and humans will again have a chance to understand the balance of existence and live within that balance...

It is still possible to impact this civilization in a way to avoid such chaos and destruction but the opportunities grow less and outcomes are unknown...It may be possible to identify and unite people on

Earth who understand reality and who realize the changes and chances which must be taken to really make a difference but it's an unknown...A dangerous world needs cooperating nations to build smart, unassailable defenses to protect non-violent populations and governments who can work together to remove dictators using military finesse in intelligent restricted ways with unlimited support for those who recognize the necessity and advantages of a cohesive Earth, demonstrating the real possibilities of a progressive, inclusive global society through example and transparency...Nations that will protect and accept millions of humans who want equal rights and do not wish to live under seventh century edicts or another oppressive violent social structure and nations who support organizations focused on every baby born to provide nutrition, stimulation, sleep and love without conditions...Nations who recognize inequality and can change the economic structure on a world with an underground economy measured in trillions of dollars...Nations with programs to recycle everything humans use to create a sustainable ecosystem while giving everyone enough resources to shelter, feed and educate their children...A global system of wealth based on a shared existence, not money...*This will not happen...*

There are millions of ignorant males who will never accept that woman are the same creature they are and there are billions of humans dominated by racial and ethnic hatred who cannot understand a reality that proves all humans are the same...Millions of brains unable to balance intellect and reason who really do believe humans kill for gain and pleasure because it's human nature, believe in territorial conquest and a world ready to plunder, believe in gods, myths and legends that control their destiny and absolve them of responsibility and believe that any behavior which does not dominate and control for personal gain is weak and unacceptable...Despite this, the human species is growing in intelligence but it's unrealistic to imagine any parallel development of infant care and reality awareness equal to the technological advances and the addition of six billion humans in 200 years...When humans developed the technology to war on an enormous scale they fought two World Wars killing millions of innocent people...Now humans wage many wars with multiple rules to allow the destruction of entire societies and the murder and rape of thousands of women and children while edging closer every decade to a chaos and destruction that will threaten the planet...There is no utopia and there's no magic bullet...The key is a focus on early development, the existence of every human born and find a way to communicate with the billion(s) who do understand the nature of human reality and the necessity of maintaining a stable ecological balance on a single world...With exponential growth of population and the entrenched powers with no ability to recognize limits, it's increasingly clear this will not happen in time...

A Reality with No Boundaries...When you go past the quantum foam there is more...something else...When you go out to the edge of this reality (46 billion light years) there is more, something beyond, something before...No matter how many times you consider the nature of human existence, it's unsolvable...unexplainable...no answers...but it exists...it's something...Why is it so big and why is a single species on a single planet of life curious about such a large reality where 99.999% of all activity has no impact on our existence?

The BOSS Great Wall (BGW) is the largest structure found in the universe, a long string of active galaxies (830) within four superclusters a billion light years long believed to be about five billion light years from Earth...

Questions

What is existence? Why does existence exist and where did it come from? Was it an accident or by design and who or what designed it? These are the biggest questions and the mind considering these questions about eternity, forever and infinity will have to return to the reality we know with questions unanswered and an awareness of humanity's inability to conceptualize the true nature of existence...An expanding accelerating void that is mostly empty and very young with a beginning just 13.7 billion years ago and an unknown future that will continue for trillions of years...And in the midst of this incomprehensible emptiness a replicating consciousness has evolved to understand the size, structure and properties of this void in a way that staggers the imagination...Vast clusters of millions of galaxies arranged in recognizable patterns of filaments and nodes that resemble the connective networks in the human brain, impossible not to think about symmetry, universality and coincidence...

Stranger still is that each and every one of us is moving through this emptiness at millions of miles per hour with no sensation of that movement, a violent, active universe seemingly frozen in time, changing in extreme slow motion from the three-dimensional perspective of a species extremely short-lived and very diminutive in size...Has another consciousness evolved to recognize the structure of this reality and in this one galaxy with billions of stars and trillions of planets, where is life? One possible answer is the age of this universe...Our journey from single-celled life to the complexity of the human brain took over three billion years, a very long time in a universe this young and it's possible we may be the first civilization to reach this level of awareness...In 50 billion years this galaxy could be filled with electromagnetic signals from emerging civilizations or it might still be silent...There is no way to know and what the human species seeks more than anything else is to know...After three billion years of evolution what we do know is the size of this universe, the mystery of reality and the challenge of survival promises this curious species a future of exploration and discovery for millions of years..

Science has determined the universe "started" 13.7 billion years ago (a very young universe) and with laws of physics and the value of the speed of light, 13.7 billion light years is the observational limit of any device humans can develop...However, because of the rate of expansion within that 13.7 billion year time frame the objects observed at this observational limit are now 46 billion light years in distance...Using an expansion rate measured from the time of the Big Bang verified in a multitude of ways, this "emptiness", a Universe that seems endless in all directions, is estimated to have an actual diameter of more than 93 billion light years growing bigger, ever faster, every second...In 2016 scientists believe the universe is "flat" and will continue to expand for trillions of years to an infinite size, scientific conjecture based on a measurement of the "angles" between the ripples in the CMB (cosmic background radiation) and our human need to define the size and structure of our existence...The nature of reality beyond this observational limit will always remain unknown unless something unimaginable happens in a future too distance to predict...In local space the nearest star is 24,687,600,000,000 miles (4.2 ly) away, close enough for a spaceship going 40,000 miles per hour to reach in 70,456 years...

The energy event at the beginning that initiated the movement of everything is unimaginable...Our galaxy is still moving at 1.3 million mph billions of years after the "event" that started everything...Estimates are the energy scale of this 'event' was over 1,000 trillion degrees Celsius with the initial acceleration (after inflation) at the speed of light...Without mass, energy particles continued to move at the speed of light until interaction with the Higgs Field (everywhere in the Universe) determined the mass of all elementary particles with a measurable mass...

Did "time" exist before a consciousness evolved that could measure the relative motion of anything/everything?

Time

Time...Every thought experiment concerning time barriers and restrictions of human motion through reality indicate time travel is simply not possible in one's own time...Traveling "back" in time makes no sense given the *entropy* of the universe, (everything evolving from a more ordered state to a less ordered state) and the absence of visitors from some future to see the 'dawn of space travel'...Also there's the "grandfather paradox" with an absurd possibility of you killing your grandfather before your mother was conceived eliminating the possibility of you when there clearly is a you...It seems clear by any 21st century conjecture that going backward in time is simply not possible...That leaves the present (now) and the future...The present is a never-ending journey, a continuous movement through multiple events with no ability to stop or slow the time you exist in and no way to reach back even one second and change any event that has happened...What is now is instantly past and it continues forever without cessation...The future is more interesting with a clear possibility of acceleration and return to a future timeline you *used* to exist in but you can't travel forward in your own time meaning *you cannot go forward in a timeline you exist in*...You can accelerate to enormous speeds, slow down and return to a timeline you used to exist in, but the time that has passed since you left, time measured by an aging of the planet and the aging of the population, did not exist for you...You can only go forward in time to someone else's time since you left that time to travel at relativistic speeds...And you can never meet you in any paradox since there is only one you...Continuous motion through infinite space...

Every object with mass in motion experiences it's own time and until you get to extremely high velocities this means very little to creatures like us who do not move in any way fast enough to notice relativistic effects...It is unknown at this early era of scientific discovery if time is a fundamental part of nature but the fact that biological processes can be slowed relative to a 'control population' simply by moving extremely fast suggest a complex causality between time and everything in existence...*Time is an individual measurement of mass in motion* and it's unique for every single creature and every object with mass at extreme differences unnoticed by humans living in a relatively stable existence around a single star...Human's have found precision in the rhythms and movements of the very small (the length of a second is determined by vibrations of the Cesium atom – 192,774,892 vibrations = one second) and in precise rhythms and movements of the very large (Earth's orbit and rotation / Pulsars)...Using the only natural spatial references we know our species gave these periods of motion labels (second, day, month, year) and created a precise measurement of time...A simple measurement of motion...We measure a pulsar spinning 40x's a second and another intelligent species seeing the same pulsar would measure it moving at exactly the same speed because that's the reality of it's movement, but they would have different labels for their time based upon their local spatial reference and maybe the vibrations of a different kind of atom that corresponded to the measurement of movement in their star system...Time is individual to every creature or object in motion and since all humans live in the same gravitational indentation created by the solar system and our continuous movement through space, every human experiences time the same way with measurements and labels for all human interactions with motion (aging, motion around the sun, etc)...If nothing moved (absolute zero, nothing moves, not electrons, not galaxies) there would be no time...There would be no aging, no passage, no measurement, nothing...

If the universe were fixed and static as believed for most of human history it's likely we would not have this existential question of origins, eternity and forever...It would be easy to believe it had always been and it would always be and there would be comfort(?)in that thought...The fact that we know the universe had a beginning believed to have started from an infinitesimal, unimaginable size called a singularity means something existed before this universe...This measurable reality came from somewhere and something and that is the biggest mystery...

Black Holes, which exist in the center of almost all galaxies, were discovered only 45 years ago...

NASA's new black-hole-hunter spacecraft, NuSTAR, has detected it's first 10 supermassive black holes, all at the center of distant galaxies between 0.3 and 11.4 billion light-years from Earth...

Black Holes

The strangest reality in this universe is a black hole, an immense gravitational indentation in space existing at the center of nearly every galaxy observed...Recent research predicted there were over 30 million black holes spread across the universe within the first billion years after the Big Bang and it's believed black holes may be fundamental in the creation and stabilization of every known galaxy in existence excepting for a few of the irregular galaxies or dwarf galaxies...

Science recently confirmed the existence of black holes at the center of dwarf galaxies with masses several hundred thousand times that of the Sun...The black hole at the center of our galaxy is 4 million times the mass of the Sun...

These unseen "objects" vary in size and mass and spin with an effect that theoretically "drags" space-time, distorting and curving the fabric of space around the black hole in severe ways to a point where light (and everything else too near) disappears into the curvature and is never seen again...This boundary that surrounds every black hole where light becomes trapped is called the *event horizon,* essentially the edge of a black hole...Stephen Hawking, one of Earth's resident experts on black holes, has explained how black holes can dissipate over time through a quantum effect emitting extremely small amounts of radiation (Hawking radiation) but it would take trillions of years for a large black hole to completely disappear...The real mystery is this: If mass at a certain level turns into a singularity, a hole in space, and this singularity is the destination of all matter flowing into the black hole, then regardless of how much material was ingested it would remain a singular, unchanging size...It's mass would be determined by the original event creating the singularity and it could not "grow"...Theory insists that matter is completely obliterated' upon entering a black hole and if it's true that matter is destroyed entering the black hole (vanishing into a singularity beyond the grasp of science) how would black holes grow larger in size and mass over time? This growth suggests black holes may not be just holes but also a kind of exotic matter dense enough to trap light, impossible to image...(Hawking's newest paper in 2016 suggests the charged particles being sucked into a black hole leave behind a kind of two-dimensional holographic imprint of their information on the event horizon, a quantum effect)

The idea of a billion black holes throughout the universe ripping the fabric of space and creating singularities where all science ceases to make sense is very challenging to explain mathematically and impossible to grasp in any logical way...And a little alarming...What would be the nature of a universe with billions, maybe trillions of black holes that are essentially holes in space? And if all black holes are holes in the fabric of space, holes to where?

In 2014, scientists using the Chandra X-ray telescope and Europe's XMM-Newton confirmed the rotational speed of a supermassive black hole 6 billion light years from Earth...The black hole (an extremely active quasar) is perfectly located on a line of sight beyond a large elliptical galaxy which magnifies the black hole's activity through gravitational lensing allowing the scientists to read the quasar's X-ray output across a known spectrum at different energies...After measuring the radius of the black hole's event horizon from the changes in the observed X-ray energies, they estimated the spin of this unimaginably dense object at more than half the speed of light...

In an ambitious effort to take technology to the edge of what is possible, scientists are combining multiple radio telescopes in a network controlled by extremely precise atomic clocks with the goal of directly imaging the event horizon of the Milky Way's central black hole...To create such an image would require a telescope the size of Earth and by connecting telescopes in Arizona, Mexico, Hawaii, Spain and Antarctica with the Atacama Large Millimeter Array in Chile, scientists will create the equivalent of an Earth sized telescope that they hope will image the glowing ring of gas surrounding the black hole...The Event Horizon Telescope is expected to begin functioning in 2017...

Imagine a continuous fabric of reality so diffuse and spread out that we exist in this fabric without noticing it's existence...Humans are so tiny that we flow through the fabric without even recognizing it as anything (we call it a vacuum, but it's real, it's something) and all things within this fabric, including stars and galaxies, can move through this bent, curved fabric without impedance, without a sensation of it's existence...But it has form and a structure recognized by the particles of light that follow its contours as well as planets and stars...It does exist and it is real...

Gravity

(A distinction of considerable debate)...Gravity is a fundamental PROPERTY of the universe, not a FORCE...Even Einstein understood this when working out his theory of general relativity but gravity continues to be viewed as a fundamental force intrinsically linked with the search for a unified theory of everything when gravity is simply a conceptual term humans use to identify a fundamental property that determines the contour and shape of the fabric of space...Unfortunately, humans have no idea what the 'fabric of space' is...Suggestions are it's a quantum foam of unknown exotic particles popping in and out of existence (hard to visualize) or vibrating strings in ten dimensions (really hard to visualize), a combination of these or 'something' yet to be realized...What is curious about our reality is there are no size boundaries in any direction with the smallest 'realities' discovered or predicted made of something smaller and the largest reality we know, a 'space' now estimated to be 93 billion light years in diameter, contained within a larger, unknown reality...Human's exist in a three-dimensional universe where space itself (empty by human standards with less than 9 atoms per cubic centimeter) is an unknown reality but it is *something,* a substance that bends and stretches when you place objects with mass in it...The Sun, the largest object in this region of space for several light years, sits like a large ball in the center of a vast three-dimensional trampoline with a slope that gradually gets less as it bends space in every direction for an unknown distance...*At what point does the slope level out and another slope starts toward another heavy object in space?* Two light years? Maybe three light years? Anything 'rolling' down this slope toward our Sun is captured in this gravitational indentation of space and very few objects escape...Take this idea and start *everything* moving thousands of miles per hour in continuous motion for billions of years...This is human reality...

"Planck length" is believed to be the smallest size any particle can be in this universe, theoretically the shortest measurable length (although no one has measured anything this small) because the "uncertainty principle" predicts the universe is indeterminate and probabilistic at smaller scales...The dividing line between quantum mechanics and general relativity...A single atom of hydrogen is more than 10 trillion trillion Planck lengths wide...
Planck length = .0000000000000000000000000000000016

Gravity is not a force that 'pulls' on objects, it is a fundamental property of human reality that's seemingly consistent everywhere in the universe (*this universal consistency of gravity everywhere is now questioned by new research into dark matter/dark energy*)...Space is flexible and reacts with every object that has mass, any mass, and creates an indentation and a gravitational slope even on the tiniest scale...The gravity 'waves' science continues to try and measure are simply reactionary ripples through space caused by the movement of matter in this three dimensional substance, ripples that travel through a reality so tenuous many believed it would be impossible to detect them but this may have changed in 2015 when two identical detectors (LIGO) operating in unison 3,000 kilometers apart detected the gravitational waves created by the merger of two black holes 1.3 billion light years away...Another example of our species ability to build incredible tools with extraordinary purpose capable of detecting a change in distance 1/10,000[th] the width of a proton measuring human reality in extraordinary ways...

If you stopped all motion the three fundamental forces would not exist (electro-magnetic force, weak nuclear force and strong nuclear force) but the warping of space would continue determined by the mass and distance of every object in existence...As a fundamental property of reality gravity will never be 'controlled' but it may be possible to create devices that interact with magnetic fields to counter gravity's effect enabling a flight technology unlike anything in existence today...

Human Travel to the Stars

Human science has concluded in many, many ways it's impossible for any object with mass to travel faster than the speed of light...299,792,458 meters per second...670 million miles per hour...In a reality containing trillions of objects with the nearest star (Proxima Centauri) 4.24 light-years away requiring over 70,000 years to reach in the fastest spaceship ever launched from Earth, it seems a cruel joke that an intelligence who can measure the universe is forever limited to travel within it's own star system...Warp drives used frequently in science fiction seem to require some kind of multidimensional element to work and wormholes large enough and predictable enough to enter and exit without damage are very problematic and unknown...But in a reality still not understood, insights and recent discoveries continue to provide glimpses of what might be possible someday...

The nature of quantum reality and extreme physics only recently considered by science hint at a possibility of interstellar travel in a future not yet imagined as this species begins to understand there may be extreme processes fundamental to human reality that exist without regard to the limits of mass and acceleration...The quantum world has exhibited an odd phenomenon called *quantum entanglement,* a quantum reality in which the quantum states of two or more objects are entangled in such a way that the state of one object is instantly reactive or reflective to the state of the other object, a correlation between entangled pairs that works instantaneously even when these pairs are vast distances from one another...Repeated experiments continue to verify this reactive correlation occurring at many times the speed of light and in the 21st century with human science, experimentation, exploration and technology still in it's infancy, the science behind this phenomenon is unknown...

That we can observe any action happening at many times the speed of light means this is not an immutable barrier...

In trying to understand the fabric of space scientists have proposed space may be a quantum foam of some kind with particles popping in and out of existence...Predictable reality at a scale that is currently unmeasurable and unimaginable today *but what if quantum foam were a repeatable pattern,* a quantum field of an unknown dynamic that could be precisely measured and predicted...The entirety of human existence is composed of repeated patterns and the very definition of intelligence is an ability to recognize and understand increasingly complex patterns of reality...Suppose that all of space was identifiable by predictable and recognizable patterns of quantum activity unique to location and it was possible to measure, model and replicate a quantum field of certain dimensions with technology yet to be developed...Future humans could model this quantum foam activity with extremely powerful precise computers (quantum computers?) to a degree where they could predict a quantum energy field at an exact time within very precise dimensions...Dimensions large enough to contain a spacecraft...We could choose a measured dimension 20,000 miles above the surface of Jupiter and with advanced technology unrealistic for thousands (millions?) of years, humans could replicate the exact properties of a known quantum field beside Jupiter in a device on Earth of the same exact dimensions containing a spaceship and instantly the spaceship would appear in the quantum field beside Jupiter...

The replicated quantum field on Earth and the predicted quantum field in space would have precisely the same properties and in a future with unknown possibilities there could be instantaneous travel through space...Using extreme quantum computers, perhaps the only way to model a dynamic quantum field, all spaceships sent out this way would be able to replicate a quantum field around their spaceship that corresponded with a known quantum field of exact dimensions at an orbital point around the Earth and they could return...In a far distant future humans might understand this well enough to model predictable quantum fields around stars light years away and by replicating the exact nature of that field at a precise time, transfer instantaneously through interstellar space...

This is only the dream of a small human on a single planet with no way to escape but knowing there are possibilities beyond the speed of light gives hope humans will one day travel to other stars...

Dark Energy

Is it possible that dark energy is not an accelerating force within the Universe pushing it apart but instead is a vacuum force outside of this Universe pulling it apart? Science would still measure an accelerating expansion with no way to detect the cause or mechanism...

In another unique collaboration of human tool builders with purpose, an extremely sensitive 570 megapixel digital camera has been built and mounted on the four meter telescope at the Cerro Tololo Observatory in Chile to survey the southern sky for five years focusing on 200 million galaxies, 4,000 supernovae and 100,000 galactic clusters...It will map the three-dimensional distribution of the clusters measuring the *baryon acoustic oscillations* believed to be "sound echoes" from the Big Bang, it will study the same kind of supernovae that led to the original discovery of dark energy to 'read' the expansion history of the universe and it will study the *gravitational lensing* of individual galaxies and clusters...The camera has 62 charge coupled devices (CCD) which can image light from more than 8 billion light years with an extreme level of sensitivity to the specific wavelength of red light to pinpoint distant objects...The Dark Energy Survey (DES) officially began on August 31, 2013...

In a previous study the BOSS (Baryon Oscillation Spectroscopic Survey) team using the 2.5 meter Sloan Foundation Telescope said their data suggests dark energy is a property of space that increases with time...As the universe expands and creates more space it creates more dark energy which at some point overwhelms gravitational deceleration and accelerates the expansion of the universe at an ever increasing rate...The team observed 50,000 closely spaced quasars building a three dimensional map of the distribution of hydrogen gas clouds at a distance of 11 billion light years and found patterns of clustering even at this distance...They intend to map the locations of 1.5 million galaxies and more than 160,000 quasars over the next few years...November 2012

Current estimates are that Dark Energy represents 70% of the Universe (it's unknown what it is) and Dark Matter represents 25% of the Universe (it's unknown what it is) and "normal matter" is less than 5% of the Universe...

In April 2014 the BOSS team reported the universe has expanded at an increasing rate of 1% every 44 million years for the past 10 billion years, the most precise measurement of dark energy to date...Viewing 140,000 distant quasars, two independent studies provided the data with one study mapping quasar distribution and the other study measuring the positions of hydrogen clouds using the quasar's light to map distribution ...

Dark Matter

The theoretical presence of the unknown substance *dark matter* was "discovered" by Jan Oort and Fritz Zwicky with Oort measuring the perpendicular motions of stars relative to the galactic disk and Zwicky measuring galactic clusters...Both scientists found that the visible mass observable and measurable was far less than the mass needed to account for the movement and gravitational attraction predicted by modern physics...The most powerful evidence of dark matter is implied by the observed rotational velocities of millions of spiral galaxies, a rotational velocity that is constant from the center of the galaxy out to the edge of the visible disk of stars...To explain this observational reality within the framework of the accepted understanding of gravity, an invisible "mass" of some kind would have to envelope the galaxy accounting for as much as 10x's the measurable mass of the galaxy...Continued observations of various star clusters, galactic clusters and elliptical galaxies all indicate some unknown substance is required to explain the movements and gravitational lensing measured by astronomers...

A group of radio interferometers, four 20-meter radio telescopes installed in Japan at various distances to create a single radio telescope some 2,300 kilometers in diameter comprise a joint project called VERA (VLBI Exploration of Radio Astronomy) which has been in operation since 2007 and has verified the galactic rotational velocity of the Milky Way between 10,000 and 50,000 light years from the galactic center...This is the first analysis of the Milky Way using more than 50 objects and from this analysis researchers report a galactic rotational velocity of 240 kilometers per second, a velocity that's applicable to almost any star at any location within the galaxy...

Current theories suggest dark matter is composed of massive neutrinos, stellar remnants, brown dwarfs, weakly interacting massive particles or some sort of cold, dark matter of unknown origin...

Almost eighty years after these discoveries predicted dark matter must be pervasive throughout the universe no one has verified the discovery of a "piece" of dark matter and no one has been able to detect dark matter...In South Dakota an enormous underground detector (Large Underground Xenon = LUX) was built deep in the Earth (4,850 feet) in the latest effort to detect predicted interactions of dark matter with regular matter...Researchers expected to identify dark matter by detecting interactions with WIMP's (Weakly Interacting Massive Particles) tiny elusive particles that fill the universe, billions of WIMP's passing through everything continuously without interacting with the 'normal' matter familiar to science but after three years and the analysis of half a million gigabytes of data, no evidence has been found...As the search for dark matter continues a growing number of researchers investigating the properties of gravity suggest that gravity may not be the stable, predictable, inflexible and universal attractor current theory holds but may be flexible in differing acceleration frames, new ideas that could explain the observations currently attributed to the existence of dark matter without dark matter...And it continues...Exploring alternative theories is what humans do and it's very possible that one day all the mysteries determining properties and motion in this observable measurable existence will be known if the human species can survive this critical transition to a science based intelligence...A new detection facility being built (LUX-ZEPLIN) will be 70x's more sensitive when it starts taking data in 2020...

Recently suggested is dark matter may result from billions of tiny black holes spread throughout the galaxy (no one has observed a hint of evidence supporting this) but imagine if dark matter is instead the product of all the matter/energy flowing into a central galactic black hole, disappearing from this dimension of reality and interspersing in a uniform way to surround the galaxy in an unknown, undetectable dimension existing in the same space...

A new mission to map the galaxy in unprecedented detail (GAIA) was successfully launched in 2013 by the European Space Agency designed to chart a three-dimensional map of the galaxy with precise positional and radial velocity measurements of a billion stars, less than 1% of the galactic total...Gaia will profile these stars with precise measurement of distance, motion, brightness, temperature and composition...Technology to track any objects that pass in front of it's camera is expected to discover thousand's of objects not yet detected by ground telescopes...GAIA entered a stable, operational orbit in January 2014 around L2, a Lagrange point (there are five) where the Earth, the Sun and the spacecraft balance in a predictable, continuous orbit...L2 is on the night-side of Earth which will keep GAIA in constant darkness about 1.5 million kilometers from Earth, a perfect location for continuous observations...

On the temperature scale of reality just 400+ degrees colder than the operating temperature of a human being stops all motion at the atomic level while there seems to be no limit to how hot it can be (1,000 trillion degrees Celsius?)...Humans are extremely cold creatures on this scale, existing very close to the bottom...

Will humans ever know exactly how many stars are in our galaxy? Incredible technologies would be needed...

What's fascinating is this universe is just 13.7 billion years old with the human species circling a ordinary third generation star at very nearly the beginning of time and our species may be one of the pioneering life forms in a newly created cosmos...We have survived to a critical point in the evolution of intelligence with the knowledge that fate is not predetermined or manipulated by deities or forces not understood...Our decisions as a species, not as nations, not as cultures or ethnicities, but as a species will determine the future of humanity...

Maybe there is a connection between the patterns of neurons and pathways in the human brain to the nodes, pathways and structure that define this Universe...The similarities between the physical structure and symmetry of the most complex object known (brain) and the Universe at it's largest scale is striking...All science believes this Universe came from a singularity, a location (?) in space and time with no size...We came from somewhere...There is something outside of this reality and this symmetry between the overall structure of the Universe and the physical reality of the brain may communicate to sentient life that this reality is 'structured' and life can survive here for billions of years...A reality that we exist in and a reality that exists in us...

"The extinction event that is happening right now is the first one in history that is the responsibility of a single species...There's no meteorite this time, no exceptional volcanic eruptions, no 'Snowball Earth,' just us, prospering at the expense of other species" Richard Fortey

EPILOGUE
The Briefest Moment in Time...Now

Think of a pyramid sticking out of the ocean with just the tip visible, a tip representing 1 to 1.5 billion humans who understand the nature of existence, the fundamental properties of a proven reality and can now understand what no human has been able to understand in the short history of this species (200,000 years)...Underwater, a majority of the rest (6 billion humans) believe in fate, destiny and mysterious deities who determine the life of all living creatures on Earth, persistent beliefs despite almost four centuries of scientific discoveries that verify our human reality...The only way to turn this equation around is to increase the intelligence of an entire species and the only way to increase Earth's IQ is to focus on the needs of every baby...It sounds easy, a formula of care with multiple caretakers, adequate nutrition and an awareness of cause and affect but in reality this change will take hundreds, even thousands of years and corrupted political structures existing in this century cannot wait hundreds of years...Thousands of variables make it impossible to predict a future Earth but the time to act is now with empathy, unlimited humanitarian aid, inclusion, self-realization and technology by a consortium of nations who will confront 'leaders' who control populations with murder, torture and rape, terrorists groups who commit atrocities in the name of a self-serving god, cultures who vilify and dominate females out of fear and ignorance and those who starve their populations and steal their wealth...

45 million centuries after Earth's creation the 21st century may be the most critical one of all...

With a history of just 200,000 years in a stable reality that's existed for billions of years, the human species has no real understanding of what is possible or what may be possible...In a continuum of staggering size with no limits in any direction we are just beginning to recognize realities that define us and there are mysteries and unknowns which may take not thousands of years, but millions of years to recognize, explore and understand...A new understanding of the science that defines our existence has swept away centuries of ignorance and for the first time in history we have the tools, understanding and knowledge to manipulate a known reality and manage our planet...Two centuries ago there was no radio, no phones, no television and no power grids in a world of isolated populations with many people doing many things but without any global awareness of human possibilities, activities, dangers or new discoveries...On a planet with less than a billion people, individuals could live a life and never know anyone outside their own group...Not anymore...This century there is a global awareness of everything, everywhere and everybody...21st century humans live in a reality saturated with thousands of designed frequencies in a bubble of existence that pulses with energy...You can turn on a device almost anywhere on this planet that decodes millions of speed of light particles into language or data and can do it safely because of an evolutionary history allowing us to co-exist with all electromagnetic frequencies below the frequency of visible light with no risk...We exist in an atmospheric sea of frequencies, billions of wave particles passing through everything every second accessing information and real-time events instantly with an improving technology to message and communicate with anyone on Earth...We can finally see our world...We can see territorial priorities and political division...We can see degradation of habitat and a growing pollution that increases every year...We can see chronic hunger and billions of people living in desperate ways...We can see political coalitions, autocrats, despots and ignorant leaders responsible for a growing inequality and we can see thousands of babies and very young children in extreme negative situations...With a global vision unimagined 200 years ago we can see the stress of overpopulation, conflicts that seemingly never end and the violence of discrimination and ignorance...

Is there any nation on Earth where there are not hungry children or humans living in desperate situations? The legacy of a sentient self-aware intelligence who believe themselves to be the pinnacle of logical planning and thought, the perfect creation of deity...If one analyzes the unrest and violence that continues unchecked in so many places there is a clear pattern of poverty and desperation which drives violence and chaos in every nation, people who year after year see political leaders with military power and corporations with influence taking for themselves without any inclusion of the humans who live in desperation...A blueprint for disaster in any civilization but on a scale unprecedented in a world civilization...With the unchecked population growth of Earth's most destructive creature on a planet just 7,926 miles in diameter, the options and solutions to implement real change and effect global practices grows less each year and despite technology that could make a huge difference and a global knowledge base that grows more every year, nothing will change long-term until we address the human condition, a growing chasm of exclusion in thousands of places on an Earth that will not 'grow' in size...With the growing fragmentation of politics, ineffective policies and unilateral priorities that continue to define this 1st World Civilization, consensual action without conflict seems impossible...

There are thousands of humans on Earth in multiple societies who write and craft very sensible regulations governing all aspects of a working civilization but there are millions of humans who work just as hard to find every way possible to avoid regulations or solutions in pursuit of profit...Legal systems worldwide allow private lawyers representing corporations to actively and intentionally do harm to our species and our planet for short-term profit...

It's imperative that the human species set a bottom standard of existence available to anyone and everyone on Earth...Every adult human must be included in this vast global system of economics and trade and every person should have four basic rights as a human being: Healthcare, Nutrition, Shelter and Education...When humans do this we will be a global community...When we do this we will be a intelligent, sentient species with priorities and the highest priority is to care for every baby and provide a standard, a bottom for every human on Earth...In 2016 the human species is nowhere close to this...

The impact of inclusion (giving all adult humans on Earth $1,000 annually costing around $7 trillion out of a global GDP of $77.6 trillion in 2014) would result in rising markets and corporate profits as four billion humans gain predictability and inclusion in a world where they finally have a voice and a reason to care about their impact on the environment and the future...A global policy dismissed by every financial power on the planet when opined...

We have technology to make all chemical storage tanks leak-proof, technology to filter waste runoff from agriculture and industry, technology to build nuclear, coal-fired and natural gas power plants to a standard of quality and safety without compromise, technology to prevent a Chernobyl or a Fukushima and technology to recycle almost everything on the planet...What's lacking is a consensual awareness among a majority to recognize causality and consequence in a fragmented, unstructured hierarchy with a priority on profit and territory...Everywhere on Earth are politicians, corporate leaders, foundations and pundits who will justify this global negligence with cries about the "lack of money" to do things that should be done while supporting corrupt policies and corporate structures who continue to do anything and everything to avoid paying taxes required by law or their 'fair' share while posting enormous profits year after year...With intelligence and technology we could find way to exist within this ecosystem without the enormous damage being done to the planet today but there are powerful opponents to any action that would diminish the burning of Earth and the profit of waste...The common denominator in all of these problems is an absence of *reality intelligence* and the absence of an intuitive awareness that would recognize reality and use innovative, practical and creative solutions to advance civilization without corruption and mismanagement...The most obvious indicator of the primitive state of human development is represented by the billions of weapons on Earth with more constructed every year, more deadly, easier to operate, and enough for every man, woman and child on Earth...Weapons with the single purpose of facilitating the killing of humans by humans...

In many situations the only option will involve the military, a reality of 21ˢᵗ century civilization, but how nations use military force to advance solutions makes all the difference in all cases...First, and this is a big first, there must be a consortium of nations with equal realizations and equal standing who can work in unison to address severe situational realities anywhere on Earth...There should be temporal governments organized by an improved United Nations able to coordinate infrastructure responsibility and population security under a strict time limit of reorganization without foreign powers taking control and profiting from a nation's resources...Al-Assad should have been taken out when he dropped the first bombs on his citizens but he has killed hundreds of thousands over five years and is as entrenched as ever...Mugabe has destroyed Zimbabwe and he's still in power through deceit and violence while the leaders of North Korea have shown little awareness of anything for more than 60 years limiting brain growth, inclusion and informational awareness in millions of people while still protected and tolerated by China, a situational powder-keg on an Earth where nations war over lines drawn in the sand...

To many who would question how any human can judge the actions of others, there's a standard by which all activities on Earth can be measured...*The sanctity and singularity of a human life*...Any society, government or culture who will kill innocents to control resources, sexually mutilate millions of women and children for any reason, starve infants and entire populations to enrich themselves (while still 'accepted' by the world to represent that population), war for expansion and territorial domination and enact restrictive laws to terrorize and murder thousands because they can, should be challenged, disassembled and removed from power and responsibility by a world organization with the power of an intelligent majority and an awareness of reality...An effective world coalition could manage an efficient reorganization of any enclave with an strict intolerance of violence eliminating punitive policies and sanctions to include everyone in the global economy and taking measures to prevent the rise of another oppressive, violent regime...During reorganization humanitarian aid would be unlimited with a focus on women and children and every action designed to stabilize and benefit indigenous populations, not foreign corporations or control by any powerful agency...A very simplistic view of change but the only way this could succeed is with a cohesive multinational military in concert with global organizations who are transparent with no hidden agendas...Such cooperation is unlikely in this century...

The financial "wizards' who create derivatives and high-speed trading while manipulating interest rates for personal gain should be capable of designing a transparent global financial structure to recognize every human being in the global economy...They have shown no interest in doing this...In such a structure every human would have access to a standard necessary to provide for health, education, shelter and nutrition...If an individual wishes to do nothing more than sit in a chair for the rest of their lives they would still be afforded this standard of care as long as they do not violate the first law of existence but most humans will not sit...Billions of humans would be a vibrant part of the world's economy living fulfilling lives without worry of desperation, sickness or poverty and there would still be many layers of wealth and activity determined by desire, ambition and creativity as humans trade, buy and sell as they always have...An alternative to the primitive economic structure of 2016 which has every nation on Earth holding enormous debt while corporate structures and individuals create vast wealth for themselves, "gaming" the system while billions of people continue to exist without hope in dangerous and desperate ways...

The First Law of Existence
Every human has the right to live their life as they choose as long as they do not interfere with any other human's right to live their life as they choose and do not negatively effect humanity's only habitat...

The Second Law of Existence
No human has the right to force another human to participate in their existence under any circumstance...

The Third Law of Existence
Every human has the right to change the defined structure of their existence in any way at any time for any reason if that change does not violate the first or second law...

The rise of ISIS is not new...This has been going on for thousands of years as ignorant humans without an ability to understand reality or human singularity seek dominance with no understanding of why...What is perplexing is the inability of the most powerful militaries on Earth to eliminate the core of this radical terrorist group who are allowed to burn an innocent man alive and post a video of it on the internet...They can capture entire villages, kidnap and murder the inhabitants while acquiring huge amounts of money in some way and nations surrounding Syria are so dysfunctional and distrustful that cooperation requires days of 'talks' and 'deals' before any real response is considered while the powerful nations in the west all agree it will take "years" to rid the world of ISIS...why?

It's increasingly clear that a primitive species who let millions of babies die and give life saving drugs to only those who can pay are nowhere close to understanding the power and potential of a civilization that provides a basic level of healthcare and subsistence for every member of their species...

Global Patterns

This growing paradigm of excess and ignorance is apparent in the Middle East and Asia, across all of Africa and increasingly in western nations who believe themselves to be politically stable, well organized and impervious to internal unrest...What they do not understand is the perversity of money when the focus is wealth without reason and the unchecked growth of populations who exist without hope in a global capitalistic tiered system that does not recognize limitations and the needs of billions of humans on a shared planet...Socialism is attacked as a system of financial equality that prohibits or endangers profit, innovation and growth when in reality socialism is a definable level of existence that insures health, sustainability and safety...In this century with eyes turned inward and the threat of chaos at their door many governments will use military force on domestic and foreign populations to protect excess preventing a majority of humans from attaining any level of sustainability...Wealthy nations worldwide continue to turn a blind eye to this obvious reality while supporting despots and human monsters with financial agreements and aid to continue a status quo that allows oppressive regimes to increase personal wealth while destroying millions of lives...What's different is a new technological global environment where "arrangements" of convenience are no longer opaque...The world watches and unrest will continue to grow as the wealthy get richer and more militaristic to protect their wealth while proposing solutions unworkable and doing little to understand the real cause of suffering and desperation in the daily lives of billions of humans...

Corporations and politicians everywhere are increasingly indistinguishable with an indifference to science or causality, an indifference which is destroying the planet instead of using technology and a global workforce to engineer realistic systems of effective renewable non-polluting sustainable energy sources...After twenty years of global discussions emphasizing the responsibility of industrial nations to reduce carbon pollution, the U.S. actually increased it's greenhouse gas emissions in 2013 with an entrepreneurial unregulated oil and gas industry that fails to recognize any limitations on the recovery and burning of fossil fuels...The result of America's unregulated energy industries is an explosion of *methane* (one of the most potent greenhouse gases) into the atmosphere as industries cut costs for profit leaving hundreds (thousands) of leaking pipes, fittings, disposal sites and refineries actively leaking gas while polluting drinking water with underground "injections" that are responsible for thousands of earthquakes in places where there were none before...And the United States has effectively spread this questionable work model around the world lobbying multiple nations to obtain access for America's gas & oil corporations in developing nations that do not have serious regulation...Earth in 2016...Human population continues to grow and this world will explode one day as billions rebel against 'systems' that create unthinkable wealth for a minority while billions have no food, no electricity and no sanitation...

Clear in any in-depth reporting on the priorities of governments worldwide is the enormous amount of money and resources being spent on military forces instead of education, feeding the poor, setting up renewable energy grids and making the world a sustainable environment for thousands of years...

Increasingly obvious with mass migration, warring factions, financial crisis and ongoing global pollution is how incompetent almost all governments are at recognizing reality, dealing with causality and managing their territorial societies...The almost singular cause of this is money and a focus on profit, enrichment and greed above everything else...The pattern across South America, African, Europe and North America countries is gross, gross mismanagement of almost everything...Even in northern European nations with reputations of fairness, social justice and equality well established, governments are now using private (for profit) organizations to deal with social problems and people are falling through the cracks as services are limited and cut in pursuit of profit...An intelligent species should pride itself on efficiency & effectiveness but money falls through the cracks in chaos so it continues...

The Ostrich Syndrome

The Kyoto Protocol is an international agreement through the United Nations with a singular focus on reducing the amount of greenhouse gas emissions that result from the processes humans use to burn the planet and create energy (protocol adopted in 1997 and entered into force in 2005)...The first commitment period by treaty participants to bind developed countries to a reduction of greenhouse gas emissions was 2008-2012 with a non-binding commitment by 37 industrial nations and Europe to reduce emissions 5% against 1990 levels...*Some nations were successful and some were not*...The 2nd commitment period (2013-2020) required a reduction of emissions to 18% below 1990 levels...*A new agreement (COP 21) in Paris (2015) has been signed to replace Kyoto in 2020*...A report in April 2014 by the Intergovernmental Panel on Climate Change (IPCC) demonstrates that many nations on planet Earth have their heads buried firmly in the sand...

The IPCC is composed of a panel of climate experts who won the Nobel Prize in 2007 for their efforts to inform the world of the consequences of the continued burning of this world for energy...

The report shows greenhouse gas emissions *grew* by more than 2% in the first decade of this century with nations worldwide subsidizing the acquisition and burning of fossil fuels to a far greater extent than any efforts aimed at developing renewable clean energy for the future (global emissions dropped slightly in 2014 and 2015 giving new hope that these targets are attainable, but new studies foretell a grim reality even if the pollution were stopped immediately and this civilization will continue to emit greenhouse gases for the rest of this century)...Citing population growth and economic growth as primary drivers of greenhouse gas emissions, the report predicts current technologies will be unable to help if nations are unable to limit emissions in a significant way over the next 15 years...The original goal of limiting global warming to 3.6 degrees Fahrenheit is already unattainable and an effort to keep concentrations of CO_2 in the atmosphere below 500 parts per million will fail in just a few decades (400 parts per million have now been detected globally)...Already there is growing speculation on new technology to remove the CO_2 from the atmosphere but unknowable are the many consequences to the plant and animal life on Earth that depend on the Sun's energy and predictable cycles of light to continue an ecosystem that has been essential in the development of life on Earth...New reports out in 2016 show sea levels rising, temperatures increasing and over 35% of Earth's coral reefs experiencing trauma that might never be recovered...Coral reefs support the existence of many millions of humans worldwide providing nutrients and shelter to over 25% of all marine creatures in what is arguably the most critical food chain on the planet...It can take up to 10,000 years for a coral reef to grow from scratch and millions of years for barrier reefs and atolls to fully develop (NOAA)...Billions of humans don't understand consequence and the viability of human life in the future grows increasingly uncertain as corporations and political bodies continue to push for energy without regard for consequence caring little about a future beyond their own existence...A defining characteristic of a brain without the ability to understand reality...

In 2014, OxFam reported the 85 richest humans on Earth possess wealth equal to 3.5 billion of the poorest humans on the planet...One half of Earth's population have less money than the 85 with the most money...

July 2017

In the time it's taken to write this book little progress has been made in any of the critical areas that will determine the survivability and stability of the human species...The "Panama Papers' have shown thousands (million?) of humans who control the wealth, resources and militaries of Earth hiding, stealing and misappropriating trillions of dollars while millions of children go hungry and millions more die from an inability to pay for health services...Generations of children are being lost in the many wars and conflicts that continue to bring huge profits for the military industrial corporations in a profit from death scenario that grows more as the world grows more 'connected'...In 2014, America reported $36.2 billion in weapon sales (NYT's) $10 billion more than 2013 giving deadly weapons to Qatar and Saudi Arabia as well as South Korea and Iraq where these weapons are being used by ISIS militants to wage war on innocents...The congressional report that provided these numbers also noted **"the international arms market is not likely growing over all"** because of **"the weakened state of the global economy" "A number of weapons-exporting nations are focusing not only on the clients with which they have held historic competitive advantages due to well-established military-support relationships but also on potential new clients in countries and regions where there have not been traditional arms suppliers"**...*A renewed focus on supplying 'new clients' who do not have 'suppliers'*...A stunning lack of intellect and awareness by those who have the power and position to change this ignorance but greed has long been a priority over empathy on this world...

Possible is a thousand year stalemate with powerful nations and corporations hiding behind military's and huge walls protecting their geographical sovereignty with tactical nuclear weapons and the technology to produce food and necessities for their populations while billions try to survive in the chaos and anarchy of tribal Earth...

Crisis in Nigeria with 244,000 children to suffer from 'severe acute malnutrition' with an estimate from UNICEF that one in five (50,000 children) will die without aid...Children under age five...Already damaged by a babyhood without nutrients, sleep or security with no chance to grow the dense neural nets necessary for awareness and intelligence...

Crisis in Yemen with Saudi Arabia (backed by the U.S.) bombing infrastructure, civilians, children, funerals, weddings and medical facilities with missiles and bombs supplied by America's top defense contractors...Violence, trauma and a severe lack of nutrition guarantee another generation with little empathy and no ability to understand possibilities...A bill introduced in the U.S. Senate to block arms sales to Saudi Arabia for this indiscriminate killing failed 71 to 27...

Crisis in Syria with Soviet jets and Syrian jets destroying entire cities and targeting children, schools and hospitals in a madman's quest to absolutely destroy any human who had the audacity to question his authority...Nothing is done...Five years...The most powerful military in the world cannot stop this stunningly ignorant 'leader' from killing Syrian children on a massive scale because of a fear of 'involvement' or 'territorial sovereignty'? This does not get better hiding behind closed doors... How many times can these "leaders of the free world" walk up to a podium in their sharply pressed suits and 'condemn' these actions while neatly folding up their repetitive speeches and disappearing back into safe houses...Why does NATO, Europe and America have a military if you can't save the children...In 2017 a treaty was agreed to that essentially gave Al-Assad his victory and control of the nation...

Crisis and famine in South Sudan...Crisis and suffering in Libya...America is focused on their celebrities and politicians (often indistinguishable) in a bizarre election involving Russians, billionaires and corruption in a process transparent everywhere in the world except America...Donald Trump will not change...Britain is focused on their 'Brexit' and migrants across the English Channel and all of Europe is fixated on refugee allocation while thousands die trying to cross a small sea...The leaders of all these nations wear nice clothes, eat great food, have state dinners and sleep in safe beds somehow without dreams of four year old children floating face down until they wash up on a beach to become a small sensation in a 24 hour news cycle, disappearing from the screen in a week, sometimes just days...

The lowest level of existence tolerated by a species defines the empathy, morality and intelligence of that species...

A study published in January 2016 in the journal Pediatrics estimates that as many as half of the world's 2 billion children experienced physical, sexual or emotional violence in the previous year...In response seven months later, the World Health Organization announced the first coordinated plan to end violence against children that includes a seven-point strategy consisting of practical measures like implementing and enforcing laws that limit access to firearms and trying to change beliefs and values around gender roles which would target nations where girls have fewer rights and less freedom...Also creating safe environments by improving housing, increasing parent and caregiver support, strengthening economies and funding support services such as treatment programs for juvenile offenders and educating children in life and social skills...How they might do this is unknown...

There has been little progress on the proliferation of plastics in the oceans (it's impossible to walk within any environment in America without seeing plastic bottles thrown away in every trashcan visible destined for the landfill) and every CEO of every energy company vows to fight any regulation that will reduce pollution if it reduces their profit margin...Reports from the Arctic estimate over 1.7 billion metric tons of carbon in the permafrost that will be released when this once permanently frozen part of Earth cycles through thaws and refreezing as the planet heats up (an enormous amount of toxic mercury will also be released by this process) and the Paris Climate Talks in 2015 have no binding deadlines, no binding targets and years before any nation needs to show progress with many countries using sleight of hand and technicalities to burn this world as aggressively as they have in the past...

2016...A study published in Scientific Reports concludes the goal agreed to just 7 months earlier in Paris (COP21) to keep global warming below a 1.5 degree Celsius increase over Earth's pre-industrial averages will fail even if global emissions were stopped immediately...Research shows that temperature averages reported in 2016 are already above a one degree Celsius increase and burning this planet for energy will not cease for many years...

UNICEF
(Report published in October 2016)

Around 300 million children in the world live in areas with outdoor air so toxic it can cause serious health damage including harming children's developing brains...Satellite imagery confirms that around 2 billion children live in areas where outdoor air pollution caused by factors such as vehicle emissions, heavy use of fossil fuels, dust and burning of waste exceeds minimum air quality guidelines set by the World Health Organization...South Asia has the largest number of children living in these areas (620 million) with Africa next at 520 million children...East Asia and the Pacific region have 450 million children living in areas that exceed guideline limits..."Air pollution is a major contributing factor in the deaths of around 600,000 children under five every year" said UNICEF Executive Director Anthony Lake. "Pollutants don't only harm children's developing lungs – they can actually cross the blood-brain barrier and permanently damage their developing brains – and, thus, their futures."

Conflict with North Korea is inevitable before the end of this decade with an ignorant ruling class focused on nuclear war and an arrogant president in America incapable of understanding actions and policies necessary and effective on a global stage...Trump's belligerence toward Iran also threatens to re-ignite a nuclear arms race in the Middle-East...

America's response to Kim Jong-un's aggression is to increase advanced arms sales to surrounding allies in the region increasing profits for U.S. national defense corporations...

It is surreal to exist on this primitive planet and listen to this species rationalizing and justifying total chaos, unimaginable suffering and death...Listen to the news reports and analysis worldwide and it's stunning how everyone seems to believe that America arming Israel, Egypt and Saudi Arabia is a very"normal" state of affairs...Nobody questions the primitive aspect of this existence while watching a world civilization in turmoil with millions of babies and children dying...After talk of be-heading's, millions in refugee camps and millions more hungry and poor, the news returns to politics and celebrity...An emerging sentient species killing without conscience, strung out on money and discovering technology without any idea as to how to improve the human animal...Surreal

It continues...

About the Author

I was born in Paris a million years too early at a critical time in the evolutionary history of a single planet where the actions of this entire species will determine if there is an advanced civilization in a million years, a brief time in the history of Earth but a perilous time in the history of the only self-aware, sentient species known to exist...The progress of this species is undeniable but the recent discovery of advanced technology (for this era) without a parallel and critical realization of what is necessary to create a human with the empathy and intelligence to use that technology for the advancement of every member of this species has initiated a series of events (millions) that are evolving in an exponential sequence that threatens the existence of every creature on this world...It is unknown how many humans on this planet understand the reality and nature of human existence but the identification and unity of those who do may be the only variable with a chance to change this ignorance, maybe 1.3 billion who recognize this reality and understand what is necessary for an 'advanced' species to survive in a star system for millions of years...This means over 6 billion humans will mostly ignore the reality around them, waiting for salvation, existing in a 'normal' where uncounted children die from conflict, famine and disease while war is just something humans do because they always have...This book seeks to find the 1.3 billion who understand the absurdity of political policies without direction, the destructive instincts of military's that dominate every territory on Earth, the ignorance of wealth without reason and the primitive mindset that can rationalize the death of billions because they do not have the resources to buy food or medicine from those without empathy who will never recognize their excess...Those who do understand this reality can go to *primitiveplanet.org* and simply click so we know you exist...If there were unity and a shared knowledge by those who understand there may still be hope...

The author resides on planet Earth by accident or design with no memories of origins or of those who nurtured him in his first years of existence...After years of studying human behavior in every aspect of existence in relation to human empathy, morality and behaviors impossible to explain in any conventional sense, the essential importance of the first years after birth are undeniable...His best friend and constant companion is a five year old emissary who will make a real impact on this equation of ignorance vs intelligence one day...Thank you Clinton...And I was fortunate enough to find a caretaker who would change the nature of our entire species if she cared for all of Earth's children in their first and most determinant years of life...Every child you touch recognizes an instinctual, embracing love without conditions...Thank you Jeni...This book was written for all the babies of Earth in hope the next world civilization will understand and never let any baby suffer violence, hunger, indifference or neglect in the first years of life so they can become adults with the empathy, morality and intelligence to change a world...

Thank you Josina Maria for what you did to make this possible...

Earth's IQ seeks a priority standard of care for all the children of Earth during the first three years of life so they are never hungry without someone to feed them, never lonely without someone to hold them, never hurt without someone to fix them and never awake without someone aware...Every child born on Earth...The first thousand days...